Springer Series in Reliability Engineering

Series Editor

Hoang Pham, Department of Industrial and Systems Engineering, Rutgers University, Piscataway, NJ, USA

More information about this series at http://www.springer.com/series/6917

Anu G. Aggarwal · Abhishek Tandon · Hoang Pham
Editors

Optimization Models in Software Reliability

Springer

Editors
Anu G. Aggarwal
Department of Operational Research
University of Delhi
New Dehli, India

Hoang Pham
Department of Industrial and Systems Engineering
State University of New Jersey
Piscataway, NJ, USA

Abhishek Tandon
Shaheed Sukhdev College of Business Studies
University of Delhi
New Delhi, India

ISSN 1614-7839 ISSN 2196-999X (electronic)
Springer Series in Reliability Engineering
ISBN 978-3-030-78918-3 ISBN 978-3-030-78919-0 (eBook)
https://doi.org/10.1007/978-3-030-78919-0

This Springer imprint is published by the registered company Springer Nature Switzerland AG
The registered company address is: Gewerbestrasse 11, 6330 Cham, Switzerland

Preface

In this age of Internet technology, software has become an indispensable part of our everyday life. With the paradigm shift of businesses, education, defence, healthcare, etc. from physical to digital platforms, our reliance on software has grown in leaps and bounds. Development of highly reliable software at reasonable prices with quick deliveries has emerged as the prime objective of IT firms. Optimization techniques/tools have gained popularity among IT managers and software developers for decision-making while meeting multiple needs of the users. Optimization is a set of mathematical techniques where the constraints affecting the decisions are identified and included in the model while maximizing gains or minimizing costs and risks. The use of optimization techniques for the judicious allocation of resources and strategic decisions pertaining to software release schedules, software versions management, and warranty/maintenance policies is a relatively new field of study. The main focus of this book is the evaluation and optimization of key decisions related to the software development process. This book consists of 16 chapters featuring a broad range of topics including Software Reliability Modeling, Estimation and Prediction, Optimal Allocation and Selection Decisions, and up-gradations problems. Each chapter is written by well-known researchers and IT practitioners to present the recent trends and research opportunities in the area of software reliability engineering. More specifically,

Chapter "Software Reliability Modeling and Methods: A State of the Art Review" provides a comprehensive review of probabilistic software reliability models with different groups with major focus on Nonhomogeneous Poisson Process (NHPP)-based software reliability models which address various aspects related to software development practices, such as testing efficiency, testing coverage, multiple fault types, time-delay fault removal, and environmental factors. In chapter "Software Reliability Growth Models Incorporating Software Project/Application's Characteristics as a Power Function with Change Point," it discusses a framework based on error-removal phenomenon model by incorporating the software project/application characteristics as a parameter. This chapter also describes a software reliability growth model developed under a change point scenario, which is then utilized to develop release time policy balancing reliability and expected cost incurred during software application development. Chapter "Robust Multi-Response Based Software Reliability Modeling" discusses robust multi-response-based software reliability modeling and also illustrates the methodology on a large-scale communication software system. The two responses under study are time-between-detection of defects and effort. Optimization using response surface methodology is performed on overall desirability value.

Chapter "Multi-criteria Decision Making in Optimal Software Testing-Allocation Problem" presents optimal testing-resource allocation for the module testing in the software testing phase based on multi-attribute utility theory. The authors have optimized the utilities for the reliability, testing-resource, and testing-cost for resource allocation. Chapter "Release Planning Analysis Through Testing Coverage and Fault Reduction Factor Based Models with Change Point Perspective" presents software reliability growth model and associated software release time problem after the incorporation of testing coverage and Fault Reduction Factor (FRF) simultaneously. The release planning problem consists of minimization of the development cost subject to the reliability aspiration level. Chapter "Understanding Interactions Among Software Development Attributes and Release Planning Problem Through ISM and MAUT" presents a study to analyze the relationship among different attributes of the software development process and its importance from the customer's perspective. A hierarchical model is described to evaluate the importance of attributes at various levels along with their interrelationships using Interpretive Structural Modeling (ISM). An optimal release problem with multiple objectives is developed using Multi-attribute Utility Theory (MAUT). Chapter "Software Reliability Modeling and Assessment Integrating Time Dependent Fault Reduction Factor in Random Environment" describes a software reliability model with the consideration that the operating environment of software is different from the controlled testing environment and affects software execution and its reliability significantly. It also discusses optimal planning of software release time with cost and reliability criteria. Chapter "Multi-objective Release Time Problem for Modular Software using Fuzzy Analytical Hierarchy Process" deals with optimal release time problem for a modular software system considering the relative importance of the modules. To obtain the modular weights, Fuzzy Analytical Hierarchy Process (FAHP) is used.

The planning and implementation of software development processes go hand in hand with the user's expectations. Chapter "Neutrosophic AHP Approach for Budget Constrained Reliability Allocation Among Modules of Software System" describes a reliability allocation model by integrating analytical hierarchical process and reliability maximization model based on the budget and reliability constraints. The comparisons at different levels of hierarchy are performed under a Neutrosophic environment. Chapter "Testing Resource Allocation for Software System: An Approach Integrating MEMV-OWA and DEMATEL" deals with testing-resource allocation for a modular software system based on DEMATEL (Decision-Making Trial and Evaluation Laboratory). The module weights have been determined by using Maximal Entropy Minimum Variance Ordered Weighted Averaging (MEMV-OWA) method. Chapter "Modeling Allocation Problem for Software with Varied Levels of Fault Severity" describes a resource allocation problem with the objective of maximization of removal of fault content from a modular software. The allocation problem takes into account the different levels of severity of faults and module-wise cost is considered to be dependent on its severity level.

Chapter "Integration of FAHP and COPRAS-G for Software Component Selection" includes a structural decision-making mechanism for selecting Commercial Off-the-Shelf (COTS) components based on multiple attributes. Multiple Criteria Decision-Making (MCDM) techniques have been implemented to determine the critical weights of the attributes using Fuzzy Analytical Hierarchy Process (FAHP) followed by Complex Proportional Assessment of alternatives with Grey Relations (COPRAS-G). In chapter "Estimation and Testing Procedures for the Reliability Functions of Exponentiated Generalized Family of Distributions and a Characterization Based on Records," characterization based on record values for exponentiated generalized family of distributions is provided. Two measures of reliability are considered. Point as well as interval estimates for unknown parameter(s) of reliability based on records are discussed. A number of parameter estimation methods are compared through simulation.

Chapter "Modelling of Non-linear Multi-objective Programming and TOPSIS in Software Quality Assessment Under Picture Fuzzy Framework" presents nonlinear multi-objective programming and TOPSIS-based software quality assessment under picture fuzzy framework. The quality of open-source software system is assessed by taking into account performance, cost-based criteria, and on the basis of feedback gathered from the users and the software experts. Chapter "Requirement Barriers to Implement the Software Projects in Agile Development" deals with the study and analysis of different requirement barriers, which causes problems in agile implementation. The study is based on responses gathered through interviews of developers and testers and presents the roadmap for the software managers to take appropriate steps for effective software implementation. Chapter "Ranking of Multi-release Software Reliability Growth Model Using Weighted Distance-Based Approach" consists of weighted distance-based approach to rank the multi-release SRGMs using the Maximum Deviation Method (MDM) and Distance-Based approach (DBA). The models are ranked based on selection criteria having different priority weights and composite distance values.

This book will be an important resource of information for the postgraduate students, researchers, academicians, and software industry experts. Using optimization modeling, IT consultants, SRE research community, and system analysts can definitely make efficient software performance assessment and related strategic decisions related to software project budget utilization, delivery planning, software versioning to name a few.

We acknowledge Springer for this opportunity and professional support. Importantly, we would like to thank all the chapter authors and reviewers for their time and efforts for this work.

New Dehli, India — Anu G. Aggarwal
New Dehli, India — Abhishek Tandon
Piscataway, USA — Hoang Pham
April 2021

Contents

About the Editors

Dr. Anu G. Aggarwal is Professor in Department of Operational Research, University of Delhi with more than 20 years of teaching and research experience. She is an active researcher with more than 150 research publications in international peer-reviewed journals, book chapters, and international and national conference proceedings. She has guided more than 20 M.Phil and 5 Ph.D. students; served as technical chair, co-chair for a number of international conferences; undertaken a number of research projects funded by government agencies; delivered invited talks at international conferences; and served as resource person at faculty development programs. She has been awarded Young Researcher Award for Applications of Soft Computing Techniques in Software Reliability at 4th International Conference on Quality, Reliability and Infocom Technology (Trends and Future Directions) in 2009. Her key research interests include reliability growth modeling for software systems and applications of soft computing techniques; mathematical modeling and optimization in consumer buying behavior; innovation-diffusion modeling; and multi-criteria decision-making techniques.

Dr. Abhishek Tandon is currently working as a Senior Assistant Professor in the Department of Financial Studies at Shaheed Sukhdev College of Business Studies, University of Delhi, Delhi and is a Visiting Faculty at Department of Business Economics, University of Delhi. He has a teaching, research, and corporate experience of more than 10 years. He holds a Ph.D. Degree in Marketing and Reliability from University of Delhi and has published extensively in journals of repute. His core areas of interests are interdisciplinary research, application of mathematical modeling in marketing, reliability, Internet marketing, mobile commerce, quantitative social science, etc. He was awarded as Best Teacher by the Government of NCT of Delhi in 2016. He was awarded prestigious UGC BSR meritorious students' scholarship by University Grants Commission, Ministry of Education and was also awarded by Society of Reliability, Quality and Operations Management for Exemplary promise and potential in Research in 2009. He has successfully completed a major research project fully funded by Indian Council of Social Science Research, Ministry of Education and two Innovation projects from the University of Delhi.

Dr. Hoang Pham is a Distinguished Professor and former Chairman (2007–2013) of the Department of Industrial and Systems Engineering at Rutgers University, New Jersey. Before joining Rutgers, he was a Senior Engineering Specialist with the Boeing Company and the Idaho National Engineering Laboratory. He has served as Editor-in-Chief, Editor, Associate Editor, Guest Editor, and board member of many journals. He is the Editor of *Springer Book Series in Reliability Engineering* and has served as Conference Chair and Program Chair of over 40 international conferences. He is the author or coauthor of 8 books and has published over 200 journal articles, 100 conference papers, and edited 15 books including *Springer Handbook in Engineering Statistics* and *Handbook in Reliability Engineering*. He has delivered over 40 invited keynote and plenary speeches at many international conferences. His numerous awards include the 2009 IEEE Reliability Society *Engineer of the Year Award.* He is a Fellow of the Institute of Electrical and Electronics Engineers (IEEE) and the Institute of Industrial Engineers (IIE).

Software Reliability Modeling and Methods: A State of the Art Review

Mengmeng Zhu and Hoang Pham

Abstract The increasing dependence of our modern society on software systems has driven the development of software products to be more competitive and time-consuming. At the same time, large-scale software development is still considered a complex, effort-consuming, and expensive activity, given the influence of the transitions in software development, which are the adoption of software product lines, software development globalization, and the adoption of software ecosystems. Hence, the consequences of software failures become costly and even dangerous. In this chapter, we thus review probabilistic software reliability models with different groups. Since nonhomogeneous Poisson process (NHPP) based software reliability models have been successful tools in practical software reliability engineering, this chapter mainly focuses on the review of NHPP based software reliability models that address various concerns in software development practices, such as testing efficiency, testing coverage, multiple fault types, time-delay fault removal, and environmental factors.

Keywords Nonhomogenous Poisson process · Software reliability growth model · State of the art review

1 Introduction

In today's technological world, almost everyone is directly or indirectly in contact with computer software. Computers have been rapidly expanding to a wide array of complex machinery and equipment applied in our everyday safety, security, infrastructure, transportation system, and financial management. Since software product

M. Zhu (✉)
Department of Textile Engineering, Chemistry and Science, North Carolina State University, Raleigh, NC 27606, USA
e-mail: mzhu7@ncsu.edu

H. Pham
Department of Industrial and Systems Engineering, Rutgers University, Piscataway, NJ 08854, USA
e-mail: hopham@rci.rutgers.edu

A. G. Aggarwal et al. (eds.), *Optimization Models in Software Reliability*, Springer Series in Reliability Engineering,
https://doi.org/10.1007/978-3-030-78919-0_1

is extensively involved in various industries and service-based applications, the increasing dependence of our modern society on software-driven systems has led the development of software product to be very competitive and time-consuming (Febrero et al. 2016). Unlike hardware systems, software cannot break down or wear out during its life cycle but can fail or malfunction under certain configurations within specific conditions (Hartz et al. 1997). Hence, the development, measurement, and qualifying of software are challenging yet critical in such a fast-growing technological society.

In 1980, Lehman (1980) summarized the *Laws of Program Evolution*. The first law, *Continuing Change*, expressed the universally observed fact that large programs are never completed. They just continue to evolve until the more cost-effective updated version replaces the systems. The second law, *Increasing Complexity*, could also be viewed as an instance of the second law of thermodynamics. As an evolving program is continually changed, its complexity, reflecting deteriorating structure, increases as well, unless the mission is done, or maintenance is needed. The third law, *The Fundamental Law of Program Evolution*, is subject to dynamics that make the programming process, measures system attributes and collaborative projects, and self-regulating with statistically proven trends and invariances. The fourth law, *Conservation of Organizational Stability (Invariant Work Rate)*, and the fifth law, *Conservation of familiarity*, both lead to the third law. The fourth law more focuses on the steadiness of multiloop self-stabilizing systems. A well-established organization is good at avoiding dramatic change and particularly discontinuities in the increasing growth of an organization. Especially in the past two decades, the complexity of the task that the software system performs has grown dramatically, faster than hardware due to the fast-paced high technology development (Catelani et al. 2011).

A modern software product is prone to include a large number of modules, system components, and Lines of Code (LOC) (Catelani et al. 2011; Han et al. 2012; Chang et al. 2014; Chatterjee et al. 1997; Chatterjee and Singh 2014). The size of a software product is no longer measured in terms of thousands of LOC, but millions of LOC. The latest investigation states that more than 10 Microsoft commercial software products could have more than 600 million LOC (Dang et al. 2011). In view of such a great amount of LOC, the complexity of software product, domain knowledge of programmer/tester, testing methodologies, testing coverage, and testing environment should be all carefully taken into account in software development.

Since the inception of electronic computing in the late 1940s, the development race of the computer industry has led to an unprecedented process (Patterson and Hennessy 2013). Powerful, inexpensive computer workstations replaced the drafting boards of circuits and computer designers. Moreover, an increasing number of design steps were automated. Computer and communication industries have grown into the largest, amongst the new rising industries in the twentieth century (Moravec 1998). Hardware advances have allowed software programmers to create wonderful coding and develop new features and functionality (Moravec 1998). However, there exists uneven progress between software and hardware in the computer revolution in the past few decades. Based on the latest technology review, hardware is leaving software behind. As a matter of fact, software is relied on a less firm foundation, at the same time, carries a larger burden than hardware in operation. Given the current technology in manufacturing and electrical engineering, software has more potentials to allow designers to contemplate more ambitious systems in consideration of a broader multidisciplinary scope (Lyu 1996, 2007).

The nonperformance and failures of a software system are inconvenient, sometimes can lead to severe consequences, especially in the application of aerospace engineering and national defense systems. In March 2015 (COMPUTERWORLD 2020), a software glitch carried in the software package of Lockheed Martin F-35 Joint Strike Fighter aircraft had made the aircraft could not correctly detect the target. The sensor on the plane cannot distinguish the difference between singular and multiple threats. Additionally, different F-35 aircraft provide different detection information even they are aiming at the same threat, which depends on the angles they are aiming at and what their sensors have received. The delivery date had to be postponed as well because of this issue.

Software failures can also cause serious consequences in an automobile. Toyota had to recall almost 2 million Prius hybrid vehicles, in order to fix a software glitch along with its engine control units (ECUs) in February 2014 (COMPUTERWORLD 2020). A malfunction within the car's hybrid drive system caused by a software glitch could, in certain circumstances, cut the system's power and cause the car to an unscheduled halt. A software glitch affecting the ECUs controlling the motor/generator and the hybrid system could put extra thermal stress on certain transistors under certain conditions. The same software issue recurred in July 2015, which has resulted in the recall of 625,000 Prius cars globally. Software failures have affected the healthcare system as well. Emergency services were unavailable for around six hours across seven U.S. states in April 2014 (COMPUTERWORLD 2020). The incident had a major impact on 81 call centers, meaning about 6,000 people who made 911 calls that were not able to connect in these seven states. There is a study announced by the Federal Communications Commission found that the cause of service unavailable was an entirely preventable software error.

The nonperformance and failures of software are expensive. A study carried by the National Institute of Standards & Technology in 2002 found that inadequate infrastructures for fixing software bugs cost the U.S economy $59.5 billion every year. What about the global cost of fixing software bugs every year? This study also estimated that more than a third of software bugs could be eliminated by improving software testing scheduling and methodology (Tassey 2002).

Hence, developing reliable software is a major challenge to the software industry, information technology (IT) industry, and other related industries, which leads to the fact that *Software Reliability Engineering* is popular in both academia and industry. We thus discuss the recent trends in software development and the importance of developing software reliability models in Sect. 2. Indeed, the deterministic and probabilistic software reliability models are reviewed in Sect. 3. Since nonhomogeneous Poisson process (NHPP) based software reliability models have been successful tools in practical software reliability engineering, Sect. 4 thus focuses on the review of NHPP based software reliability models that address various concerns in the software development practices. Section 5 concludes this chapter.

2 Software Reliability Engineering

2.1 Trends in Software Development

Large-scale software development is a very complex, effort-consuming, and expensive activity. Even though many innovations and improvements have been proposed on software architecting systems and development approaches, large-scale software development is still largely unpredictable and error-prone (Bosch and Bosch-Sijtsema 2010).

Bosch and Bosch-Sijtsema (2010) discussed three trends in software development in 2010, which will further accelerate the complexity of large-scale software development. The first trend is the increasing adoption of software product lines. A software product line consists of a software platform shared by a group of products. Each software product can select and configure components in the platform and extend the platform with desirable functionality. At the same time, the platform consists of many components with the associated team. Each team takes charge of one product or several products. Software development is taken place within many teams in the organization. During the development cycle, the interactions and communications among teams are much than traditional software development teams. Some research identified the adoption of software product lines allows 50–75% of development expenses reduction and decreases the defect density if the adoption is successful (Hallsteinsen et al. 2008; Clements and Northrop 2002). However, the adoption of software product lines also brings a new level of dependency on organizations in software development. The second trend is software development globalization. Companies often have multiple software development sites globally or partnered with other remote companies especially located in India and China. There are many advantages in terms of software development, e.g., cycle time reduction, travel cost reduction, fewer communication issues about user experiences, and faster response to customers (Cascio and Shurygailo 2003). Nevertheless, software development globalization also brings challenges given the culture difference, time zone, software engineering maturity in every country, and technical skills between different countries. The third trend is the adoption of software ecosystems. A software ecosystem is defined as a set of businesses functioning as a unit and interacting with a shared market for software and services. There are relationships amongst those units, which are supported by a technological platform, operating through the exchange of information, resources, and artifacts (Messerschmitt and Szyperski 2005; Jansen et al. 2007). Software ecosystem also takes external developers, domain experts, and users. Hence, community-centric collaboration and coordination are very important, which are similar to the adoption of software product lines (the first trend in software development). Thus, the dependencies between components will increase and the complexity of software development will increase accordingly (Bosch and Bosch-Sijtsema 2010).

2.2 *Importance of Software Reliability Model*

Given the complexity of large-scale software development, how we can assure software quality is one of the challenging problems in the industry. One of the fundamental quality characteristics is reliability. It is generally accepted that reliability is the key factor in software quality since it quantifies failures and misbehaviors of the product. As recognized in both industry and academia, reliability is an essential measurement metric for developing a robust and high-quality software product (Febrero et al. 2016; Lyu 1996, 2007). According to the definition given by ANSI/IEEE, software reliability is defined as the probability that a software system can perform its designed function without failure during a specified time on a given set of inputs under defined environments (Lyu 1996).

On the other hand, the increasing complexity and shortened iteration cycle of software products bring in a decrease in average market life expectancy (Tassey 2002). Thus, since the 2000s, there is a great attention shift from hardware development and testing to improve software quality and reliability with the purpose of winning more market share. Moreover, high reliability is desirable if a software company plans to reduce the total cost of software products from an economic point of view. It is undoubtedly that lower reliability software product not only results in the negative impact regarding customer satisfaction but also brings in the additional cost occurred during the operation phase because fixing a software fault in the operation phase costs more resources compared with in-house testing. Since the fixing cost for a software fault in the operation phase is much higher than the in-house testing phase, most organizations try to minimize these expenditures occurred in the operation phase. That is why many high technology organizations need to release multiple versions for a software product instead of fixing software faults in the operation phase, to improve product reliability and introduce new features to improve the user experience.

It is necessary to develop a practical and applicable model that can capture the software failure growth trend, predict the number of failures and software reliability given a specific period of operating time, propose the optimal release time of new products, and schedule the delivery time for the next release based on the predetermined level of reliability. Thus, software reliability models are applied to evaluate software reliability and capture the failure growth trend in the past few decades. There are several ways to measure software reliability. A practical and common one is model software reliability by utilizing the past failure behaviors obtained from the testing phase.

3 Software Reliability Models

Software failure data, collected mostly in the testing phase, are applied to study the behavior of software systems, such as software reliability given a specific time interval, failure growth rate, the number of remaining faults in the system, and the optimal release time. Hence, a great number of studies have been focused on the development of software reliability models in the past four decades with various assumptions, such as testing methodology, testing coverage, fault removal efficiency, fault dependency, and time-delay debugging.

The classification of software reliability models was presented by different researchers (Bastani and Ramamoorthy 1986; Goel 1985; Musa et al. 1990; Mellor 1987). One of the widely utilized classification methodologies categorizes software reliability models into two types: the deterministic models and the probabilistic models (Pham 2000). The deterministic models are used to study: (1) the element of a program by counting the number of distinct operators, operands, errors, and instruction; (2) the control flow of a program by counting the branches; (3) the data flow of a program (data sharing and passing). There are two well-known models: Halstead's software metric (Halstead 1977) and McCabe's cyclomatic complexity metric (McCabe 1976). These models provided an innovative and pioneering quantitative approach to analyze and measure the performance of software systems at that time; however, a random event is not involved, hence, these models are not suitable to apply in modern software. The probabilistic models take into account failure detection and failure removal as probabilistic events during software development. The classification of the probabilistic software reliability models is given by Pham (2000, 2007), Xie (1991): (1) error seeding; (2) failure rate; (3) curve fitting; (4) reliability growth; (5) Markov structure; (6) time series; (7) NHPP.

In Sect. 3.1, we review the research articles regarding the probabilistic software reliability models with groups stated as follows: error seeding, failure rate, curving fitting, reliability growth, Markov structure, and time series. Since NHPP based software reliability models have been successful tools in practical software reliability engineering, this chapter mainly focuses on the review of NHPP based software reliability models. Therefore, Sect. 3.2 introduces the general theory of NHPP and Sect. 4 reviews a great number of NHPP software reliability models with different considerations, such as testing effort, testing coverage, fault removal efficiency, fault dependency, time-delay fault removal, environmental factor on affecting software reliability, and multiple-release software.

3.1 Probabilistic Software Reliability Models

In the error seeding based software reliability model, the number of errors in a program is estimated by applying the multi-stage sampling technique (Mills 1972; Cai 1998; Tohma et al. 1991). Errors are categorized as indigenous errors and induced (seeded) errors. The number of indigenous errors, which is unknown, is estimated from the number of induced errors and the ratio of these two types of errors obtained from software debugging data. We list three models in the group of errors seeding based software reliability models.

Mills (1972) proposed an error seeding method to estimate the number of errors in a program by introducing seeded errors into the program. If the probability of detecting both indigenous errors and induced errors are equal, then the probability of k induced errors in r removed errors follows a hypergeometric distribution, given as

$$P(k; N, n_1, r) = \frac{\binom{n_1}{k}\binom{N}{r-k}}{\binom{N+n_1}{r}}, \quad k = 1, 2, \ldots, r$$

where N is the total number of indigenous errors, n_1 is the total number of induced errors, r is the total number of errors removed during debugging, k is the total number of induced errors in r removed errors, and $r - k$ is the total number of indigenous errors in r removed errors. Given the parameters n_1,r, and k are known, the maximum likelihood estimation (MLE) of N can be shown as (Huang 1984) $\widehat{N} = \lfloor N_0 \rfloor + 1$, in which $N_0 = n_1(r-k)/k - 1$. Note that N_0 and $N_0 + 1$ are both MLEs of N if N_0 is an integer. Tohma et al. (1991) introduced a reliability model based on the hypergeometric distribution to estimate the number of errors in the program. Later, Cai (1998) modified Mills' model by dividing software into two parts, Part 0 and Part 1.

In the failure rate class, these studies (Jelinski and Moranda 1972; Schick and Wolverton 1978; Moranda 1981) focused on how failure rates change at the failure time during the failure intervals. The number of faults in the program is a discrete function, thus, the failure rate of a program is a discrete function as well. The Jelinski-Moranda model (Jelinski and Moranda 1972) is one of the earliest software reliability models, which states the program failure rate at the ith failure interval is given by $\lambda(t_i) = \emptyset[N - (i - 1)],\ i = 1,\ 2, \ldots,\ N$, in which $\emptyset$ is a proportional constant representing the contribution of one fault makes to the overall program, and N is the number of initial faults in the program. The Schick-Wolverton model (Schick and Wolverton 1978) modified the Jelinski-Moranda model by assuming the failure rate at the ith time interval increases with time t_i since the last debugging. Later, Moranda (1981) proposed a reliability model considering the program failure rate function as initially a constant D and decreases geometrically at failure times.

In the curve fitting class (Belady and Lehman 1976; Miller and Sofer 1985), the models use statistical regression analysis to illustrate the relationship amongst software complexity, the number of faults, and failure rate in the software. Linear regression analysis, nonlinear regression analysis, or time series approach is applied between the dependent and independent variables. Estimation of errors, complexity, and failure rate are investigated in the modeling. Belady and Lehman (1976) introduced a model by applying time series approach to estimate software complexity. Miller and Sofer (1985) also proposed a model to estimate software failure rate by assuming the failure rate is a monotonically non-increasing function.

In the reliability growth class (Coutinho 1973; Wall and Ferguson 1977), the improvement of program reliability is measured and predicted via the testing phase by reliability growth models. The failure rate is a function of time or the number of testing cases in this group of models. Coutinho (1973) pointed out that the failure rate is a function of the cumulative number of failures and testing time. Wall and Ferguson (1977) proposed a model that is similar to Weibull model to predict software failure rate during testing.

In the group of Markov structure models (Goel and Okumoto 1979a; Littlewood 1979; Yamada et al. 1998; Goseva-Popstojanova and Trivedi 1999; Dai et al. 2005), the assumption is that the failure of the modules is independent of each other. Goel and Okumoto (1979a) proposed a linear Markov model with imperfect debugging. Meanwhile, they gave the transition probability of the model. Littlewoods (1979) developed a reliability model incorporating the transitions between modules while operating. Two types of failures are considered: failure from each module, modeled as a Poisson failure process, and failure from the interface between modules. Yamada et al. (1998) performed a software safety model to illustrate software's time-dependent behavior using the Markov process. Goseva-Popstojanova and Trivedi (1999) proposed a software reliability modeling framework that can consider the phenomena of failure correlation and further studied its effects on the software reliability measures based on the Markov renewal process. Dai et al. (2005) proposed a software reliability model based on a Markov renewal process for the modeling of the dependence among successive software runs, in which four types of failures are allowed in the general formulation. Meanwhile, the cases of restarting with repair and without repair are considered.

In the time series model group (Chatterjee et al. 1997; Singpurwalla and Soyer 1985; Ho and Xie 1998; Xie and Ho 1999), autoregressive integrated moving average method is applied to study software reliability. Singpurwalla and Soyer (1985) introduced several ramifications into a random coefficient autoregressive process of order 1 to describe software reliability. Besides, several research papers (Chatterjee et al. 1997; Ho and Xie 1998; Xie and Ho 1999) also used time series approach to address software reliability prediction in the testing phase and operation phase.

3.2 General Theory of NHPP

Let $N(t)$ be the cumulative number of software failures by time t. The counting process $\{N(t),\ t \geq 0\}$ is said to be a NHPP with the intensity function $\lambda(t),\ t \geq 0$. The probability of exactly n failures occurring during the time interval $(0,\ t)$ for the NHPP is given by

$$P\{N(t) = n\} = \frac{[m(t)]^n}{n!} e^{-m(t)} \text{ for } n = 0,\ 1,\ 2, \ldots$$

where $m(t) = E[N(t)] = \int_0^t \lambda(s)ds$. $m(t)$ is the expected number of failures up to time t, which is also known as the mean value function (MVF).

Note that the forms of the MVF vary with different assumptions. In NHPP, the stationary assumption is relaxed compared with the Poisson process. In other words, $N(t)$ is Poisson-distributed with a time-dependent failure intensity function $\lambda(t)$, while the Poisson process holds the stationary assumption, $m(t) = \lambda t$.

A general NHPP model includes the following assumptions: (1) the failure process has an independent increment; (2) the failure rate of the process is given by $P\{N(t + \Delta t) - N(t) = 1\} = \lambda(t)\Delta t + o(\Delta t)$; (3) during a small interval Δt, the probability of more than one failure is negligible, that is $P\{N(t + \Delta t) - N(t) \geq 2\} = o(\Delta t)$, in which $o(\Delta t)$ represents a quantity that tends to be zero for a small Δt. The instantaneous failure intensity function $\lambda(t)$ is defined as

$$\lambda(t) = \lim_{\Delta t \to 0} \frac{R(t) - R(\Delta t + t)}{\Delta t R(t)} = \frac{f(t)}{R(t)}$$

where $R(t) = P[N(t) = 0] = e^{-m(t)}$.

The MVF is expressed as $m(t) = \int_0^t \lambda(s)ds$. One of the main objectives of NHPP software reliability model is to derive appropriate $m(t)$. The failure intensity function is equivalent to the derivative of MVF, which is $\lambda(t) = m'(t)$. Different assumption on the fault detection and fault removal process lead to different failure intensity function $\lambda(t)$. Reliability and other related measurements can be obtained by solving the differential equation $m'(t)$. The least-square estimate or MLE are commonly applied to estimate unknown parameters.

Software reliability $R(t)$ is defined as the probability that a software failure does not occur in $(0, t)$, that is

$$R(t) = P[N(t) = 0] = e^{-m(t)}$$

In general, during time interval $(t, t + x)$, software reliability can be described as

$$R(x|t) = P[N(t + x) - N(t) = 0] = e^{-[m(t+x) - m(t)]}$$

4 NHPP Software Reliability Models

NHPP has been successfully applied to model software reliability since Goel and Okumoto (1979b) firstly proposed their innovative model in 1979. Based on the development of Goel-Okumoto model, many NHPP software reliability models have been proposed in the past four decades to address different scenarios in software fault detection and fault correction processes, such as testing coverage, fault removal efficiency, fault dependency, time-delay fault removal, environmental factor on affecting software reliability, and multiple-release software. The failure processes are described by NHPP property with the MVF at time t, $m(t)$, and the failure intensity of the software program, $\lambda(t)$, which is also the derivative of MVF. Most existing NHPP software reliability models are developed based on the model below

$$\frac{dm(t)}{dt} = h(t)[N(t) - m(t)] \tag{1}$$

where $m(t)$ denotes the expected number of software failures by time t, $N(t)$ denotes the total number of fault content by time t, and $h(t)$ denotes the time-dependent fault detection rate per unit of time. The underlying assumption of Eq. (1) is the failure intensity is proportional to the residual fault content in the software. Depending on the model considerations, $N(t)$ and $h(t)$ can be modeled as a constant or a time-dependent function. Given many NHPP software reliability models are developed based on Eq. (1), we thus review these models depending on the focused scenarios in software development process. In the following sections, we rewrite the reviewed models based on the format of Eq. (1), and the coefficients of the reviewed models are nonnegative unless specified. The basic assumptions for the reviewed models are as follows unless specified: (1) software fault detection follows the NHPP; (2) the number of failures detected at any time t is proportional to the remaining faults in the software program; (3) when a failure is detected, the error that caused the failure is immediately removed; (4) all faults in a program are mutually independent from the perspective of failure detection.

4.1 NHPP Exponential Models

The Goel-Okumoto model (Goel and Okumoto 1979b) assumed that the isolated faults are removed prior to future test occasions and no new errors are introduced. The Goel-Okumoto model thus has $h(t) = b$ and $N(t) = a$ in Eq. (1) to obtain the MVF. Musa (1975) proposed a similar model to the Goel-Okumoto model by considering the relationship between execution time and calendar time, which assumed that $h(t) = c/nT$ and $N(t) = a$ in Eq. (1) to obtain the MVF, in which a is the number of failures in the program, c is the testing compression factor, T is the meaning time to failure at the beginning of the test, and n is the total number of possible failure during the maintained life of the program.

Ohba (1984) proposed the hyper-exponential growth model in consideration of different clusters of modules in a program. Each module contains a different initial number of errors and different failure rates, which are all assumed as constants in the software reliability model. It is well-known that the sum of an exponential distribution is a hyper-exponential distribution. Thus, the system software reliability model is more like the summation of each module's reliability model. Similar to the model proposed by Ohba (1984), Yamada and Osaki (1985) also proposed a software reliability model that considered software can be divided into K modules. The probability of faults for each module will be taken into consideration. The fault detection rate is the same within modules; however, it is various between modules. The total number of errors in the software is assumed as a constant and there are no new errors will be introduced during the fault detection process.

4.2 NHPP S-Shaped Models

In the NHPP S-shaped models, software reliability growth curve behaves as S-shaped. The curve crosses the exponential curve from below and the crossing occurs only once (Pham 2007). Time-dependent fault detection rate is applied in modeling software reliability growth trends. The concept of the S-shaped model is proposed to describe the changes of fault detection rate. Fault detection rate can be changed because of the difficulty level to detect different types of faults or the working experience that involves a learning process of software testers. Ohba et al. (1982) discussed a general NHPP model with S-shaped MVF in which $h(t)$ in Eq. (1) is treated as a time-dependent function. Ohba and Yamada (1984) proposed the NHPP model with the S-shaped MVF and considered the cumulative number of detected faults often seems to perform S-shaped. They stated that some of the faults are not detectable before some other faults are removed. The model proposed by Ohba and Yamada (1984) is called the inflection S-shaped model, which assumes $h(t) = b/(1+\beta e^{-bt})$ and $N(t) = a$ in Eq. (1), in which b and β represent failure detection rate and inflection factor, respectively.

Around the same time, Yamada et al. (1983, 1984, 1986) proposed several software reliability models considering the software fault detection process as a learning process. Specifically, when software testers get more familiar with the testing environment, specifications, and requirements, the fault detection rate will be higher. For example, Yamada et al. (1984) proposed a model, called the delayed S-shaped model, which assumed that $h(t) = b^2t/(bt + 1)$ and $N(t) = a$ in Eq. (1), in which b is the error detection rate per error in the steady state. Nakagawa (1994) developed the connective NHPP model with S-curve forms. A group of modules called, main route modules, are tested first, followed by other modules. Even the failure intensity in the main route modules and other modules are similar, the failure growth curve performs as S-curve since the detection starts at different time points. Afterward, S-shaped reliability models are further developed in many studies (Chatterjee et al. 1997; Chatterjee and Singh 2014; Pham 1993; Pham and Zhang 1997; Pham et al. 1999). For example, Pham et al. (1999) developed the PNZ model in consideration of both the imperfect debugging and the learning effects, which assumed $h(t) = b/(1 + \beta e^{-bt})$ and $N(t) = a(1 + \alpha t)$ in Eq. (1), in which faults can be introducted during the debugging process at a constant rate of α, a is the total number of initial faults, and b and β have the same meanings as the model proposed by Ohba and Yamada (1984). Chatterjee and Singh (2014) incorporated a logistic-exponential testing coverage function in developing software reliability model, which assumed that $h(t) = c'(t)/(1 - c(t)) - d(t)$, $c(t) = \left(e^{bt} - 1\right)^k / \left[1 + \left(e^{bt} - 1\right)^k\right]$, $d(t) = \beta/(1 + \beta t)$, and $N(t) = a$ in Eq. (1), in which k is the positive shape parameter, b is a positive scale parameter, $d(t)$ is the fault introduction rate, modeled as a decreasing function of time.

4.3 NHPP Imperfect Debugging Models

NHPP perfect debugging models often assume when a failure occurs, the fault that caused the failure can be immediately removed, and no new faults are introduced (Goel and Okumoto 1979a; Ohba and Yamada 1984; Yamada et al. 1983; Hossain and Dahiya 1993), which means $N(t) = a$. Many models described in NHPP exponential models and S-shaped models are also NHPP perfect debugging models, in which $N(t)$ is modeled as a constant.

The concept of imperfect debugging is based on the assumptions (Pham 2000, 2007): (1) when the detected errors are removed, it is possible to introduce new errors; (2) the probability of finding an error in a program is proportional to the number of remaining errors in the program. Many reliability models are proposed based on NHPP imperfect debugging concept (Pham 2007, 1993; Yamada et al. 1984, 1991, 1992; Pham and Zhang 1997; Pham et al. 1999; Pham and Pham 2000; Inoue and Yamada 2004; Jones 1996; Kapur et al. 2007, 2011; Teng and Pham 2006; Tokuno and Yamada 2000; Fang and Yeh 2016; Xie and Yang 2003; Pham and Normann 1997). In the 1990s, Yamada et al. (1992) proposed two imperfect debugging models considering two types of fault content functions $N(t)$, which are $N(t) = \alpha e^{\beta t}$ and $N(t) = \alpha(1 + \gamma t)$, respectively, in which α is the number of initial fault content in the program prior to software testing, β and γ are the increasing rates of the number of the introduced faults to the program.

Software reliability models can belong to multiple categories, such as perfect debugging, imperfect debugging, S-shaped, exponential, testing effort, testing coverage, fault dependency, environmental factors, and software multiple-release. As an example, we review models that belong to both the categories of NHPP imperfect debugging and S-shaped models. S-shaped models were initially proposed to focus on the change of fault detection rate considering the difficulty level of detecting different types of software faults and the efficiency of detecting faults based on software testers' learning process, we therefore name the models that belong to both the categories of NHPP imperfect debugging and S-shaped models as NHPP imperfect debugging fault detection (IDFD) model. Besides the general assumptions of NHPP software reliability models, the generalized NHPP IDFD models also include the following assumptions: (1) the error detection rate differs among faults; (2) new faults are introduced during debugging.

Pham and Normann (1997) provided a generalized solution of Eq. (1). Specific MVF can be obtained by substituting different fault detection functions. The PNZ model (Pham et al. 1999) described in Sect. 4.2 also belongs to the category of the NHPP IDFD model. Moreover, Pham and Zhang (1997) proposed a model with $N(t) = c + a(1 - e^{-\alpha t})$ and $h(t) = b/(1 + \beta e^{-bt})$ in Eq. (1). Pham (2000, 2007) proposed a model by considering the fault introduction rate is an exponential function of the testing time, and the error detection rate follows a learning process, which assumed that $N(t) = \alpha e^{\beta t}$ and $b(t) = b/(1 + ce^{-bt})$ in Eq. (1). Later, Kapur et al. (2011) proposed two general frameworks for developing NHPP software reliability model in the presence of imperfect debugging and error generation. The first framework was formulated based on the assumption that there is no differentiation between failure observation and fault removal process. $h(t)$ and $N(t)$ in Eq. (1) are modeled as $h(t) = pF'(t)/(1 - F(t))$ and $N(t) = A + \alpha m(t)$, respectively, in which p is the probability of perfect debugging, $F(t)$ is the failure time distribution, A is the initial number of faults, and α is a constant fault introduction rate. The second framework is thus extended based on the assumption that there is a differentiation between failure observation and fault removal process.

4.4 NHPP Software Reliability Models on Software Testing

The common way to improve software reliability is to focus on in-house testing. Myers et al. (2011) defined software testing as a process of executing a program with the intent of finding errors. There are two fundamental rules in software testing. Firstly, it is intended to detect as many faults as possible during the in-house testing phase and remove the detected faults from the software system. Secondly, software failure data will be collected to predict system reliability, estimate the remaining faults, and schedule the product delivery date.

Owing to the fact that software debugging, testing, and verification are accounted for 50–70% of a software product's development cost. Indeed, software testing is always defined as a difficult and expensive section in software development (Ohmann and Liblit 2017; Hailpern and Santhanam 2002). Software debugging cost even goes higher if debugging is carried out in the operation phase. In practice, it is unlikely to release bug-free software products owning to its natural characteristics. Post-deployment failures are inevitable in complex software.

It is generally accepted that the longer time spent on software testing, the fewer faults that software will carry and the more reliable the software will be. However, this is not a practical approach. Exhaustive testing to execute all possible inputs unlikely to happen since too many possible combinations result in little improvement in system reliability (Weyuker 2004; Kaner et al. 2000). Moreover, full execution tracing is usually impractical for complex software programs due to the limitation of cost and resources (Ohmann and Liblit 2017). Furthermore, after software reaches a certain level of refinement, any further effort on removing faults will cause an exponentially increase in the total development cost but not much increase in reliability assessment (Pham and Zhang 1999a, b). Thus, how to test software efficiently and meet the predetermined reliability is a challenging task for both researchers and practitioners. In this section, we review NHPP software reliability models considering different scenarios in software testing, including testing coverage, testing efficiency, testing effort, time-delay fault removal, and multiple fault types.

Testing Coverage Models—Software systems have been widely applied in numerous safety-critical domains; however, large-scale software development is still considered a complicated and expensive activity. Since the testing phase plays an essential role in software development, a great number of software reliability models focus on the specific scenarios in the software development process, such as testing coverage, testing efficiency, testing resource allocation, and so on. Testing coverage is a measure that enables software developers to evaluate the quality of the tested software and determine how much additional effort is needed to improve the quality and reliability (Pham 2007). At the same time, the information on testing coverage can provide customers with a quantitative confidence criterion for software products. Pham and Zhang (2003) thus introduced a generalized model incorporating the measurement of testing coverage into software reliability assessment, in which $h(t) = c'(t)/(1 - c(t))$ in Eq. (1). This model indicates that the failure intensity depends on both the rate, the coverage rate *c'(t)*, and the percentage of the code that has not yet been covered by testing by time *t*, expressed as *1-c(t)*. Note that different functions of $N(t)$ and $c(t)$ can be plugged into Eq. (1) to obtain the MVF, given the formula of $h(t)$. One of the examples for the expressions of $c(t)$ and $N(t)$ is $c(t) = 1 - (1 + bt)e^{-bt}$ and $N(t) = a(1 + \alpha t)$. The model developed in Chatterjee and Singh (2014), reviewed in Sect. 4.2, is based on the model developed in Pham and Zhang (2003) by considering the fault introduction rate into $h(t)$, expressed as $h(t) = c'(t)/(1 - c(t)) - d(t)$, in which $d(t)$ is the fault introduction rate.

Inoue and Yamada (2004) proposed an alternative evaluation metric for the testing coverage in their study and further proposed a software reliability model by formulating the relationship between the alternative testing coverage evaluation function and the number of detected faults. The testing coverage measures are classified into several types, such as statement coverage, branch coverage, and path coverage. The measure of testing coverage is defined as the proportion of the number of statements that have been executed in the total number of statements. The software reliability proposed by Inoue and Yamada (2004) assumed that $h(t) = sc(t)$ and $N(t) = a$ in Eq. (1), in which $c(t) \equiv dC(t)/dt, C(t) = \alpha(1-e^{-b_{sta}t})/(1+ze^{-b_{sta}t}), z = (1-r)/r$, $r = b_{ini}/b_{sta}$, α is the target value of testing coverage to be attained, b_{ini} is the initial testing skill factor of the test case designers, and b_{sta} is the steady-state testing skill factor. Later, Li et al. (2008) incorporated logistic testing coverage function to develop a software reliability model. The time-varying test coverage function is expressed as $C(t) = C_{max}/(1 + Ae^{-at})$, in which C_{max} is the ultimate testing coverage that can be achieved by testing, a is the parameter of testing coverage increasing rate, and A is a constant. The proposed reliability model assumed that $N(t) = N$ and $h(t) = C'(t)/(1 - C(t))$ in Eq. (1).

Testing Efficiency Models—Section 4.3 reviewed software reliability models that addressed *new faults are introduced into debugging* based on the concept of imperfect debugging. Moreover, imperfect debugging can also be understood as the detected faults are removed at a certain rate instead of 100%. Jones (1996) stated that the faults removal efficiency (FRE) is an important factor in software quality and process management, which can provide software developers with the estimation of testing effectiveness and the prediction of additional effort. Note that FRE usually ranges from 15 to 50% for unit test, 25–40% for integration test, and 25–55% for system test (Pham 2007). Most software reliability models (Pham 2007) assumed that the detected faults are removed 100%. However, fault removal is not always 100% in practice. Some represented models are reviewed below. Zhang et al. (2003) proposed a generalized software reliability model based on imperfect debugging considering new faults can be introduced while debugging and the detected faults may not be removed completely. They defined the FRE in the study, which presented a new idea for later research. The FRE is defined as the percentage of bugs eliminated by reviews, inspections, and tests. Incorporating FRE into software reliability analysis will not only improve the prediction accuracy of software metrics but also define a tangible and quantifiable factor. The model proposed in Zhang et al. (2003) is expressed as $dm(t)/dt = b(t)[a(t) - pm(t)]$ and $da(t)/dt = \beta(t)dm(t)/dt$, in which p is the FRE, which means p percentage of detected faults can be completely eliminated during the debugging. Note that Zhang et al. (2003) provides a general solution for their proposed model and a specific solution with $b(t) = c/(1 + \alpha e^{-bt})$ and $\beta(t) = \beta$.

Kapur et al. (2007) proposed a software reliability model that incorporates testing efficiency regarding testing efforts in the testing phase and usage function in the testing phase. They (Kapur et al. 2007) assumed that: (1) when a software failure occurs, an instantaneous repair effort starts with the fault content is reduced by one with probability p and remains unchanged with probability $1 - p$; (2) the number of failures during the operation phase is dependent upon the usage function. Thus, their proposed model considers $h(t) = \left[pb/(1 + \beta e^{-bW(t)})\right]dW(t)/dt$ and $N(t) = a + \alpha m(t)$ in Eq. (1), in which $W(t)$ represents the cumulative testing effort in the time interval $(0, t]$. Base on the model proposed in Zhang et al. (2003), Li and Pham (2017) further proposed a software reliability model by incorporating FRE with $dm(t)/dt = h(t)[N(t) - pm(t)]$, $h(t) = \beta c'(t)/(1 - c(t))$, and $N(t) = a + \alpha m(t)$, in which β is proportionality constant and p is the FRE (same meaning as defined in Zhang et al. (2003)). Later, Zhu and Pham (2016) proposed a new way to formulate a software reliability model that addresses non-removed errors due to the experience of software testers, expressed as $dm(t)/dt = b(t)m(t)[1 - m(t)/L] - c(t)m(t)$, in which $b(t)$ is the fault detection rate per unit of time, L is the maximum number of faults existed in the program, and $c(t)$ is the non-removed error rate per unit of time. Zhu and Pham (2016) provided a general solution for the proposed model and a specific model with $b(t) = b/(1 + \beta e^{-bt})$, and $c(t) = c$.

Testing Effort Models—Yamada et al. (1991) proposed a software reliability model by using exponential and Rayleigh curves to describe the behavior of the amount of test effort spent on software testing. The proposed model (Yamada et al. 1991) is expressed as $dm(t)/dt = rw(t)[a - m(t)],\ 0 < r < 1$, and $w(t) = \alpha\beta e^{-\beta t}$ or $w(t) = \alpha\beta t e^{-\beta t^2/2}$, in which $w(t)$ is the test effort function representing the current test resource expenditures at testing time t, α and β are the coefficients associated with exponential and Rayleigh function. Huang and Kuo (2002) investigated a software reliability model based on the NHPP by incorporating a logistic testing effort function. They used the same base model (Yamada et al. 1991) and further proposed a new logistic testing effort function $w(t) = NA\alpha e^{-\alpha t}/\left[1 + Ae^{-\alpha t}\right]^2$, in which N is the total testing effort eventually consumed, α is the consumption rate of testing effort expenditures in the logistic testing effort function, and A is a constant parameter.

Huang and Lyu (2005) studied the impact of the testing effort and testing efficiency on modeling software reliability and the cost for optimal release time. The model proposed in Huang and Lyu (2005) used the same basic model (Yamada et al. 1991) and further considered a generalized logistic testing effort function $w(t) = N/\left[((k+1)/\beta)/(1 + Ae^{-\alpha kt})\right]^{1/k}$, in which N is the total amount of testing effort eventually consumed, k is the structuring index whose value is larger for better-structured software development efforts, A is the constant parameter in the logistic testing effort function, β is a normalized constant, and α is the consumption rate of testing effort expenditures in the logistic testing effort function. Huang (2005) further proposed a software reliability model incorporating the testing effort function in Huang and Lyu (2005) and the concept of change-point. Later, Lin and Huang (2008) incorporated the concept of multiple change-points into Weibull-type testing effort functions to propose a new software reliability model. Peng et al. (2014) proposed software reliability models in terms of fault detection process and fault correction process by incorporating testing effort function and imperfect debugging. Peng et al. (2014) assumed that $h(t) = b(t)w(t)$ for fault detection process, in which $b(t)$ is the fault detection rate per unit of testing effort at time t and $w(t)$ is the current testing effort expenditure at time t, and provided a general solution. Peng et al. (2014) also proposed the MVF for fault correction process with debugging delay $m(t) = \int_0^t \lambda_d(y)F(w(t) - W(y))dy$, in which $F(W(t) - W(y))$ is the probability that the fault detected at time y is corrected before time t, and $\lambda_d(t)$ is the fault intensity function of the fault detection process.

Time-delay Fault Removal Models—Time-delay fault removal models are also discussed in many studies. Xie and Zhao (1992) generalized Schneidewind's model by assuming a continuous time-dependent delay function which quantifies the expected delay in correcting the detected faults. Delay is treated as an increasing function of time t. The faults are easy to be corrected in the early stage of testing and become difficult to detect as time goes by. The MVF, $m_d(t)$, proposed in the fault detection process, is similar with Goel-Okumoto model, which is $m_d(t) = (\alpha/\beta)(1 - e^{-\beta t})$. The MVF, $m_c(t)$, proposed in the fault correct process is formulated as $m_c(t) = m_d(t - \Delta t),\ t \geq \Delta_t$.

Hwang and Pham (2009) developed a generalized NHPP software reliability model by considering quasi-renewal time-delay fault removal. They assumed that: (1) time-delay is defined as the interval between fault detection and fault removal; (2) time-delay is considered as a time-dependent function, described by a quasi-renewal process with parameter α and the first interarrival time s_1. This model provides a more relaxed assumption in software testing and debugging, which is very close to the practical testing and debugging process. Note that the testing resource allocation during the testing phase, which is usually depicted by the testing effort function, is affected not only by the fault detection rate but also the time to correct a detected fault. Moreover, Peng et al. (2014) not only incorporated the testing effort function and fault introduction into the fault detection process but also considered the debugging delay in the fault correction process.

Multiple Fault Types Models—Many studies stated that there exists more than one type of software fault in the program (Grottke et al. 2010; Laprie et al. 1990; Avizienis 1985; Grottke and Trivedi 2005, 2007; Shetti 2003; Alonso et al. 2013; Yazdanbakhsh et al. 2016; Deswarte et al. 1998; Laprie 1995). Different fault classes are categorized by practitioners and researchers to describe the characteristics of software faults that cause failures during the testing and operation phase (Grottke et al. 2010; Laprie et al. 1990; Yazdanbakhsh et al. 2016; Deswarte et al. 1998; Laprie 1995). The limits and challenges in the dependability of computer systems in terms of the fault class, such as physical faults, design faults, and interaction faults, are discussed in Yazdanbakhsh et al. (2016), Deswarte et al. (1998) as well. Ohba (1984) discussed two types of software faults, mutually independent faults, and mutually dependent faults. Tokuno and Yamada (2000) proposed an imperfect debugging software reliability model with two types of software failures involved. The first type is caused by the fault latent in the system, which is described by a geometrically decreasing function; the second type fault is randomly regenerated in the testing phase, which has a constant hazard rate. Lyu (1996) divided software failures into four groups according to the severity including catastrophic failure (a failure that may cause death or mission loss), critical failure (a failure that may cause severe injury or major system damage), marginal failure (a failure that may cause minor injury or degradation in mission performance), and minor failure (a failure that does not cause injury or system damage but may result in system failure and unscheduled maintenance).

Kapur and Younes (1995) considered the leading error and dependent error in model development. The expressions of $N(t)$ and $h(t)$ in Eq. (1) for the MVF, $m_1(t)$, for the leading error is written as $N(t) = q_1$ and $h(t) = b$. The expressions of $N(t)$ and $h(t)$ in Eq. (1) for the MVF, $m_2(t)$, for the dependent error is written as $N(t) = q_1$ and $h(t) = cm_1(t-T)/q$, in which c is the dependent error removal rate, T is the time-delay between the removal of the leading errors and the removal of the dependent errors, and $m_1(t-T)/q$ represents the ratio of the leading error removed to the initial error content at time t. Note that the cumulative number of software failures detected and removed by time t will be the summation of $m_1(t)$ and $m_2(t)$.

Pham (1996), Pham and Deng (2003) also studied multiple failure types with different detection rates. Three different types of errors are defined in Pham (1996), Pham and Deng (2003) including critical errors, which are very difficult to detect and remove; major errors, which are difficult to detect and remove; minor errors, which are easy to detect and remove. The $N(t)$ and $h(t)$ in Eq. (1) defined in Pham (1996) are $N(t) = n_i(t), h(t) = b_i$, and Eq. (1) is reformulated as $dn_i(t)/dt = \beta_i dm_i(t)/dt$, in which i represents different types of errors defined and β_i represents the type i error introduction rate that satisfies $0 \leq \beta_i \leq 1$. Later, Pham and Deng (2003) further considered the expression of $h(t)$ can be modeled as a non-decreasing S-shaped model, in which $h(t) = b_i/(1+\theta_i e^{-b_i t})$. Huang and Lin (2006) incorporated fault dependence and delay debugging in the software reliability growth model. Huang and Lin (2006) considered all detected faults can be categorized as either leading faults or dependent faults. For the leading faults, the expressions of $N(t)$ and $h(t)$ in Eq. (1) are $N(t) = a_1$ and $h(t) = r$, in which a_1 is the total number of leading faults. For the dependent faults, the expressions of $N(t)$ and $h(t)$ in Eq. (1) are $N(t) = a_2$ and $h(t) = \theta m_1(t - \varphi(t))/a$, in which $m_1(t)$ is the expected number of leading errors, θ is the fault detection rate of dependent faults, a is the total number of initial faults, and $\varphi(t)$ is the delay-effect factor.

Grottke et al. (2010) studied the proportion of the various fault types including Bohrbugs, non-aging-related Mandel bugs, aging-related bugs, and unknown bugs and their evolvement with time based on the fault discovered in the onboard software for 18 JPL/NASA space missions. However, they did not provide a quantitative way to estimate the number of faults. Zhu and Pham (2017a) proposed a new NHPP software reliability model by considering software fault dependency and imperfect fault removal. Two types of software faults are defined, Type I (independent) fault and Type II (dependent) fault, based on the consideration of fault dependency. The assumptions in Zhu and Pham (2017a) are: (1) Type I fault is detected and removed in Phase I. Type II fault is detected and removed in Phase II. The un-removed Type I faults from Phase I are still not able to detect in Phase II; (2) in both phases, there exists a certain portion of software faults that the software development team is not able to remove. In Phase I, the MVF of Type I fault is expressed as $dm_1(t)/dt = b_1(t)[a_1(t) - m_1(t)] - c_1(t)m_1(t),\ t \leq t_0$, in which $a_1(t)$ is total software content, $b_1(t)$ is Type I fault detection rate, $c_1(t)$ is the non-removable fault rate in Phase I. In Phase II, the MVF of Type II fault is expressed as $dm_2(t)/dt = (b_2(t)/a_2(t))m_2(t)[a_2(t) - m_2(t)] - c_2(t)m_2(t),\ t > t_0$, in which $a_2(t)$ is total software content in Phase II, $b_2(t)$ is Type II fault detection rate, and $c_2(t)$ is the non-removable fault rate in Phase II.

4.5 *NHPP Multiple-Release Software Reliability Models*

As software development moves further away from the rigid and monolithic model, the importance of software multiple-release is brought to the vanguard. It is unlikely to deliver all features that customers wanted in the single release because of the limited budget, unavailable resources, estimated risk, and constrained working schedules. Staying competitive in the market and keeping profitable for a software product is difficult with having only a single release especially when rival releases a new release carrying more attractive features and satisfying more customer requirements (Saliu and Ruhe 2005). As a result of multiple releases planning, software organization will have more competitive and overwhelming advantages to balance the competing stakeholder's demands and benefits according to the available resource (Ruhe and Momoh 2005; Svahnberg et al. 2010). On the other hand, a large software system continually desires to align with the changing customer requirements for the sake of market share. In order to obtain feedback from users, figure out what customers really look for, and assign a lower software development cost, a certain portion of increments on the requirements for a multiple-release product is essential for the growth of an organization (Maurice et al. 2006; Greer and Ruhe 2004; Missbauer 2002). Thus, software organization needs to modify parts of the existing modules to extend the current functionality, usability, and understandability by adding new features and correcting the problems from the previous releases (Al-Emran and Pfahl 2007; Gorschek and Davis 2008). Additionally, agile software development is getting more attention in recent years. Agile is an iterative and team-based approach, which emphasizes the rapid delivery of an application in complete functional components (Lotz 2018). The wide adoption of agile methodology also promotes software multiple-release.

Furthermore, most software products are not introduced into the market with full capacities at their initial release. New features will be added and existing features will be enhanced after launched software for a while. Hence, software multiple-release is critical to keeping a software product stays competitive in the market. Modeling and predicting software failure behaviors for single-release software systems have been extensively studied in the past few decades. However, only a few researchers studied multiple-release software reliability and introduced prediction models to explain software fault detection process and fault removal process for multiple-release software.

Garmabaki et al. (2011) incorporated different severities level used to describe the difficulty of correcting faults in the upgrade process to develop a multi up-gradation software reliability model. Faults are classified into two categories, simple fault, and hard fault. The fault removal for the development of the new release depends on the fault from the previous releases and the fault generated in that release. Hu et al. (2011) considered the effect of multiple releases regarding the fault detection process in software development. They assumed that there is no gap between the release of the previous version and the development of the next version. In this work, in order to study the effects of multiple releases on the fault dynamics during the whole software development, they considered a scenario where a software development team develops, tests, and releases software version by version. The field test of each version continues after its release so that faults can continue to be detected and corrected until the next version is ready to be tested. In case a fault is detected in the field test of any version, it will be reported and corrected in both the current version and its subsequent version.

Kapur et al. (2012) introduced the combined effect of schedule pressure and resource limitations by the use of the Cobb–Douglas production function in software reliability modeling. The Cobb–Douglas function illustrates the total production output can be obtained by the amount of labor input, capital input, and total factor productivity. An optimal release planning problem is formulated in this study for software with multiple releases with the solution obtained by applying the genetic algorithm method. Yang et al. (2016) incorporated fault detection and fault correction process in multiple-release software reliability modeling. They considered there is a time-delay in fault repair after detecting faults. The time-delay function is explained by an exponential function or a gamma function. They also assumed the faults in a new version including both the undetected faults from the last version and the newly introduced faults during the development process of the new version. Pachauri et al. (2015) proposed a software reliability growth model by considering fault reduction factor (FRF) and extended this idea to multiple-release software systems. FRF is defined as the ratio of the total number of reduced faults to the total number of failures. FRF is not a constant, which can be affected by other factors, such as resources allocation.

The multiple-release software reliability models reviewed above mainly focused on obtaining optimal release time by optimizing the software cost model without considering the dependent relationship of software faults generated from different releases. Therefore, Zhu and Pham (2017b) focused on the development of a multiple-release software reliability model considering the remaining software faults from the previous releases and the newly introduced faults resulting from the newly introduced features in the development of the next release. Additionally, the dependent fault detection process is taken into account in this research. In particular, the detection of a new software fault for developing the next release depends on the detection of the remaining faults from the previous releases and the detection of the newly introduced faults. They further discussed the behaviors of the proposed software reliability model through mathematical proofs.

4.6 NHPP Environmental Factor Based Software Reliability Models

Software development process has gone through a great change during the past one and half decades. The rise of the Internet had led to rapid growth in the demand for international information display and email systems on the World Wide Web. Software programmers are required to handle various illustrations, maps, photographs, and other images, plus simple animations at a rate we have never seen before. The high technology has an ever-increasing impact on daily life, which drives the software release cycle to become shorter than before, for instance, many companies have shortened their software release cycle from traditional 18 months to 3 months, in order to respond to the fast-changing and competitive market (HP Applications Handbook 2012; Khomh et al. 2012). Moreover, as high technology gets more involved in our everyday life, there are a wide variety of computational devices like mobile phones, tablet PCs, laptops, desktops, and notebooks (Gallud et al. 2012), which also brings more challenges to software developers, such as application maintenance, device consistency, and dynamic version settings (Eisenstein et al. 2001). Customers also have more requirements on the specific design and functionality of the software product. A user-friendly interface, involved in the interaction amongst users, designers, hardware systems, and software systems, has been emphasized to a great extent nowadays. Furthermore, for practitioners and researchers, programming skills, programming language, domain knowledge, and even the programmer organization and team size are different compared with a decade ago. Finally, software development is distributed across multiple locations as the development of globalization (Ramasubbu and Balan 2007). However, such cross-site work patterns may take a much longer time and require much more effort, even though the work size and complexity are similar (Herbsleb et al. 2000, 2001; Herbsleb and Mockus 2003).

Given the current trends of the software development process, which are the adoption of software product lines, software development globalization, and the establishment of software ecosystems, the complicated and human-centered software development process needs to be addressed more appropriately. Meanwhile, environmental factors play significant impacts on affecting software reliability during the software development process (Zhang and Pham 2000; Zhang et al. 2001; Zhu et al. 2015; Zhu and Pham 2017c; Misra et al. 2009; Chow and Cao 2008; Clarke and O'Connor 2012; Sawyer and Guinan 1998; Roberts et al. 1998). Indeed, how to define and incorporate single/multiple environmental factors that present a significant impact on reliability into the software reliability model is critical to address modern software development in practice.

Environmental Factors in Software Development—Although no general definition has been given to defining what are the environmental factors affecting software reliability during the software development process, there have been many related works that defined different types of factors in software development from various perspectives.

Zhang and Pham (2000) defined 32 environmental factors and characterized the impacts of these environmental factors affecting software reliability during the software development process for single-release software. These 32 environmental factors are defined from the four phases of software development, general information, and the interaction with hardware systems. Software development is divided into four phases in this study: analysis phase, design phase, coding phase, and testing phase. The authors conducted a survey investigation and obtained empirically quantitative and qualitative data from managers, software engineers, designers, programmers, and testers, who participated in software development practices. This study also identified the important environmental factors in software development and analyzed the correlations between these environmental factors. Later, Zhang et al. (2001) provided an exploratory analysis to further analyze the detailed relationships of these environmental factors. Zhu et al. (2015) revisited these 32 environmental factors defined in Zhang and Pham (2000) and analyzed their impacts on software reliability during software development based on a current survey distributed to software development practitioners. As the application of agile development and the increasing popularity of multiple-release software products in many organizations, Zhu and Pham (2017c) further conducted another study to investigate the impact level of these 32 environmental factors on affecting software reliability in the development of multiple-release software to provide a sound and concise guidance to software practitioners and researchers.

Sawyer and Guinan (1998) presented the effects on software development performance that depend on the production method of software development and the social process of how people work together in the software development environment. Roberts et al. (1998) proposed five factors that are essential to implement a system development methodology, including organizational system development methodology transition, functional management involvement/support, use of models, and external support. Chow and Cao (2008) collected the survey data from 109 agile projects from a diverse group of organizations with different sizes, industries, and geographic locations to provide empirical information for the statistical analysis. Based on the multiple regression analysis, the critical success factors are identified as a correct delivery strategy, a proper practice of agile software engineering techniques, and a high-caliber team. Three other factors that could be critical to certain success dimensions are identified as a good agile project management process, an agile-friendly team environment, and strong customer involvement.

Misra et al. (2009) conducted a large-scale survey-based study to identify the success factors from the perspective of agile software development practitioners who have successfully adopted agile software development in their projects. This study identified nine out of the fourteen hypothesized factors that have statistically significant relationships with "success". The important success factors are customer satisfaction, customer collaboration, customer commitment, decision time, corporate culture, control, personal characteristics, societal culture, and training and learning. Clarke and O'Connor (2012) researched the situational factors affecting the software development process. Rigorous data coding techniques from Grounded Theory have been applied in this study. They concluded that the resulting reference framework of situational factors consists of eight classifications and 44 factors that inform the software process. On the other hand, this framework also provides useful information for practitioners who are challenged with defining and maintaining the software development process.

Environmental Factor based Software Reliability Models—Only a few studies incorporated environmental factors, the random effect of the testing/operating environments, or other factors, such as FRF that could be influenced by many environmental factors, to develop software reliability models.

Teng and Pham (2006) presented a new methodology for predicting software reliability in the field environment. A generalized random field environment (RFE) software reliability model which can cover both the testing phase and operating phase is proposed in this study by assuming all the random effect in the field environments can be captured by a unit-free environmental factor. Two specific RFE software reliability models are developed by the use of the generalized RFE software reliability model, called the γ-RFE model and the β-RFE model, to describe different random effects in the operation phase. Hsu et al. (2011) integrated the FRF into software reliability models. The FRF is proposed by Musa (1975), which is generally defined as the ratio of net fault reduction to failure experience (Musa 1980; Musa et al. 1987), which could be influenced by many environmental factors, such as fault dependency, human learning process, imperfect debugging, and delay debugging. The authors firstly studied the trend of the FRF and considered it as a time-variable function, and then incorporated the FRF in software reliability growth modeling to improve the accuracy of failure prediction. Pachauri et al. (2015) also considered the impact of FRF in developing a software reliability growth model. Pham (2014) incorporated the uncertainty of the operating environments into a software Vtub-shaped fault detection rate model. In particular, the fault detection rate in this study is represented by a Vtub-shape function, and the uncertainty of the operating environments is represented by a random variable, modeled as a gamma distribution.

With the recent investigations of the significance of environmental factors in software development, Zhu and Pham (2018) incorporated one of the top 10 significant environmental factors from Zhu et al. (2015), Zhu and Pham (2017c), Percentage of Reused Modules (PoRM), to be a random variable which has a random effect on fault detection rate. This study introduced the Martingale framework, specifically, Brownian motion and white noise process into the stochastic fault detection process, which is used to model the impact resulting from the randomness of PoRM. They further proposed a single-environmental-factor software reliability model considering the gamma-distributed PoRM and the randomness associated with PoRM. Later, considering the significance of the impacts from multiple environmental factors (Zhu et al. 2015; Zhu and Pham 2017c), Zhu and Pham (2020) proposed a generalized software reliability model with multiple environmental factors and the associated randomness under the Martingale framework. The randomness is reflected in the process of detecting software faults. Indeed, this is a stochastic fault detection process. Software practitioners and researchers are able to obtain a specific multiple-environmental-factors software reliability model according to the individual application environments from the proposed generalized multiple-environmental-factors software reliability model.

5 Conclusion

Given our modern societies are increasingly dependent on software systems, such as transportation networks, smart grids, and healthcare systems, software systems malfunction can result in cascading failures. Meanwhile, large-scale software development is still a complex, effort-consuming, and expensive activity. The consequences of software failures thus become costly and even dangerous. In this chapter, we review probabilistic software reliability models with different groups. Considering the wide adoption of NHPP based software reliability models in practical software reliability engineering, this chapter mainly focuses on the review of NHPP based software reliability models that address various concerns in software development practices, such as testing efficiency, testing coverage, multiple fault types, time-delay fault removal, and environmental factors.

References

Al-Emran A, Pfahl D (2007) Operational planning, re-planning and risk analysis for software releases. In: International conference on product focused software process improvement. Springer, pp 315–329

Alonso J, Grottke M, Nikora AP, Trivedi KS (2013) An empirical investigation of fault repairs and mitigations in space mission system software. In: Proceedings of 2013 43rd annual IEEE/IFIP international conference on dependable systems and networks (DSN). IEEE, pp 1–8

Avizienis A (1985) The N-version approach to fault-tolerant software. IEEE Trans Softw Eng 12:1491–1501

Bastani FB, Ramamoorthy CV (1986) Input-domain-based models for estimating the correctness of process control programs. Reliab Theory 321–378

Belady LA, Lehman MM (1976) A model of large program development. IBM Syst J 15(3):225–252

Bosch J, Bosch-Sijtsema P (2010) From integration to composition: on the impact of software product lines, global development and ecosystems. J Syst Softw 83(1):67–76

Cai KY (1998) On estimating the number of defects remaining in software. J Syst Softw 40(2):93–114

Cascio WF, Shurygailo S (2003) E-leadership and virtual teams. Organ Dyn 31(4):362–376

Catelani M, Ciani L, Scarano VL, Bacioccola A (2011) Software automated testing: a solution to maximize the test plan coverage and to increase software reliability and quality in use. Comput Stand Interfaces 33(2):152–158

Chang IH, Pham H, Lee SW, Song KY (2014) A testing-coverage software reliability model with the uncertainty of operating environments. International Journal of Systems Science: Operations & Logistics 1(4):220–227

Chatterjee S, Singh JB (2014) A NHPP based software reliability model and optimal release policy with logistic–exponential test coverage under imperfect debugging. Int J Syst Assur Eng Manag 5(3):399–406

Chatterjee S, Misra RB, Alam SS (1997) Prediction of software reliability using an auto regressive process. Int J Syst Sci 28(2):211–216

Chow T, Cao DB (2008) A survey study of critical success factors in agile software projects. J Syst Softw 81(6):961–971

Clarke P, O'Connor RV (2012) The situational factors that affect the software development process: towards a comprehensive reference framework. Inf Softw Technol 54(5):433–447

Clements P, Northrop L (2002) Software product lines. Addison-Wesley

COMPUTERWORLD (2020) Top software failures in recent history. https://www.computerworlduk.com/galleries/infrastructure/top-software-failures-recent-history-3599618/#22

Coutinho JDS (1973) Software reliability growth. In: Proceedings of the IEEE symposium on computer software reliability, pp 58–64

Dai YS, Xie M, Poh KL (2005) Modeling and analysis of correlated software failures of multiple types. IEEE Trans Reliab 54(1):100–106

Dang Y, Ge S, Huang R, Zhang D (2011) Code clone detection experience at Microsoft. In: Proceedings of the 5th international workshop on software clones. ACM, pp 63–64

Deswarte Y, Kanoun K, Laprie JC (1998) Diversity against accidental and deliberate faults. In: Proceedings of the computer security, dependability and assurance: from needs to solution. IEEE, pp 171–181

Eisenstein J, Vanderdonckt J, Puerta A (2001) Applying model-based techniques to the development of UIs for mobile computers. In: Proceedings of the 6th international conference on intelligent user interfaces. ACM, pp 69–76

Fang CC, Yeh CW (2016) Effective confidence interval estimation of fault-detection process of software reliability growth models. Int J Syst Sci 47(12):2878–2892

Febrero F, Calero C, Moraga MÁ (2016) Software reliability modeling based on ISO/IEC SQuaRE. Inf Softw Technol 70:18–29

Gallud JA, Peñalver A, López-Espín JJ, Lazcorreta E, Botella F, Fardoun HM, Sebastián G (2012) A proposal to validate the user's goal in distributed user interfaces. Int J Hum Comput Interact 28(11):700–708

Garmabaki AH, Aggarwal AG, Kapur PK (2011) Multi up-gradation software reliability growth model with faults of different severity. In: Proceedings of 2011 IEEE international conference on industrial engineering and engineering management (IEEM). IEEE, pp 1539–1543

Goel AL (1985) Software reliability models: assumptions, limitations, and applicability. IEEE Trans Software Eng 12:1411–1423

Goel AL, Okumoto K (1979a) A Markovian model for reliability and other performance measures of software systems. In: National computer conference. IEEE, pp 769–774

Goel AL, Okumoto K (1979b) Time-dependent error-detection rate model for software reliability and other performance measures. IEEE Trans Reliab 28(3):206–211

Gorschek T, Davis AM (2008) Requirements engineering: in search of the dependent variables. Inf Softw Technol 50(1–2):67–75

Goseva-Popstojanova K, Trivedi K (1999) Failure correlation in software reliability models. In: Proceeding of the 10th international symposium on software reliability engineering (ISSRE). IEEE, pp 232–241

Greer D, Ruhe G (2004) Software release planning: an evolutionary and iterative approach. Inf Softw Technol 46(4):243–253

Grottke M, Trivedi KS (2005) A classification of software faults. J Reliab Eng Assoc Jpn 27(7):425–438

Grottke M, Trivedi KS (2007) Fighting bugs: remove, retry, replicate, and rejuvenate. Computer 40(2)

Grottke M, Nikora AP, Trivedi KS (2010) An empirical investigation of fault types in space mission system software. In: Proceedings of 2010 IEEE/IFIP international conference on dependable systems and networks (DSN). IEEE, pp 447–456

Hailpern B, Santhanam P (2002) Software debugging, testing, and verification. IBM Syst J 41(1):4–12

Hallsteinsen S, Hinchey M, Park S, Schmid K (2008) Dynamic software product lines. Computer 41(4)

Halstead MH (1977) Elements of software science. Elsevier

Han S, Dang Y, Ge, S., Zhang D, Xie T (2012) Performance debugging in the large via mining millions of stack traces. In: Proceedings of the 34th international conference on software engineering. IEEE, pp 145–155

Hartz MA, Walker EL, Mahar D (1997) Introduction to software reliability: a state of the art review. The Center

Herbsleb JD, Mockus A (2003) An empirical study of speed and communication in globally distributed software development. IEEE Trans Softw Eng 29(6):481–494

Herbsleb JD, Mockus A, Finholt TA, Grinter RE (2000) Distance, dependencies, and delay in a global collaboration. In: Proceedings of the 2000 ACM conference on computer supported cooperative work. ACM, pp 319–328

Herbsleb JD, Mockus A, Finholt TA, Grinter RE (2001) An empirical study of global software development: distance and speed. In: Proceedings of the 23rd international conference on software engineering. IEEE Computer Society, pp 81–90

Ho SL, Xie M (1998) The use of ARIMA models for reliability forecasting and analysis. Comput Ind Eng 35(1–2):213–216

Hossain SA, Dahiya RC (1993) Estimating the parameters of a non-homogeneous Poisson-process model for software reliability. IEEE Trans Reliab 42(4):604–612

HP Applications Handbook (2012) Shorten release cycles by bringing developers to application lifecycle management. http://www.hp.com/hpinfo/newsroom/press_kits/2011/optimization2011/Lifecycle.pdf

Hsu CJ, Huang CY, Chang JR (2011) Enhancing software reliability modeling and prediction through the introduction of time-variable fault reduction factor. Appl Math Model 35(1):506–521

Hu QP, Peng R, Xie M, Ng SH, Levitin G (2011) Software reliability modelling and optimization for multi-release software development processes. In: Proceedings of2011 IEEE international conference on industrial engineering and engineering management (IEEM). IEEE, pp 1534–1538

Huang X (1984) The hypergeometric distribution model for prediction the reliability of software. Microelectron Reliab 24(1)

Huang CY (2005) Performance analysis of software reliability growth models with testing-effort and change-point. J Syst Softw 76(2):181–194

Huang CY, Kuo SY (2002) Analysis of incorporating logistic testing-effort function into software reliability modeling. IEEE Trans Reliab 51(3):261–270

Huang CY, Lin CT (2006) Software reliability analysis by considering fault dependency and debugging time lag. IEEE Trans Reliab 55(3):436–450

Huang CY, Lyu MR (2005) Optimal release time for software systems considering cost, testing-effort, and test efficiency. IEEE Trans Reliab 54(4):583–591

Hwang S, Pham H (2009) Quasi-renewal time-delay fault-removal consideration in software reliability modeling. IEEE Trans Syst Man Cybern Part A Syst Hum 39(1):200–209

Inoue S, Yamada S (2004) Testing-coverage dependent software reliability growth modeling. Int J Reliab Qual Saf Eng 11(04):303–312

Jansen S, Brinkkemper S, Finkelstein A (2007) Providing transparency in the business of software: a modeling technique for software supply networks. Establishing the foundation of collaborative networks. Springer, pp 677–686

Jelinski Z, Moranda P (1972) Software reliability research. Stat Comput Perform Eval 465–484

Jones C (1996) Software defect-removal efficiency. Computer 29(4):94–95

Kaner C, Falk J, Nguyen HQ (2000) Testing computer software, 2nd edn. Dreamtech Press

Kapur PK, Younes S (1995) Software reliability growth model with error dependency. Microelectron Reliab 35(2):273–278

Kapur PK, Gupta A, Jha PC (2007) Reliability analysis of project and product type software in operational phase incorporating the effect of fault removal efficiency. Int J Reliab, Qual Saf Eng 14(03):219–240

Kapur PK, Pham H, Anand S, Yadav K (2011) A unified approach for developing software reliability growth models in the presence of imperfect debugging and error generation. IEEE Trans Reliab 60(1):331–340

Kapur PK, Pham H, Aggarwal AG, Kaur G (2012) Two dimensional multi-release software reliability modeling and optimal release planning. IEEE Trans Reliab 61(3):758–768

Khomh F, Dhaliwal T, Zou Y, Adams B (2012) Do faster releases improve software quality? An empirical case study of Mozilla Firefox. In: Proceedings of the 9th IEEE working conference on mining software repositories. IEEE, pp 179–188

Laprie JC (1995) Dependable computing: concepts, limits, challenges. In: Special issue of the 25th international symposium on fault-tolerant computing, pp 42–54

Laprie JC, Arlat J, Beounes C, Kanoun K (1990) Definition and analysis of hardware-and software-fault-tolerant architectures. Computer 23(7):39–51

Lehman MM (1980) Programs, life cycles, and laws of software evolution. Proc IEEE 68(9):1060–1076

Li Q, Pham H (2017) A testing-coverage software reliability model considering fault removal efficiency and error generation. PloS One 12(7):e0181524

Li H, Li Q, Lu M (2008) Software reliability modeling with logistic test coverage function. In: Proceedings of the 19th international symposium on software reliability engineering (ISSRE). IEEE, pp 319–320

Lin CT, Huang CY (2008) Enhancing and measuring the predictive capabilities of testing-effort dependent software reliability models. J Syst Softw 81(6):1025–1038

Littlewood B (1979) Software reliability model for modular program structure. IEEE Trans Reliab 28(3):241–246

Lotz M (2018) Waterfall vs. Agile: which is the right development methodology for your project? https://www.seguetech.com/waterfall-vs-agile-methodology/

Lyu M (1996) Handbook of software reliability engineering. IEEE Computer Society Press, McGraw-Hill

Lyu MR (2007) Software reliability engineering: a roadmap. In: 2007 future of software engineering. IEEE Computer Society, pp 153–170

Maurice S, Ruhe G, Saliu O (2006) Decision support for value-based software release planning. Value-based software engineering. Springer, pp 247–261
McCabe TJ (1976) A complexity measure. IEEE Trans Softw Eng 4:308–320
Mellor P (1987) Software reliability modelling: the state of the art. Inf Softw Technol 29(2):81–98
Messerschmitt DG, Szyperski C (2005) Software ecosystem: understanding an indispensable technology and industry. MIT Press
Miller DR, Sofer A (1985) Completely monotone regression estimates of software failure rates. In: Proceedings of the 8th international conference on software engineering. IEEE Computer Society, pp 343–348
Mills H (1972) On the statistical validation of compute programs. IBM federal systems division report, 72-6015
Misra SC, Kumar V, Kumar U (2009) Identifying some important success factors in adopting agile software development practices. J Syst Softw 82(11):1869–1890
Missbauer H (2002) Aggregate order release planning for time-varying demand. Int J Prod Res 40(3):699–718
Moranda PB (1981) An error detection model for application during software development. IEEE Trans Reliab 30(4):309–312
Moravec H (1998) When will computer hardware match the human brain. J Evol Technol 1(1):10
Musa JD (1975) A theory of software reliability and its application. IEEE Trans Softw Eng 1(03):312–327
Musa JD (1980) The measurement and management of software reliability. Proceedings of the IEEE 68(9):1131–1143
Musa JD, Iannino A, Okumoto K (1987) Software reliability: measurement, prediction, application, McGraw-Hill
Musa JD, Iannino A, Okumoto K (1990) Software reliability. Adv Comput 30:85–170
Myers GJ, Sandler C, Badgett T (2011) The art of software testing. Wiley
Nakagawa Y (1994) A connective exponential software reliability growth model based on analysis of software reliability growth curves. IEICE Trans 77:433–442
Ohba M (1984) Software reliability analysis models. IBM J Res Dev 28(4):428–443
Ohba M, Yamada S (1984) S-shaped software reliability growth models. In: Proceedings of the 4th international colloquium on reliability and maintainability, pp 430–436
Ohba M, Yamada S, Takeda K, Osaki S (1982) S-shaped software reliability growth curve: how good is it? In: Proceedings of COMPSAC'82, pp 38–44
Ohmann P, Liblit B (2017) Lightweight control-flow instrumentation and postmortem analysis in support of debugging. Autom Softw Eng 24(4):865–904
Pachauri B, Dhar J, Kumar A (2015) Incorporating inflection S-shaped fault reduction factor to enhance software reliability growth. Appl Math Model 39(5–6):1463–1469
Patterson DA, Hennessy JL (2013) Computer organization and design: the hardware/software interface. Morgan Kaufmann Publishers
Peng R, Li YF, Zhang WJ, Hu QP (2014) Testing effort dependent software reliability model for imperfect debugging process considering both detection and correction. Reliab Eng Syst Saf 126:37–43
Pham H (1993) Software reliability assessment: imperfect debugging and multiple failure types in software development. EGandG-RAAM-10737. Idaho National Engineering Laboratory
Pham H (1996) A software cost model with imperfect debugging, random life cycle and penalty cost. Int J Syst Sci 27(5):455–463
Pham H (2000) Software reliability. Springer Science & Business Media
Pham, H. (2007). System Software Reliability. Springer Science & Business Media.
Pham H (2014) A new software reliability model with Vtub-shaped fault-detection rate and the uncertainty of operating environments. Optimization 63(10):1481–1490
Pham H, Deng C (2003) Predictive-ratio risk criterion for selecting software reliability models. In: Proceedings of the 9th international conference on reliability and quality in design, pp 17–21

Pham H, Normann L (1997) A generalized NHPP software reliability model. In: Proceeding of 3rd ISSAT international conference on reliability and quality in design, Aug 1997
Pham L, Pham H (2000) Software reliability models with time-dependent hazard function based on Bayesian approach. IEEE Trans Syst Man Cybern Part A Syst Hum 30(1):25–35
Pham H, Zhang X (1997) An NHPP software reliability model and its comparison. Int J Reliab Qual Saf Eng 4(03):269–282
Pham H, Zhang X (1999a) Software release policies with gain in reliability justifying the costs. Ann Softw Eng 8(1–4):147–166
Pham H, Zhang X (1999b) A software cost model with warranty and risk costs. IEEE Trans Comput 48(1):71–75
Pham H, Zhang X (2003) NHPP software reliability and cost models with testing coverage. Eur J Oper Res 145(2):443–454
Pham H, Nordmann L, Zhang Z (1999) A general imperfect-software-debugging model with S-shaped fault-detection rate. IEEE Trans Reliab 48(2):169–175
Ramasubbu N, Balan RK (2007) Globally distributed software development project performance: an empirical analysis. In: Proceedings of the 6th joint meeting of the European software engineering conference and the ACM SIGSOFT symposium on the foundations of software engineering. ACM, pp 125–134
Roberts TL, Gibson ML, Fields KT, Rainer RK (1998) Factors that impact implementing a system development methodology. IEEE Trans Softw Eng 24(8):640–649
Ruhe G, Momoh J (2005) Strategic release planning and evaluation of operational feasibility. In: Proceedings of the 38th Annual Hawaii international conference on system science. IEEE, pp 313b-313b.
Saliu O, Ruhe G (2005) Software release planning for evolving systems. Innov Syst Softw Eng 1(2):189–204
Sawyer S, Guinan PJ (1998) Software development: processes and performance. IBM Syst J 37(4):552–569
Schick GJ, Wolverton RW (1978) An analysis of competing software reliability models. IEEE Trans Softw Eng 2:104–120
Shetti NM (2003) Heisenbugs and Bohrbugs: why are they different. Techn. Ber. Rutgers, The State University of New Jersey
Singpurwalla ND, Soyer R (1985) Assessing (software) reliability growth using a random coefficient autoregressive process and its ramifications. IEEE Trans Softw Eng 12:1456–1464
Svahnberg M, Gorschek T, Feldt R, Torkar R, Saleem SB, Shafique MU (2010) A systematic review on strategic release planning models. Inf Softw Technol 52(3):237–248
Tassey G (2002) The economic impacts of inadequate infrastructure for software testing. National Institute of Standards and Technology, RTI Project, 7007(011)
Teng X, Pham H (2006) A new methodology for predicting software reliability in the random field environments. IEEE Trans Reliab 55(3):458–468
Tohma Y, Yamano H, Ohba M, Jacoby R (1991) The estimation of parameters of the hypergeometric distribution and its application to the software reliability growth model. IEEE Trans Softw Eng 17(5):483–489
Tokuno K, Yamada S (2000) An imperfect debugging model with two types of hazard rates for software reliability measurement and assessment. Math Comput Model 31(10–12):343–352
Wall JK, Ferguson PA (1977) Pragmatic software reliability prediction. In: Annual reliability and maintainability symposium, pp 485–488
Weyuker EJ (2004) How to judge testing progress. Inf Softw Technol 46(5):323–328
Xie M (1991) Software reliability modelling. World Scientific
Xie M, Ho SL (1999) Analysis of repairable system failure data using time series models. J Qual Maint Eng 5(1):50–61
Xie M, Yang B (2003) A study of the effect of imperfect debugging on software development cost. IEEE Trans Softw Eng 29(5):471–473

Xie M, Zhao M (1992) The Schneidewind software reliability model revisited. In: Proceedings of the 3rd international symposium on software reliability engineering (ISSRE). IEEE, pp 184–192
Yamada S, Osaki S (1985) Software reliability growth modeling: models and applications. IEEE Trans Softw Eng 12:1431–1437
Yamada S, Ohba M, Osaki S (1983) S-shaped reliability growth modeling for software error detection. IEEE Trans Reliab 32(5):475–484
Yamada S, Ohba M, Osaki S (1984) S-shaped software reliability growth models and their applications. IEEE Trans Reliab 33(4):289–292
Yamada S, Ohtera H, Narihisa H (1986) Software reliability growth models with testing-effort. IEEE Trans Reliab 35(1):19–23
Yamada S, Hishitani J, Osaki S (1991) Test-effort dependent software reliability measurement. Int J Syst Sci 22(1):73–83
Yamada S, Tokuno K, Osaki S (1992) Imperfect debugging models with fault introduction rate for software reliability assessment. Int J Syst Sci 23(12):2241–2252
Yamada S, Tokuno K, Kasano Y (1998) Quantitative assessment models for software safety/reliability. Electron Commun Jpn (Part II: Electron) 81(5):33–43
Yang J, Liu Y, Xie M, Zhao M (2016) Modeling and analysis of reliability of multi-release open source software incorporating both fault detection and correction processes. J Syst Softw 115:102–110
Yazdanbakhsh O, Dick S, Reay I, Mace E (2016) On deterministic chaos in software reliability growth models. Appl Soft Comput 49:1256–1269
Zhang X, Pham H (2000) An analysis of factors affecting software reliability. J Syst Softw 50(1):43–56
Zhang X, Shin MY, Pham H (2001) Exploratory analysis of environmental factors for enhancing the software reliability assessment. J Syst Softw 57(1):73–78
Zhang X, Teng X, Pham H (2003) Considering fault removal efficiency in software reliability assessment. IEEE Trans Syst Man Cybern Part A Syst Hum 33(1):114–120
Zhu M, Pham H (2016) A software reliability model with time-dependent fault detection and fault removal. Vietnam J Comput Sci 3(2):71–79
Zhu M, Pham H (2017a) A two-phase software reliability modeling involving with software fault dependency and imperfect fault removal. Comput Lang Syst Struct 53:27–42
Zhu M, Pham H (2017b) A multi-release software reliability modeling for open source software incorporating dependent fault detection process. Ann Oper Res. https://doi.org/10.1007/s10479-017-2556-6
Zhu M, Pham H (2017c) Environmental factors analysis and comparison affecting software reliability in development of multi-release software. J Syst Softw 132:72–84
Zhu M, Pham H (2018) A Software reliability model incorporating martingale process with gamma-distributed environmental factors. Ann Oper Res. https://doi.org/10.1007/s10479-018-2951-7
Zhu M, Pham H (2020) A generalized multiple environmental factors software reliability model with stochastic fault detection process. Ann Oper Res. https://doi.org/10.1007/s10479-020-03732-3
Zhu M, Zhang X, Pham H (2015) A comparison analysis of environmental factors affecting software reliability. J Syst Softw 109:150–160

Software Reliability Growth Models Incorporating Software Project/Application's Characteristics as a Power Function with Change Point

Shinji Inoue, Abhishek Tandon, and Prarna Mehta

Abstract Software Reliability is a high-handed aspect to ascertain the quality of the system, which leads to development of tools that incorporates real time set-up. One of the real time concepts is change point, which highlights on the fact that a characteristic of the model changes during the testing time duration and it is significant to incorporate its effect in the model developed to ameliorate the reliability of the system. Moreover, faults are assumed to be independent and are incurred at any arbitrary time but, in practical world, faults may occur due to many factors like the testing environment, resource allocation, code complexity, testing team skill-set etc. This rate tends to change with time and assuming it to be constant may not reflect upon an actual output. Another conundrum that is taken care of is the release time of the software project/application. The weightage that is implied by this optimization planning is due to the fact that over testing may incur high cost to the firm whereas under testing may lead to release of a project/application with high fixing cost faults affecting the manufacturer by an elevated post-release cost. In this chapter, a framework is proposed that extends error-removal phenomenon model by encapsulating the software project/application characteristics as a parameter. A realistic software development situation is also taken in account by considering the parameter not only as a constant but a time dependent function. The suggested SRGMs are also monitored under a change point scenario, which gives real-time edge to the problem, and are then utilized to develop release time policy balancing reliability and expected cost incurred by a project/an application. The models are validated using Tandem dataset and performance measures are compared quantitatively with the standard models.

S. Inoue
Faculty of Informatics, Kansal University, 2-1-1 Ryozenji-cho, Takatsuki-shi, Osaka 569-1095, Japan

A. Tandon
Department of Management Studies, Shaheed Sukhdev College of Business Studies, University of Delhi, Delhi, India
e-mail: abhishektandon@sscbsdu.ac.in

P. Mehta (✉)
Department of Operational Research, University of Delhi, Delhi, India

A. G. Aggarwal et al. (eds.), *Optimization Models in Software Reliability*, Springer Series in Reliability Engineering,
https://doi.org/10.1007/978-3-030-78919-0_2

On comparison of the results, the proposed model outperforms other extant models with and without change point.

Keywords Software reliability · Software reliability growth model · Change point · Power function · Release policy · Software testing phase

1 Introduction

The ambitious invention of a high performing and a high-defined quality software system has given software engineers the liability to envisage a more complex system. Software brings forth automation to a wide range of domain like medicine, banking system, business houses, academics, security etc. thus, becoming an indispensable element of any commercial or non-commercial field. With the increase in its application, there is a need of improvising software at a higher speed to balance with the requirements of the consumers, to keep up with the competitors in the market and most significantly to maintain quality. Thus, reliability models are entailed to assess and forecast the reliability of the software embedded in the system quantitatively, elevating system's reliability (Huang et al. 1999) and are also known as Software Reliability growth models (SRGM). A SRGM that helps in rendering a balanced amalgamation in terms of expense, reliability, productivity and performance can be considered to be generic in nature and applicable to sundry software project/application.

Many versions and forms of SRGM has been developed considering different assumptions and objective of the underlying study, under perfect or imperfect debugging conditions (Okumoto and Goel 1979). However, in the real world, there are many other factors that influence the quality of the software with time for example testing environment, complexity of the code, running environment, line of codes, testability etc. Lai et al. (2011) stated past SRGMs fail to incorporate these factors into past simulation. Thus, fault-detecting rate is influenced by external factors and it would be favorable to take them into account in order to increase the performance rate of a SRGM developed using past failure rate. In this study, these external factors are included in the model developed and coined as software application/project characteristic parameter. The given parameter is not only a function of software characteristics but can also be a time dependent power function. The parameter measuring the influence of the software project/application's characteristic may not necessarily be constant over the testing time period and is subject to change due to alteration in testing plan at some time-point. This instance of time is termed as change-point. In this chapter, SRGMs are formulated under the postulation of change point to create a real-time problem.

The main concern of the management is scheduling the testing phase and release of the project but are unaware of the latent faults remaining in the system even after the completion of testing phase (Kapur et al. 2008a). However faults in a software project/application are extensively eliminated during the testing phase of the system but excessive testing is uneconomical and impractical, thereby making the optimal release planning as a powerful decision-making aspect in a business plan (Lai et al. 2011). SRGM provides enumeration of faults that might be informative to the management to curtail development cost or to anticipate the appropriate release time of a software project/application (Huang 2005a). To précis, software release planning is a vital managerial optimization problem that involves determination of the optimal release time of the software such that the cost is minimized subject to a reliability constraint. Kapur et al. (2009a) have discussed the significance of release planning that if a product is subject early to or late release, then it may lead to either high operational cost or high testing cost which is a loss for the business. In this study, the expected number of faults obtained through SRGMs with and without change point scenarios are employed to develop optimal release plans for the respective models.

The objective of this study is condensed in three-fold. Firstly, a model is simulated taking into account of change point concept and software project/application as a power function. Secondly, models are assessed for its performance and efficiency based on a real software fault dataset and comparison is made with some extant models from the past literature. Thirdly, an optimization scheme is planned for the release of the product using the models developed. The scheme specifies an optimal combination of cost and reliability for a given software project/application, thereby focusing on a rational release policy for the same.

This chapter is divided into sections, namely, in Sect. 2 a brief summary of the past literature has been captured, following which in Sect. 3 the suggested model has been explained, in Sect. 4 the proposed model is justified with a numerical illustration, a release plan has been discussed in Sect. 5 using the developed models and lastly, Sect. 6 concludes the given study.

2 Literature Survey

In this section, an abridgment of the related literature is furnished for each research topic alluded in this study.

2.1 Software Reliability Growth Models

A SRGM follows NHPP distribution that gives an estimated count of faults for both calendar as well as running timeline. In the past literature, a vast number of SRGMs are available differentiating from each other by marginal changes in assumptions set describing or tackling a testing related problem (Lai and Garg 2012). Yamada et al. (1983) proposed an S-shaped model due the non-uniformity in reliability growth that was also a succession to Okumoto and Goel (1979) model. Kareer et al. (1990) proposed a modified S-shaped model that was based on the severity level of the faults manifesting a real time situation. Jeske and Zhang (2005) gave a realistic approach by bridging a gap between theoretical and practical application of SRGM. Many earlier SRGMs do not fit all failure datasets perfectly as stated by Chiu et al. (2008) however, incorporating learning effect into the model would enhance the results. Effects of testing skills, strategy and environment on discrete SRGMs were well captured by Kapur et al. (2008b). Kapur et al. (2009b) have presented a unified approach to evaluate a wide range of SRGMs on the basis of hazard rate. A SRGM framework demonstrated by Inoue and Yamada (2011) reflected the effect of rate of change in the testing setting. Softwares are developed in an ideal environment nonetheless, it is not the same while in the operating stage (Pham 2016). While Zhu and Pham (2018) considered single environment factor and later on taking in account the effect of multiple environmental factors on modeling and estimation (Zhu and Pham 2020).

2.2 Change Point

Zhao (1993) was motivated to apply change point concept in the field of software and hardware reliability by including it SRGM. Huang (2005b) integrated logistic testing effort and change point into the reliability modeling suggesting that the project head to invest in tools and manpower that would aid in improving the reliability of the software product. Singh et al. (2010) introduced S-shaped SRGM with change point that monitors the reliability of OSS. A discrete SRGM was modeled by Goswami et al. (2007) for different levels of bugs severity under change point. Kapur et al. (2009a) have discussed a general framework with respect to altering rate of fault detection with and without change point. In the study by Chatterjee et al. (2012), SRGM with fluctuating introduction and detection rates have been examined under change point scenario. Inoue et al. (2013) scrutinized the altering effect of testing environment on change point based SRGM. Parr-curve was deployed by Ke et al. (2014) to analyze reliability model with multiple change point. Nagaraju and Fiondella (2017) have exemplified the significance of change point modeling by assessing different SRGM with and without change point. Change point SRGMs depicts mathematically a real life situation, however, Inoue and Yamada (2018) utilized Markov process to enhance the depiction. Chatterjee and Shukla (2017) used am improvised approach to regulate the reliability assessment under change point effect.

2.3 Release Policy

SRGMs are implemented to enumerate uncertainty attached to a software system, thus aiding a software developer to achieve a highly reliable software project/application. But with increasing complexity in the software, it is hard for the software engineers to deliver fault free. At the same time, these probabilistic tools are utilized to estimate the optimal release time. The concept of optimal release policy was firstly implemented by Okumoto and Goel (1979). Following which, many researchers started implementing the optimal release time concept along with software reliability assessment as well as balancing cost parameters (Kapur and Garg 1989). Shrivastava et al. (2020) proposed an optimization model to establish the optimal release time as well as the testing stop time. Li et al. (2010) study not only deals with multiple change point and release policies but, they have also made an effort to do a sensitive analysis of the results. The failure data being uncertain and vague, pushed the researchers to indulge in fuzzy sets. Under the fuzzy environment, a release plan for the software was given by Pachauri et al. (2013). For a discrete SRGM, a release policy was proposed by Aggarwal et al. (2015). Release time decision-making can be sometime tricky when the past records are inconsistent, conversely (Chatterjee and Shukla 2017) suggested a fuzzy based release schedule. Shrivastava and Sachdeva (2019) proposed a generalized release policy under different testing environment.

3 Methodology

In this study, a SRGM has been enhanced by considering a parameter as a power function of time integrating with a change point scenario.

3.1 Notations

Symbols	Description
X	Preliminary faults existing in the software project/application in the time period (0, t]
X(t)	For a given time period, say (0, t], expected number of observed faults
p	Failure rate for a software project/application
q	Residual fault detection rate
r_1	Software project/application's characteristics measuring factor before change point
r_2	Software project/application's characteristics measuring factor after change point
k	A constant
t	Time period
τ	Change point

3.2 Assumptions

Comprehending from the past literature (Huang et al. 1999; Huang 2005a; Yamada et al. 1983; Kareer et al. 1990), in this study, SRGMs are developed, based on Non-Homogenous Poisson Process (NHPP) where, $X(t)$ is the mean value function or the expected number of faults in the time interval (0, t]. The proposed models are developed under the following assumptions,

- Due to the presence of the dormant faults, the software project/application is subject to fail during the operational phase.
- A fault detected in the software project/application is removed instantaneously.
- On removal of certain faults, it is assumed that some latent dependent faults are also eliminated in the process.
- The software project/application's operational phase is considered to be as one of the phases of the lifecycle.
- The failure or fault detection phenomenon incurred by the consumer or the testing team is considered to be equivalent.
- Software project/application characteristics' effect on the model is measured by a parameter, which might change with time.

3.3 Model Development

Kapur and Garg (1992) proposed a model that was based on the assumption that faults detected can lead to observing residual faults in the system without causing the system to fail. This model has been extended by considering a parameter that characterizes software project/application as a parameter namely, r_1. Hence, the differential equation under the given assumptions for an expected number of faults in the system is given,

$$\frac{dX(t)}{dt} = r(t)[(p + \frac{X(t)}{X}q)(X - X(t))] \tag{1}$$

where, $r(t)$ can be a constant or a time-dependent function that defines the software project/application characteristics. Zhu and Pham (2020) developed a generic model including environmental factor and on the similar lines (Inoue et al. 2013) amplified fault detection rate with the help of environment factor in their cost model. Different set up for the model in Eq. (1) has been considered in the study.

Model 1a: When $r(t)$ is considered to be a constant without change point (BCP)

Let us consider,

$$r(t) = r_1 \tag{2}$$

where, r_1 is a constant value of the above defined parameter and $r_1 > 0$. Substituting (2) in (1), the differential equation for the expected number of faults in the system is given by,

$$\frac{dX(t)}{dt} = r_1[(p + \frac{X(t)}{X}q)(X - X(t))] \tag{3}$$

The constant r_1 implies that each fault is uniformly affected or influenced by software project/application characteristics parameter. Solving (3), we obtain the expected number of faults in the testing time interval (0, t], given the initial condition, $X(0) = 0$, as,

$$X(t) = \frac{X[1 - e^{-(p+q)r_1 t}]}{[1 + \frac{q}{p}e^{-(p+q)r_1 t}]} \tag{4}$$

Model 1b: When r(t) is considered to be a constant with change point (ACP)

During the testing phase, a software project/application runs in a given environment but it is not necessarily true that parameters remain uniform throughout. The instance at which a change is observed in the pattern of the failure distribution is termed as change point. In software reliability engineering, a change point occurs due to change in testing pattern observed the testing team's, change in resource allocation, improved learning skills of the system analyst, or programmed testing codes (Kapur et al. 2008a; Zhu and Pham 2020; Ke et al. 2014). In this study, software project/application characteristics are not considered to be homogenous during the testing phase, hence developing SRGM with a change point. Let the parameter be defined as,

$$r(t) = \begin{cases} r_1, & t < \tau \\ r_2, & t \geq \tau \end{cases}$$

Considering the underlying assumptions and assuming that r(t) does not remain homogeneous during the testing phase, the mean value function is given as,

$$X(t) = \begin{cases} X\frac{[1-\exp(-(p+q)r_1)]}{\left[1+\frac{q}{p}\exp(-(p+q)r_1)\right]} & t < \tau \\ X\frac{\left[1-\left(1+\frac{q}{p}\right)\left(1+\frac{q}{p}\right)\exp(-r_1\tau)\right][\exp(-(p+q)r_1\tau)+(-(p+q)r_2(t-\tau))]}{\left[1+\frac{q}{p}\exp(-r_1\tau)\right]\left[1+\frac{q}{p}\exp(-r_2 t)\right]} & t \geq \tau \end{cases} \tag{5}$$

Model 2a: When r(t) is considered as a power function BCP

Next, let us consider as a power function of testing time which is given as,

$$r(t) = r_1 t^k \tag{6}$$

The form in (6) renders the suggested models generic and adaptable in nature (Kapur et al. 2008a). Substituting (6) in (1), the differential equation is given as,

$$\frac{dX(t)}{dt} = r_1 t^k [p + q\frac{X(t)}{X}][(X - X(t))] \tag{7}$$

where, $k \geq 0$ and the first component represents the varying relation between r_1 and t with varying value of k. If $k = 0$, the model reduces to Eq. (3). Taking the initial condition as $X(0) = 0$, we get,

$$X(t) = X\frac{[1 - \exp(-r_1(p + q)\frac{t^{k+1}}{k+1})]}{[1 + \frac{q}{p}\exp\left(-r_1(p + q)\frac{t^{k+1}}{k+1}\right)]} \tag{8}$$

Model 2b: When r(t) is considered as a power function ACP

The Eq. (3) has been extended by incorporating change point and power function of r.

$$X(t) = \begin{cases} X\frac{\left[1-\exp\left(-\frac{(p+q)r_1}{k+1}t^{k+1}\right)\right]}{\left[1+\frac{q}{p}\exp\left(-\frac{(p+q)r_1}{k+1}t^{k+1}\right)\right]} & t < \tau \\ X\frac{\left[1-\left(1+\frac{q}{p}\right)\left(1+\frac{q}{p}\right)\exp(-\frac{r_1}{k+1}\tau^{k+1})\right]\left[\exp\left(-\frac{(p+q)r_1}{k+1}\tau^{k+1}\right)+\left(-\frac{(p+q)r_2}{k+1}(t-\tau)^{k+1}\right)\right]}{\left[1+\frac{q}{p}\exp(-\frac{r_1}{k+1}\tau^{k+1})\right]\left[1+\frac{q}{p}\exp(-\frac{r_2}{k+1}t^{k+1})\right]} & t \geq \tau \end{cases} \tag{9}$$

For simplification, in this study, only one change point has been considered in both the models given by (5) and (9), however in the practical world, change points are observed more than once in a testing phase.

4 Model Validation

In this section, the SRGMs given by (3), (5), (8) and (9) have been validated on the basis of fault dataset given by Wood (1996). A change point has been observed at the 8th week of testing in the dataset and change point models given by (5) and (9) are evaluated accordingly. The parameters are estimated for each model and the performances of the respective models are recorded for comparisons.

4.1 Dataset

The proposed models have been validated using Tandem dataset (Wood 1996) that contains number of faults detected over a testing period of 20 weeks and cumulative

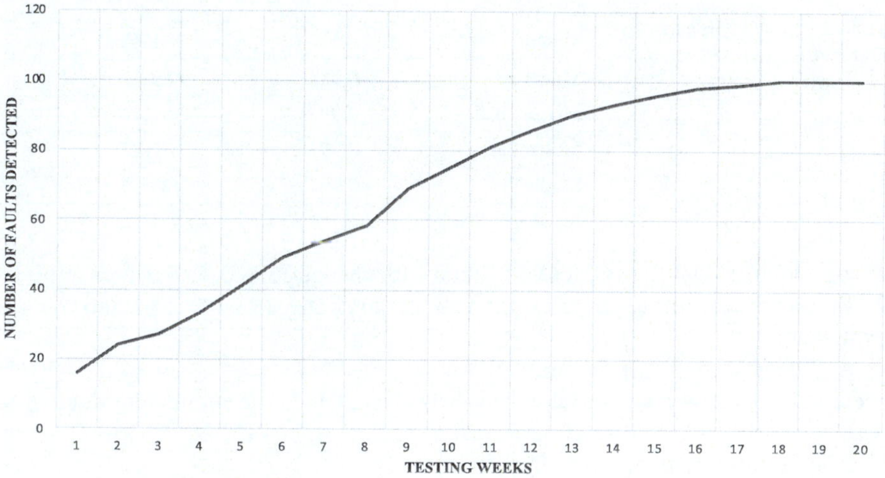

Fig. 1 A plot for Tandem Dataset

sum of number of faults being 100. This dataset has been widely implemented in the past literature for example, Park and Baik (2015), Sharma (2010), Roy et al. (2014). From the data set, a change in the number of faults detected was observed at 8th week of testing. Figure 1 depicts a graph plotting tandem data set where x-axis denote the testing weeks whereas y-axis denote cumulative number of faults detected.

4.2 Parameter Estimation

The non-linear regression option in Statistical Package for social Science (SPSS) aids in obtaining least square estimates of the parameters for a given SRGM. The parameters are valued for a given testing time in weeks denoted by t and parameters X, p, q, k, r, r_1, and r_2 delineated in Sect. 3 are estimated for respective SRGM. Table 1 represents the parameter estimation for without and with change point models.

In Table 1, "–" indicates null value of the parameter because they have no role play in the respective model. 35.1% and 93.7% software project/application characteristics are explained by the parameter r in model 1a and 2a respectively. However, without

Table 1 Parameter estimations for models before and after change point scenario

Models	X	p	q	r_1	r_2	k
Model 1a	110.829	0.222	0.268	0.351	–	–
Model 1b	116.48	0.48	0.455	0.238	0.091	–
Model 2a	114.104	0.0046	0.431	0.937	–	0.377
Model 2b	112.375	0.113	0.464	0.552	0.432	0.394

Table 2 Model comparisons with respect to performance measures

Models	R^2	MSE
Model 1a	0.989	8.971
Model 1b	0.993	5.442
Model 2a	0.991	7.183
Model 2b	0.997	2.076

change point, model 1b explains 23.8% and with change point 0.91%, whereas model 2b shows the best result, combining the effect of r_1 and r_2, 98.4% characteristics are explained.

A SRGM is said to perform well when it predicts the behavior of the system accurately by using previous fault data (Roy et al. 2014). The simulations developed were assessed with the help of performance measures, co-efficient of determination (R^2) and mean square of errors (MSE).

Looking at the R^2 and MSE values of the proposed models in Table 2, model 2b shows the best fit for the given dataset i.e. 99.7% of the variation in data is captured by the proposed model and has the lowest MSE value (2.076).

4.3 *Goodness of Fit Curves*

See Figs. 2 and 3.

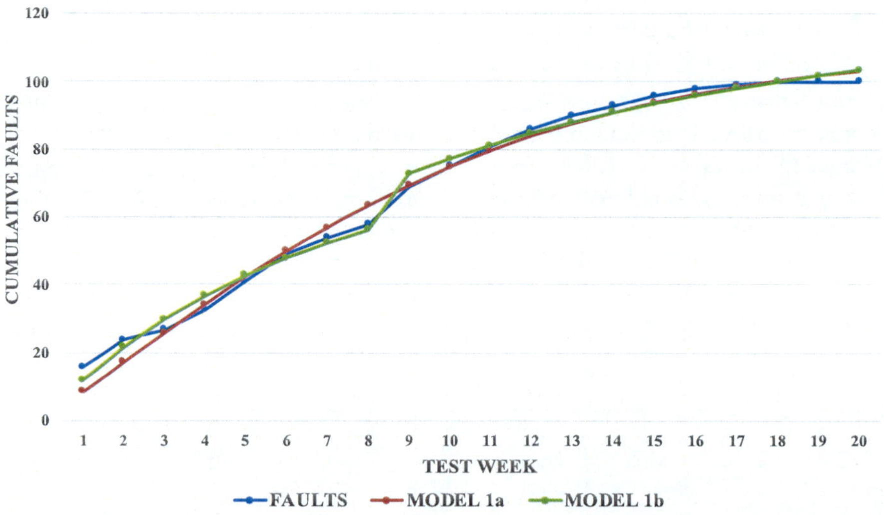

Fig. 2 Actual versus predicted values for proposed models where r is considered constant BCP and ACP

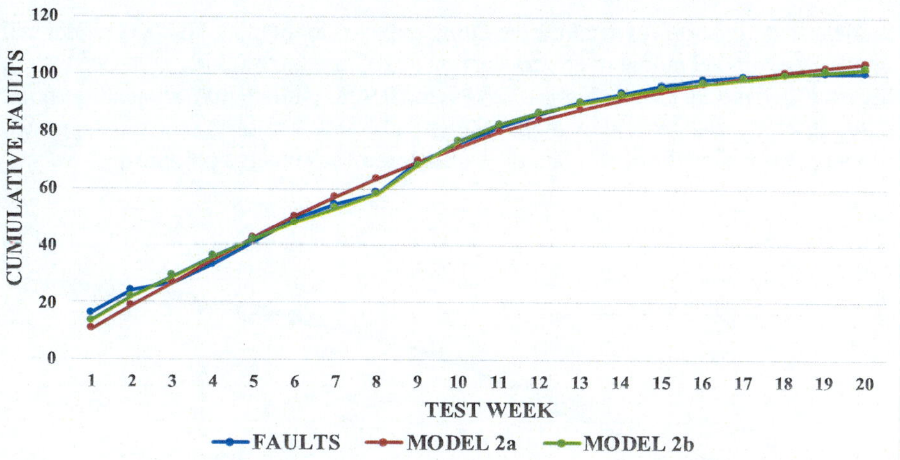

Fig. 3 Actual versus predicted values for proposed models where r is power function of time BCP and ACP

4.4 Comparison with the Previous Models

In order to validate the robustness of the proposed model, comparisons are drawn with some conventional models, namely Goel Okumoto model (Okumoto and Goel 1979), S-shaped Yamada model (Yamada et al. 1983), Kapur Garg error removal phenomenon model (Kapur and Garg 1992) respectively using the Tandem dataset. For mathematical simplifications, the proposed model 1a has been compared with the traditional models.

Table 3 depicts the comparison values for the different SRGMs with respect to change point. Nonetheless, the MSE value is lowest for the proposed model without as well as with change point than that of GO model and S-Shaped Yamada model which implies that the proposed model has the least fitting error and hence exhibit the best performance. Since the proposed model is an extension of Kapur and Garg error phenomenon model, the comparison measure is vital for this case. For the given dataset, KG Model has MSE value 18.572 and 10.937 without and with change point respectively, while those values for model 1a are nearly half of it i.e. the suggested

Table 3 Models' performance comparison with respect to coefficient of determination

	R^2		MSE	
	BCP	ACP	BCP	ACP
Model 1a	0.989	0.993	8.971	5.442
GO model	0.965	0.988	28.503	9.392
Yamada model	0.969	0.991	25.266	7.57
KG model	0.977	0.987	18.572	10.937

model has the capability to reduce the fitting error to half of that of the KG model. All in all, the extended model is a better version of KG model. Looking at the R^2 values, the proposed model has the highest (0.989 and 0.993) without and with change point cases in comparison to other models. Figures 4 and 5 reflects that the suggested model is the closest fit of the data set without and with change point respectively.

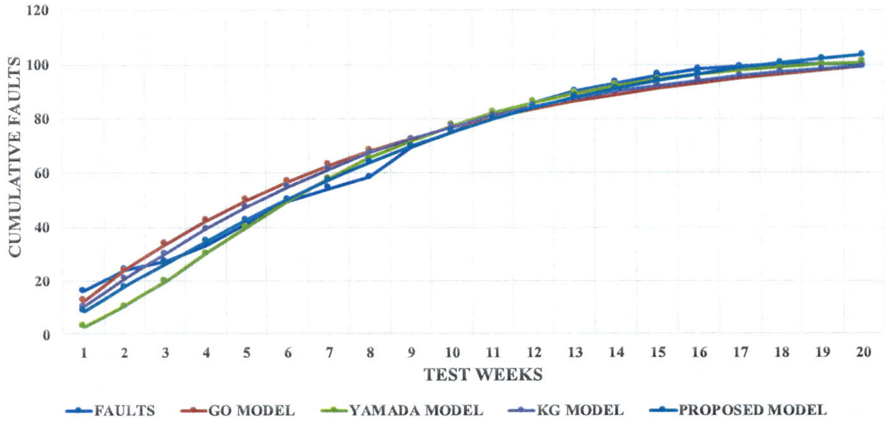

Fig. 4 A fitted curve for observed and estimated faults BCP given by proposed models and compared with some existing model

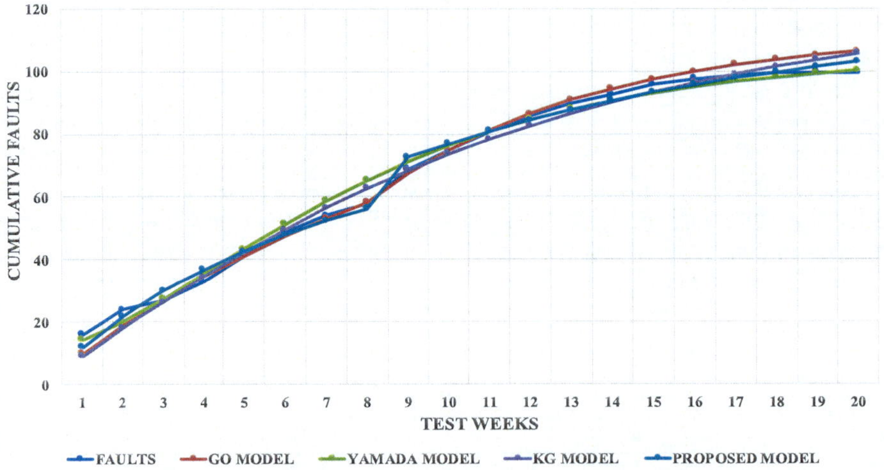

Fig. 5 A fitted curve for observed and estimated faults ACP given by proposed models and compared with some existing model

The result of the data analysis suggests that the extended version of the KG model is an effective model that enriches the data set in terms of predicting accuracy and beautifully incorporates the effect of change point under perfect debugging conditions.

5 Release Policy

With the completion of software testing phase, the software project/application is prepared for its release into the market rising to problem of optimal release policy. Protraction of testing phase implies detection of multiple faults, thus increasing the development cost, increase in the quality and delay in the release of the software project/application. The lag in release of the software may make the market restive, at the same time cause a loss to the business. On the other hand, if the testing phase is cut short, one can infer detection of a smaller number of faults implying decrease in development cost, poor quality and rapid release of the software. The accelerated release of the software may lead to loss of customers' trust towards the firm that might incur post-release high cost. In this study, one of the objectives is to minimize software outlay with respect to a desired level of reliability, thereby obtaining the optimal software release time. The outlay parameters used to illustrate release policy is given as follows,

O_1: Outlay of fault detection BCP in the testing environment ($O_1 > 0$)

O_2': Outlay of fault detection ACP in the testing environment $O_2' > 0$)

O_2: Outlay of fault detection in the operational environment ($O_2 > O_2'$)

O_3: Outlay of testing at arbitrary time point

T: stopping or release time of software project/application.

The expected outlay function without change point comprises of expense acquired by the software firm in order to detect fault BCP, during the operating phase and lastly, an overall expense per unit time before the software project/application's release time, which is given by,

$$O(T) = O_1 * X(T) + O_2 * (X - X(T)) + O_3 * T \tag{10}$$

The expected outlay function with change point is composed of expense acquired by the software firm in order to detect faults BCP, ACP, during the operating phase and lastly, an overall expense per unit time before the software project/application's release time, which is given as,

$$O(T) = O_1 * X_1(\tau) + O_2' * (X_2(T) - X_1(\tau)) + O_2 * (X - X_2(T)) + O_3 * T \tag{11}$$

where, $X_1(t)$ and $X_2(t)$ are mean value fault detection function without and with change point (τ) respectively. Henceforth, an optimization policy can be mathematically formulated as following,

$$\text{Minimise } O(T)$$
$$\text{subject to } Re(T) \geq R^* \tag{12}$$

where, Re(T) is the reliability function and R* is the desiderated level of reliability at the release time of software project/application's (0 < R* < 1). The software project developer establishes the desiderated reliability for the project in accordance to the market requirement.

5.1 *Numerical Illustration*

The formulated optimization problem is solved using MATLAB where outlay parameters are assumed on the basis of past literature (Aggarwal et al. 2015; Shrivastava and Sachdeva 2019). In order to implement the optimization problem given by (12), we have considered tandem dataset, computed parameters of the Eqs. (4), (5), (8) and (9) with the help of SPSS and given by the Table 1. The outlay parameters are assumed to be as following,

$$O_1 = 40, O_2' = 50, O_2 = 100, O_3 = 20$$

With the help of MATLAB and Eq. (12), the expected release time and reliability is computed for the four models and given by the Table 4.

The optimal release time for model 1a is 28.02 weeks with reliability 0.9823 and expected cost is 5111.01, and with change point (model 1b), the release time is 30.28 weeks at 0.9638 reliability and total expected cost being 9116.40. For model 2a, the optimal release time is 30.53, reliability attained is 0.9758 with the expected cost being 5340.80, however with change point, the release time decreases to 28.33 weeks and reliability reached is 0.9532 with the cost of 8901.80.

Table 4 Expected release time and reliability of a software project/application

Models	T	O(T)	Re(T)
Model 1a	28.02	5111.01	0.9823
Model 1b	30.28	9116.4	0.9638
Model 2a	30.53	5340.8	0.9758
Model 2b	28.33	8901.8	0.9532

5.2 Sensitivity Analysis

The expenses involved in testing phase may vary due to various external factors like skills of developers, testing efforts, resource etc. Hence, one might be interested to know which cost parameter plays an important role to improvise the reliability for a given software project/application. Sensitivity analysis is a tool that aids in the decision making of optimal input parameters in order to obtain improvised output (Li et al. 2010). The uncertainty in failure data fails to comprehend the exact value of the outlay parameters. This methodology permits the management to evaluate the change in parameters over reliability, to assess testing strategy, to condense cost values, to attain optimal resource apportionments (Castillo et al. 2008).

For the given study, software reliability growth models given by Eqs. (4), (5), (8) and (9) have been considered, whose parameters are computed using Tandem failure data sets and with the help of SPSS. Considering a desired reliability level to be fixed at 0.95, the outlay parameters (O_1, O_2', O_2, O_3) are altered by -10 and 10%, while monitoring its effect on the reliability, total expected cost as well as release time (given by Table 5). The sensitivity of outlay parameters is analyzed by using formula given by Li and Pham (2017). The relative changes observed with respect to varying outlay parameters is given by Table 6.

For different values of O_1, O_2', O_2, O_3, the altered release time, cost as well as reliability level are generated. Altering values of outlay parameters gives a guidance to the management to strike a balance between the cost and reliability and obtain the optimal release time. The sensitivity analysis of the stated release policy resulted in a very close release time, but in real time situation, the policy implemented by the management depends on the cost under consideration, SRGM adopted to describe the fault process, aim of the release plan and lastly constraints to be taken care of.

From Fig. 6a, it can be clearly observed that a 10% increase in cost O_1, release time decreases by 1.46%, overall cost increases by 8.52% and a marginal decrease of 0.1% has been noted. However, when cost O_2 is increased by 10%, release time increases by 3.26%, that has 0.3% and 0.2% effect on total cost and reliability respectively. On the other hand, with a 10% increase in O_3, the release time declines by 2%, increasing cost by 1% and reliability also declines by a mall fraction. The respective trend was observed exactly opposite when the costs O_1, O_2, O_3 were decreased by 10% for model 1(a) as depicted by Fig. 6b.

Form Fig. 6c, with an increase of 10% in cost O_1, release time, expected cost and reliability increases by 2.34%, 2.79% and 0.2% respectively. However, with an increase of 10% in cost O_2', release time and reliability declines by 2.29 and 0.2%, while the cost shoots up by 6.15%. On the other hand, with an increase of 10% in cost O_2, release time is majorly affected by 3.74%, but a marginal change of 0.4% and 0.3% on cost and reliability is observed. Lastly, with an increase of 10% in O_3, a decline of 2.7% and 0.2% is observed in release time and reliability along with increase in cost by 0.6%. The respective trends were seen exactly opposite when cost O_1, O_2', O_2, O_3 were decreased by 10% for model 1(b) as depicted by Fig. 6d except when cost O_1 decreased by 10%, the expected cost increased by 2.8%.

Table 5 Sensitivity analysis for SRGM with respect to outlay parameters

		Model 1a			Model 1b			Model 2a			Model 2b		
		T	O(T)	Re(T)	T	O(T)	Re(T)	T	O(T)	Re(T)	T	O(T)	Re(T)
O1	36	28.40	4675.27	0.9834	29.09	9372.58	0.9593	31.03	4895.10	0.9772	31.13	8656.89	0.95
	40	28.02	5111.01	0.9823	30.28	9116.40	0.9638	30.53	5340.80	0.9758	28.33	8901.80	0.9532
	44	27.61	5546.22	0.9811	30.99	9371.07	0.9661	30.00	5785.77	0.9742	31.13	9160.19	0.95
O′2	45	–	–	–	30.88	8554.45	0.9658	–	–	–	32.19	8441.06	0.9536
	50	–	–	–	30.28	9116.40	0.9638	–	–	–	28.33	8901.80	0.9532
	55	–	–	–	29.58	9676.98	0.9613	–	–	–	31.13	9509.61	0.95
O2	90	26.94	5089.57	0.9788	28.78	9071.12	0.958	29.14	5310.71	0.9713	31.13	8919.64	0.95
	100	28.02	5111.01	0.9823	30.28	9116.40	0.9638	30.53	5340.80	0.9758	28.33	8901.80	0.9532
	110	28.93	5129.10	0.9849	31.41	9156.31	0.9674	31.72	5366.52	0.979	33.24	9028.07	0.9568
O3	18	28.64	5054.36	0.9841	30.94	9055.15	0.966	31.34	5278.92	0.978	32.31	8912.47	0.954
	20	28.02	5111.01	0.9823	30.28	9116.40	0.9638	30.53	5340.80	0.9758	28.33	8901.80	0.9532
	22	27.45	5166.47	0.9805	29.65	9176.29	0.9615	29.80	5401.10	0.9735	31.13	9038.08	0.95

Table 6 Relative change observed for respective model with respect to outlay parameters

		O1		O′2		O2		O3	
		−10%	10%	−10%	10%	−10%	10%	−10%	10%
Model 1a	T	0.0137	−0.0146	0.0000	0.0000	−0.0387	0.0326	0.0223	−0.0202
	O(T)	−0.0853	0.0852	0.0000	0.0000	−0.0042	0.0035	−0.0111	0.0109
	Re(T)	0.0011	−0.0012	0.0000	0.0000	−0.0036	0.0026	0.0018	−0.0018
Model 1b	T	−0.0392	0.0234	0.0199	−0.0229	−0.0496	0.0374	0.0220	−0.0207
	O(T)	0.0281	0.0279	−0.0616	0.0615	−0.0050	0.0044	−0.0067	0.0066
	Re(T)	−0.0047	0.0024	0.0021	−0.0026	−0.0060	0.0037	0.0023	−0.0024
Model 2a	T	0.0163	−0.0173	0.0000	0.0000	−0.0456	0.0390	0.0266	−0.0239
	O(T)	−0.0835	0.0833	0.0000	0.0000	−0.0056	0.0048	−0.0116	0.0113
	Re(T)	0.0014	−0.0016	0.0000	0.0000	−0.0046	0.0033	0.0023	−0.0024
Model 2b	T	0.0988	0.0988	0.1363	0.0988	0.0988	0.1732	0.1405	0.0988
	O(T)	−0.0275	0.0290	−0.0518	0.0683	0.0020	0.0142	0.0012	0.0153
	Re(T)	−0.0034	−0.0034	0.0004	−0.0034	−0.0034	0.0038	0.0008	−0.0034

From Fig. 6e, when O_1 increases by 10%, release time decreases by 1.7%, total cost does up by 8.3% and reliability decreases by 0.2%. While O_2 increases by 10%, release time, cost and relatability also increases by 3.9%, 0.4% and 0.3% respectively. Lastly, with a 10% increase in O_3, release time and reliability decreases by 2.3 and 0.2%, on the contrary, the expected cost increases by 1.1%. When the costs O_1, O_2, O_3 decreases by 10%, an exactly opposite trend is observed which is depicted by Fig. 6f for model 2a.

From Fig. 6g, when cost O_1, O_2', O_2, O_3 increases by 10%, release time increase significantly by 9.8%, 9.8%, 17.3%, 9.8% respectively, while the expected cost also rises by 2.9%, 6.8%, 1.4% and 1.5%, but reliability declines by 0.3% for cost O_1, O_2', and O_3, on the other hand increases by 0.3% for O_2. From Fig. 6h, it is observed that when cost cost O_1, O_2', O_2, O_3 decreases by 10%, release time increases by 9.8%, 13.6%, 9.8% and 14.1% respectively, the total cost declines fpr O_1 and O_2' by 2.7% and 0.5%, however increases in the case of O_2 and O_3 by 0.2% and 0.1%. Moreover, reliability decreases by 0.3% when O_1 and O_2 decreases by 10% but increases by 0.04% and 0.08%, when cost O_2' and O_3 decrease by 10%.

6 Conclusion

In this chapter, a software reliability growth model has been comprehended with respect to software project/application's characteristics under a perfect debugging environment. The stated characteristic has been considered as a constant as well as a power function of time along with assessing the model without and with change point. The parameters are estimated with the help of SPSS and comparison has been

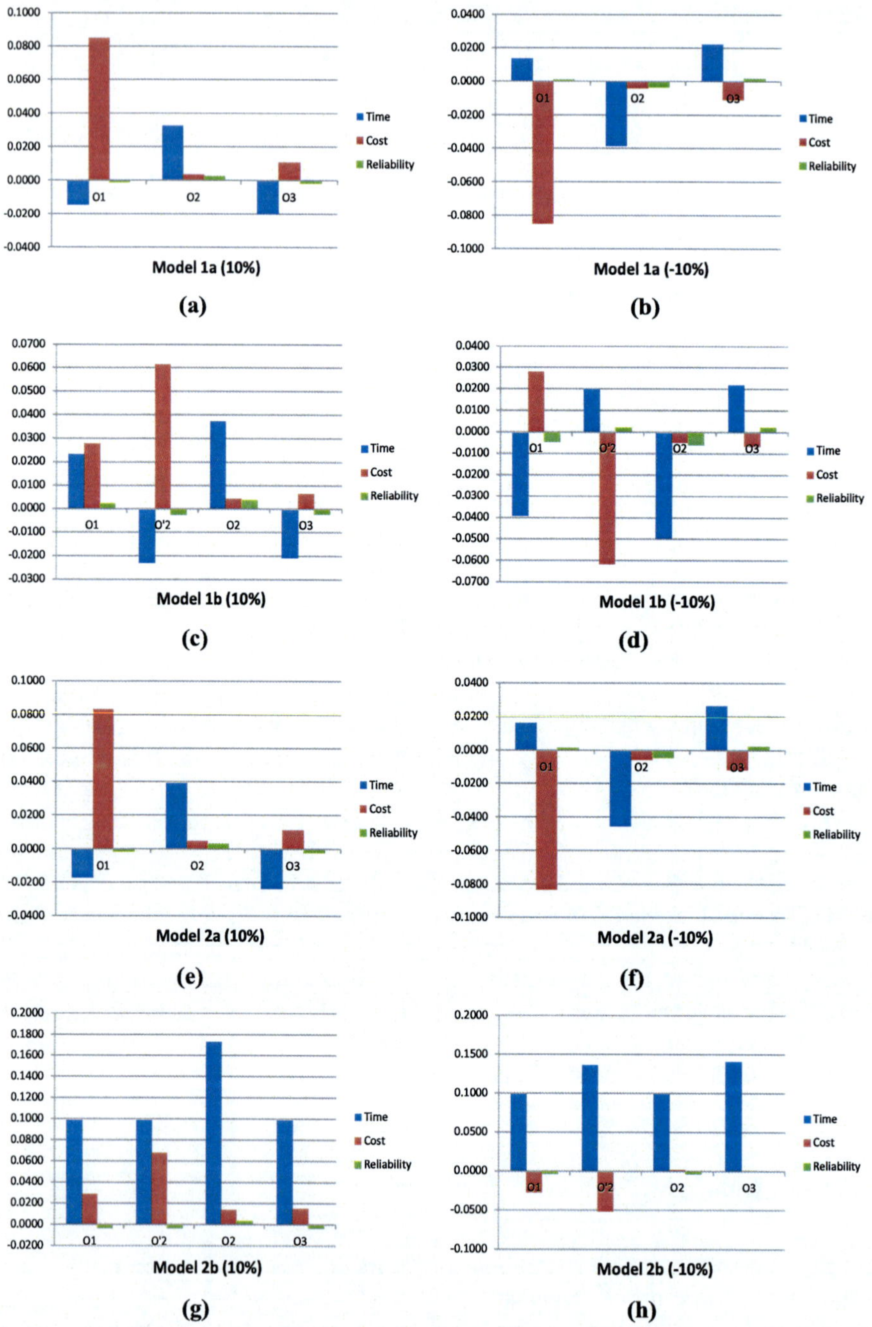
Model 1a (10%)
(a)
Model 1a (-10%)
(b)
Model 1b (10%)
(c)
Model 1b (-10%)
(d)
Model 2a (10%)
(e)
Model 2a (-10%)
(f)
Model 2b (10%)
(g)
Model 2b (-10%)
(h)
Time
Cost
Reliability
O1
O'2
O2
O3

◂**Fig. 6** Relative change in release time, cost and reliability with respect to **a** Model 1a when there is 10% increase in costs O_1, O_2 and O_3 **b** Model 1a when there is 10% decrease in costs O_1, O_2 and O_3 **c** Model 1b when there is 10% increase in costs O_1, O_2', O_2 and O_3 **d** Model 1b when there is 10% decrease in costs O_1, O_2', O_2 and O_3 **e** Model 2a when there is 10% increase in costs O_1, O_2 and O_3 **f** Model 2a when there is 10% decrease in costs O_1, O_2 and O_3 **g** Model 2b when there is 10% increase in costs O_1, O_2' O_2 and O_3 **h** Model 2b when there is 10% decrease in costs O_1, O_2' O_2 and O_3

made with extant models from literature. An optimization formulation has also been proposed for all the four models with respect to desiderated reliability and with the aim of minimizing total expected outlay. Lastly, a sensitivity analysis has been provided to validate the optimal result obtained.

Looking at the results, one can state that model with change point in both cases when the software project characteristic is constant or a power function is an efficient tool in terms of providing a realistic outlook and in comparison to other previous models.

7 Future Scope

The limitation identified for the proposed model is that it has been validated on a small dataset and the under an ideal testing environment. In order to generalize the results, using large datasets would authenticate the efficiency of the proposed model. Further research work can be implemented by considering various attributes like testing effort, imperfect debugging, resource allocation etc.

References

Aggarwal AG, Nijhawan N, Kapur P (2015) A discrete SRGM for multi-release software system with imperfect debugging and related optimal release policy. In: 2015 international conference on futuristic trends on computational analysis and knowledge management (ABLAZE). IEEE, pp 186–192

Castillo E, Mínguez R, Castillo C (2008) Sensitivity analysis in optimization and reliability problems. Reliab Eng Syst Saf 93(12):1788–1800

Chatterjee S, Shukla A (2017) An ideal software release policy for an improved software reliability growth model incorporating imperfect debugging with fault removal efficiency and change point. Asia-Pac J Oper Res 34(03):1740017

Chatterjee S, Nigam S, Singh JB, Upadhyaya LN (2012) Effect of change point and imperfect debugging in software reliability and its optimal release policy. Math Comput Model Dyn Syst 18(5):539–551

Chiu K-C, Huang Y-S, Lee T-Z (2008) A study of software reliability growth from the perspective of learning effects. Reliab Eng Syst Saf 93(10):1410–1421

Goswami D, Khatri SK, Kapur R (2007) Discrete software reliability growth modeling for errors of different severity incorporating change-point concept. Int J Autom Comput 4(4):396–405

Huang C-Y (2005a) Cost-reliability-optimal release policy for software reliability models incorporating improvements in testing efficiency. J Syst Softw 77(2):139–155
Huang C-Y (2005b) Performance analysis of software reliability growth models with testing-effort and change-point. J Syst Softw 76(2):181–194
Huang C-Y, Luo S-Y, Lyu MR (1999) Optimal software release policy based on cost and reliability with testing efficiency. In: Proceedings of the twenty-third annual international computer software and applications conference (Cat. No. 99CB37032). IEEE, pp 468–473
Inoue S, Yamada S (2011) Software reliability growth modeling frameworks with change of testing-environment. Int J Reliab Qual Saf Eng 18(04):365–376
Inoue S, Yamada S (2018) Markovian software reliability modeling with change-point. Int J Reliab Qual Saf Eng 25(02):1850009
Inoue S, Hayashida S, Yamada S (2013) Extended hazard rate models for software reliability assessment with effect at change-point. Int J Reliab Qual Saf Eng 20(02):1350009
Jeske DR, Zhang X (2005) Some successful approaches to software reliability modeling in industry. J Syst Softw 74(1):85–99
Kapur P, Garg R (1989) Cost-reliability optimum release policies for a software system under penalty cost. Int J Syst Sci 20(12):2547–2562
Kapur P, Garg R (1992) A software reliability growth model for an error-removal phenomenon. Softw Eng J 7(4):291–294
Kapur PK, Singh VB, Anand S, Yadavalli VSS (2008a) Software reliability growth model with change-point and effort control using a power function of the testing time. Int J Prod Res 46(3):771–787
Kapur P, Jha P, Singh V (2008b) On the development of discrete software reliability growth models. In: Handbook of performability engineering. Springer, pp 1239–1255
Kapur P, Aggarwal AG, Anand S (2009) A new insight into software reliability growth modeling. Int J Perform Eng 5(3):267–274
Kapur PK, Garg RB, Aggarwal AG, Tandon A (2009a) General framework for change point problem in software reliability and related release time problem. Int J Reliab Qual Saf Eng 16(06): 567–579
Kareer N, Kapur P, Grover P (1990) An S-shaped software reliability growth model with two types of errors. Microelectron Reliab 30(6):1085–1090
Ke S-Z, Huang C-Y, Peng K-L (2014) Software reliability analysis considering the variation of testing-effort and change-point. In: Proceedings of the international workshop on innovative software development methodologies and practices, pp 30–39
Lai R, Garg M (2012) A detailed study of NHPP software reliability models. J Softw 7(6):1296–1306
Lai R, Garg M, Kapur PK, Liu S (2011) A study of when to release a software product from the perspective of software reliability models. JSW 6(4):651–661
Li Q, Pham H (2017) NHPP software reliability model considering the uncertainty of operating environments with imperfect debugging and testing coverage. Appl Math Model 51:68–85
Li X, Xie M, Ng SH (2010) Sensitivity analysis of release time of software reliability models incorporating testing effort with multiple change-points. Appl Math Model 34(11): 3560–3570
Nagaraju V, Fiondella L (2017) A single changepoint software reliability growth model with heterogeneous fault detection processes. In: 2017 Annual reliability and maintainability symposium (RAMS). IEEE, pp 1–6
Okumoto K, Goel AL (1979) Optimum release time for software systems based on reliability and cost criteria. J Syst Softw 1:315–318
Pachauri B, Kumar A, Dhar J (2013) Modeling optimal release policy under fuzzy paradigm in imperfect debugging environment. Inf Softw Technol 55(11):1974–1980
Park J, Baik J (2015) Improving software reliability prediction through multi-criteria based dynamic model selection and combination. J Syst Softw 101:236–244
Pham H (2016) A generalized fault-detection software reliability model subject to random operating environments. Vietnam J Comput Sci 3(3):145–150
Roy P, Mahapatra G, Dey K (2014) An NHPP software reliability growth model with imperfect debugging and error generation. Int J Reliab Qual Saf Eng 21(02):1450008

Sharma K et al (2010) Selection of optimal software reliability growth models using a distance based approach. IEEE Trans Reliab 59(2):266–276
Shrivastava AK, Sachdeva N (2019) Generalized software release and testing stop time policy. Int J Qual Reliab Manag 37(6/7):1087–1111
Shrivastava AK, Kumar V, Kapur PK, Singh O (2020) Software release and testing stop time decision with change point. Int J Syst Assur Eng Manag 11(2):196–207
Singh V, Kapur P, Kumar R (2010) Developing S-shaped reliability growth model for open source software by considering change point. In: Proceedings of the IASTED international conference, vol 677, no 103, p 256
Wood A (1996) Predicting software reliability. Computer 29(11):69–77
Yamada S, Ohba M, Osaki S (1983) S-shaped reliability growth modeling for software error detection. IEEE Trans Reliab 32(5):475–484
Zhao M (1993) Change-point problems in software and hardware reliability. Commun Stat Theory Methods 22(3):757–768
Zhu M, Pham H (2018) A software reliability model incorporating martingale process with gamma-distributed environmental factors. Ann Oper Res 1–22
Zhu M, Pham H (2020) A generalized multiple environmental factors software reliability model with stochastic fault detection process. Ann Oper Res 1–22

Robust Multi-Response Based Software Reliability Modeling

Anusha Pai, Gopalkrishna Joshi, and Suraj Rane

Abstract In quality engineering, robustness indicates reduction in variability in responses in an environment where some of the input parameters are difficult to control. When these variations are minimized the product will be called robust product. This chapter deals with developing a framework of Robust Multi-response based software reliability modeling and later illustrates this methodology on a large scale communication software system. The two responses under study are Time-between-detection of defects; and effort which is defined as time period between defect detection and defect elimination. The inputs considered are severity, change request priority (CR Priority) and software development phase. S/N ratios are computed on these responses. Desirability values are computed on these S/N ratios and converted into single response called overall desirability value. Optimization using response surface methodology is performed on overall desirability value and a curve fitting based reliability model is developed. Reliability is defined on the basis of Time-between-detection of defects and Time-between-detection-and-elimination of defects. The approach will help the software engineers working on software development projects, in understanding the impact of large number of input variables on more responses simultaneously. The approach will also lead to a robust software product and a reliability model relating input variables and responses.

Keywords Software defects · Reliability modelling · Desirability function · Multi-response optimization · Response surface methodology · Effort · Signal-to-Noise ratio

---

A. Pai
Department of Computer Engineering, Padre Conceicao College of Engineering, Verna 403722, Goa, India

G. Joshi
Computer Science Department, KLE Technological University, Hubballi 580031, Karnataka, India

S. Rane (✉)
Mechanical Engineering Department, Goa College of Engineering, Farmagudi 403401, Goa, India
e-mail: ssr@gec.ac.in

A. G. Aggarwal et al. (eds.), *Optimization Models in Software Reliability*, Springer Series in Reliability Engineering,
https://doi.org/10.1007/978-3-030-78919-0_3

1 Introduction

One of the most important attributes of software quality is software reliability. Software has become an indispensable component of most modern systems including medical systems, household appliances, banking systems, defense organizations etc. Its failure at the customer's end has long lasting consequence on the customer's brand image. It is always a challenge for the software developers to build highly reliable software. Reliability is defined as the probability of failure free operation of a program in a specified environment for a specified time (Musa et al. 1987). Software development also has undergone changes from the traditional plan driven approach to agile software development in order to tackle the rapidly changing and evolving customer requirements. The agile development of software focuses on incremental delivery wherein each release delivers a core product with limited functionalities. Each subsequent release adds more functionality to the software product. Defects occur during software development at various stages of development.

These defects need to be identified and corrected as early as possible in the development cycle in order to reduce the cost of defect correction. Literature studies have revealed a number of factors responsible for these defects some of which are requirements and design metrics (Ma et al. 2014), static code features (Moeyersoms et al. 2015), quality metrics (Arar and Ayan 2015), defect inflow data (Rana et al. 2016), McCabe metrics (Erturk and Sezer 2015) etc. It becomes essential to identify the causes of defects in the pursuit of producing defect free software. Hence the first response studied in this work is time between the occurrences of defects. This response needs to be minimized as much as possible. The second response considered in the work is effort which is defined as the time span between when the defect was logged in the system to the time when the defect was rectified and removed from the system. This response also needs to be minimized as far as possible. The two responses are combined into a single response by using desirability function. This is done in order to study the impact of the various input factors on the combined response. In addition to studying the mean desirability value, Signal to Noise ratio (S/N ratio) on the combined D is also computed to study the robustness of the response variable to the effect of input factors. Response surface methodology (RSM) is then applied on the combined response in developing a reliability model relating the combined response to the input factors.

2 Literature Survey

This work demonstrates application of industrial engineering techniques like robust engineering, desirability function approach and response surface methodology, in software engineering. Genichi Taguchi, a Japanese engineer introduced the concept of robust engineering as variance reducing technique (Su 2013; Taguchi 1986; Phadke 1989; Ross 1996). This technique has found more applications in manufacturing and industrial engineering over the years. Software engineering has seen less application of Taguchi Methods. Applications reported in literature on Taguchi in software engineering are software quality testing (Besseris 2010), software defect cost (Lazic and Kovic 2015), information security (Jeong and Yoon 2018) and software defect management (Pai et al. 2019). The effect of interaction among parameters is important in optimization studies. The most used experimental design is the full factorial design 2^n, where each parameter is studied at two levels (Myers and Montgomery 2002). The effect of physicochemical parameters in the study of desalination of brackish water by using 2^3 full factorial design is proposed in (Gmar et al. 2017).

An important tool used in Taguchi methods called Signal-to-Noise ratio is considered as a measure of robustness. For each trial of Derringer's desirability function which is used for combining many responses into one response has been applied in multiple-response optimization in various domains of research. This technique is normally used in conjunction with other optimization techniques like Taguchi Methods (Lin et al. 2012), artificial neural networks and genetic algorithms (Elsayed and Lacor 2013), Response Surface Methodology (Islam et al. 2012). The applications cited in research are thermal engineering (Mohapatra et al. 2019), chemical engineering (Kumari et al. 2018), forestry (Hazir et al. 2018) and manufacturing (Kuram and Ozcelik 2013). Another optimization tool used in this paper is Response Surface Methodology (RSM), which has been applied by many authors to study the effect of input parameters on the response. Its application area is widely spread in various sphere of research some of which are machining (Sivaiah and Chakradhar 2018), chemical engineering (Şimşek et al. 2018), energy (Samsuri and Bakri 2017) and anthology (Chong and Gwee 2018).

3 Methodology

Full factorial design measures the response value for various combinations of input parameters. The number of trials in the experiment depend on the number of factors and their corresponding levels. Since the effect of three input factors with two factors at 4 levels and one factor at 3 levels are studied, the number of trials in the experiment are $4^2 \times 3^1 = 48$ trials as shown in Fig. 1. For each trial, mean and signal-to-noise ratio for each response is computed. S/N ratio is a metric used to measure the deviation of the quality characteristic of interest. There are three different S/N ratios depending on the type of characteristic being evaluated; Lower-is-better, Nominal-is-better and Higher-is-better. The lower-is-better S/N Ratio is chosen for our study, as the aim is to reduce the time between the occurrences of defects and to minimize the time required to eliminate these defects. The Lower-is-better S/N Ratio is given in Eq. 1 (Ross 1996).

$$\mathrm{S/N} = -10{*}\log_{10}\left(\frac{1}{n}\sum_{i=1}^{n} y_i^2\right) \quad (1)$$

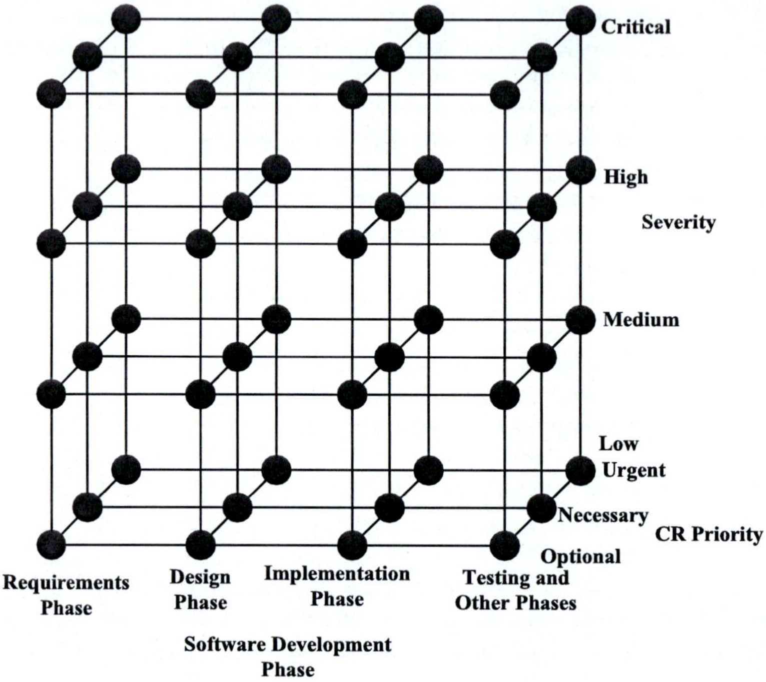

Fig. 1 Full factorial design

Since two responses are studied in this work, multi-response optimization of the two responses using desirability function is performed (Derringer and Suich 1980). Desirability function transforms the values of all responses of the trials to a scale free value between 0 and 1 which are called individual desirability values. The individual responses can be maximized using Eq. 2 and minimized using Eq. 3.

$$d_i = \begin{cases} 0 & y_i \leq L_i \\ \left(\frac{y_i - L_i}{T_i - L_i}\right)^{wt_i} & L_i < y_i < T_i \\ 1 & y_i \geq T_i \end{cases} \tag{2}$$

$$d_i = \begin{cases} 1 & y_i \leq T_i \\ \left(\frac{U_i - y_i}{U_i - T_i}\right)^{wt_i} & T_i < y_i < U_i \\ 0 & y_i \geq U_i \end{cases} \tag{3}$$

where the response variable is y_i, the upper limit is U_i, the lower limit is L_i, the target value is T_i, and $wt_i$is 1 for a linear desirability function. The Overall Desirability value, D can then be computed using Eq. 4.

$$D = (d_1.d_2.\ldots.d_m)^{1/m} \tag{4}$$

Response Surface Methodology optimizes a response under study by obtaining the optimal values of the various input parameters. RSM finds its application in design and development of new products and in the improvisation of existing processes. RSM is extensively used in studying the effect of several input factors on the response or the quality characteristic under study. The relationship between the input factors and the response can be fitted as a second-order model as shown in Eq. 5 (Myers and Montgomery 2002).

$$Y = \beta_0 + \beta_1 x_1 + \beta_2 x_2 + \beta_3 x_3 + \beta_{11} {x_1}^2 + \beta_{22} {x_2}^2 + \beta_{33} {x_3}^2 + \beta_{12} x_1 x_2 + \beta_{23} x_2 x_3 + \beta_{13} x_1 x_3 + \varepsilon \quad (5)$$

where, ε is the error term assumed to have a normal distribution with mean zero and variance σ^2 and β's is the set of unknown parameters. In order to estimate the values of β in Eq. 5, data has to be collected on the system of study. Hence the data collection phase has to be planned carefully for which Response Surface Designs are valuable. Response Surface Methodology are used to optimize a response under study, for selecting the optimal values of the various inputs in order to meet customer specifications. Figure 2 depicts the methodology used in this work.

4 Implementation

The proposed methodology was implemented on the defect data of software products delivered from a large telecommunication organization. The embedded software is installed in consumer electronic devices like IOT devices and smartphones. The organization wanted us to identify mission critical and safety critical software defects before pre-production. The company used agile software development methodology in the development of embedded software. The defect data consisted of 60,636 lines of defects logged into their database for the period June 2018 to February 2019. The total number of attributes in this system was 24. From among these 24 attributes, eight were chosen based on the domain knowledge, discussion and brainstorming with the agile team in the company. Analytic Hierarchy Process (AHP) (Saaty 1980) was used to understand the hierarchical importance of determinants of defects in software development or factors (Durmusoglu 2018). Hierarchy Process (AHP). The nature of the assessment of factors is qualitative with judgments of stakeholders which include agile coach, product owner, the programming team and the customer. A questionnaire was prepared to do the pair-wise comparison. Expert judgments have then been solicited and subsequently compiled. The factors were ranked based on the geometrical mean of the priority vectors of each questionnaire. The results of AHP were then discussed with the agile team from the company and in accordance with their recommendation three factors; Severity, CR Priority and Software development Phase were chosen for further analysis. The description and levels of factors are given in Table 1. The levels for these factors were decided by the domain experts in the company.

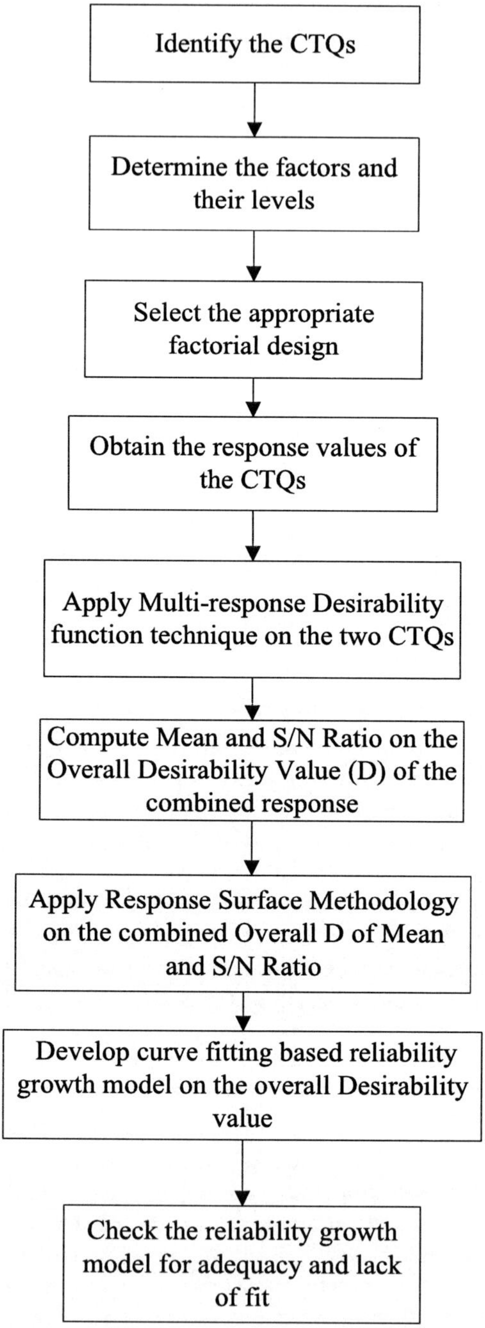

Fig. 2 Proposed methodology

Table 1 Factors, their levels and description

Input factor	Level no	Level label	Description
CR priority (A)	1	Optional	Since product acceptance will not be impacted, attention is not required
	2	Necessary	Since product acceptance may be impacted if not resolved, attention is required
	3	Urgent	Due to actual impact to internal or customer project milestones for product development, urgent attention is required
Software development phase (B)	1	Requirements	Software product requirements of the first sprint are specified
	2	Design	Detailed design of the software product is specified
	3	Implementation	Implementation of software using appropriate programming language to meet the specified design
	4	Testing and others phases	Manual and automated testing and maintenance of software after release is included in this phase
Severity (C)	1	Low	An imperfection in the implementation which does not cause a crash, functional failure or performance degradation of the product
	2	Medium	The product exhibits functional failure or performance degradation which is either not repeatable or is only associated with an internal requirement or product configuration
	3	High	The product fails to meet a product requirement due to a crash, functional failure or performance degradation but is usable

(continued)

Table 1 (continued)

Input factor	Level no	Level label	Description
	4	Critical	A crash, functional failure or performance degradation has rendered the product unusable which is consistently repeatable under normal usage conditions

The steps involved in implementing the methodology shown in Fig. 2 is listed in this section. The first step is to identify the critical-to-quality (CTQ) characteristics in the study. Time-between-defects and Time-to-eliminate-defects are the two responses that we need to optimize. The second step identifies the input factors and their levels. The input factors identified are CR priority defined at three levels: optional, necessary and urgent levels, Software development phase defined at four levels: requirements phase, design phase, implementation phase, Testing and other phases, and severity defined at four levels low, medium, high and critical. In the third step the full factorial design consisting of 48 trials for the two factors at four levels and one factor at three levels is chosen for the study. The fourth step computes the response values of the CTQs for each of the 48 trials. In the fifth step the concept of desirability is applied in order to combine the two responses into a single desirability value. The overall desirability value for the mean and S/N Ratio of the combined response is then computed. Response surface methodology is applied on the combined desirability value for the mean and S/N Ratio and a curve fitting reliability model is obtained for the robust multi-response optimization problem in sixth step. The developed model is then checked for adequacy and lack of fit. The desirability values computed by using the above stated procedure and using the full factorial design depicted in Fig. 1 are shown in Table 2.

5 Results and Discussion

As a part of statistical analysis, Analysis of Variance (ANOVA) was carried out which provides better clarity about the formulated model. Statistical measures like *P*-value and *F*-value are used for the significance testing of the prepared model. *P*-value is normally known as probability value which interprets the significance of a sample value equal to or more extreme than what was actually observed. Primarily for any statistical analysis confidence level or level of significance is decided by the analyst. For the current study, 95% confidence level with 5% of error or risk has been considered i.e. level of significance, α is set to 0.05. By comparing *P*-value with the level of significance (α), model significance is tested. If *P*-value of any model term is lesser than or equal to 0.05, then it is to be assumed to be significant, otherwise it is termed as insignificant. Thus, from the results presented in table 3 the significant

Table 2 Full factorial design

Sr no	CR priority	Severity	Software development phase	D-mean	D-S/N ratio
1	1	1	1	0.397	0.216
2	1	1	2	0.905	0.000
3	1	1	3	0.838	0.141
4	1	1	4	0.922	0.192
5	1	2	1	0.751	0.439
6	1	2	2	0.774	0.257
7	1	2	3	0.944	0.418
8	1	2	4	0.958	0.208
9	1	3	1	0.639	0.210
10	1	3	2	0.908	0.483
11	1	3	3	0.839	0.902
12	1	3	4	0.884	0.762
13	1	4	1	1.000	0.170
14	1	4	2	1.000	0.102
15	1	4	3	0.945	0.000
16	1	4	4	0.000	0.287
17	2	1	1	0.939	0.179
18	2	1	2	0.928	0.131
19	2	1	3	0.955	0.617
20	2	1	4	0.950	0.432
21	2	2	1	0.898	0.234
22	2	2	2	0.946	0.290
23	2	2	3	0.942	0.441
24	2	2	4	0.923	0.000
25	2	3	1	0.934	0.292
26	2	3	2	0.949	0.238
27	2	3	3	0.937	0.197
28	2	3	4	0.943	0.256
29	2	4	1	0.947	0.257
30	2	4	2	0.973	0.327
31	2	4	3	0.966	0.658
32	2	4	4	0.889	0.227
33	3	1	1	0.981	0.236
34	3	1	2	0.768	0.000
35	3	1	3	0.984	0.000
36	3	1	4	0.997	0.216

(continued)

Table 2 (continued)

Sr no	CR priority	Severity	Software development phase	D-mean	D-S/N ratio
37	3	2	1	0.705	0.376
38	3	2	2	0.942	0.223
39	3	2	3	0.855	0.208
40	3	2	4	0.838	0.426
41	3	3	1	0.609	0.171
42	3	3	2	0.919	0.350
43	3	3	3	0.844	0.456
44	3	3	4	0.849	0.682
45	3	4	1	0.000	0.169
46	3	4	2	0.781	0.190
47	3	4	3	0.099	0.192
48	3	4	4	0.376	0.346

Table 3 Analysis of variance for overall desirability-mean D (mean)

Source	DF	Adj SS	Adj MS	*F*-Value	*P*-Value
Model	14	1.59	0.11	3.00	0.01
Linear	5	0.49	0.10	2.59	0.04
CR priority	1	0.04	0.04	1.11	0.30
Severity	1	0.27	0.27	7.04	0.01
Software development phase	3	0.18	0.06	1.60	0.21
Square	2	0.45	0.22	5.94	0.01
CR priority * CR priority	1	0.35	0.35	9.23	0.01
Severity * Severity	1	0.10	0.10	2.66	0.11
2-Way interaction	7	0.65	0.09	2.45	0.04
CR priority * Severity	1	0.31	0.31	8.19	0.01
CR priority * Software development phase	3	0.08	0.03	0.71	0.56
Severity *Software development phase	3	0.26	0.09	2.27	0.10
Error	33	1.25	0.04		
Total	47	2.83			

factors are Severity, the quadratic term CR Priority and the interaction term CR Priority and Software Development Phase. Generally insignificant model terms have less effect on output responses; therefore, these terms are normally discarded from ANOVA table for better prediction relative to significant terms only.

The regression equation for the Overall Desirability for the mean of the combined response of time-between-defects and time-to-eliminate-defects is given in Eq. 6. To determine how well the model fits our data, goodness of fit statistics is used. S is the measure of the standard deviation of the distance between response values and the fitted values. 'S' is used to assess how well the model describes the response and is measured in the same unit as the response variable. Since the value of S is 0.19, the model describes the Overall desirability-Mean quite well. A measure of the amount of reduction in the variability of Overall desirability-Mean by using the regressor variables CR priority, severity and software development phase in the model is given by R^2. Higher the R^2 value, the better the model fits the data. The model has achieved an R^2 of 55.98%. R^2 increases when more predictors are added to the model.

$$\begin{aligned} \textit{Overall Desirability } D\ (\textit{Mean}) = {} & -0.214 + 0.907A + 0.338C \\ & - 0.094B_1 - 0.161B_2 + 0.187B_3 \\ & + 0.068B_4 - 0.1808A * A - 0.0457C * C \\ & - 0.0879A * C - 0.0253A * B_1 + 0.0141A * B_2 \\ & - 0.0618A * B_3 + 0.0730A * B_4 + 0.0239C * B_1 \\ & + 0.0856C * B_2 - 0.0143C * B_3 - 0.0953C * B_4 \end{aligned} \tag{6}$$

The fitted model is examined to make sure that it provides reasonable approximation to the true system. This is accomplished using model adequacy checking. The first among these is residual analysis. A model fits the data well if the difference between the observed values and the model's predicted values are small and not biased. Since the residuals in Fig. 3 lie approximately along a straight line, the assumption of normality is satisfied.

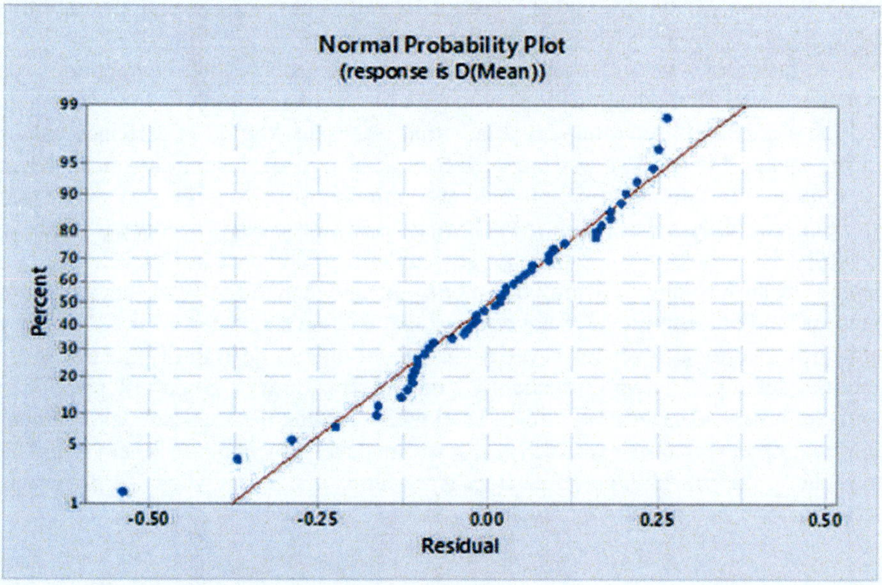

Fig. 3 Normal probability plot of residuals for D (Mean)

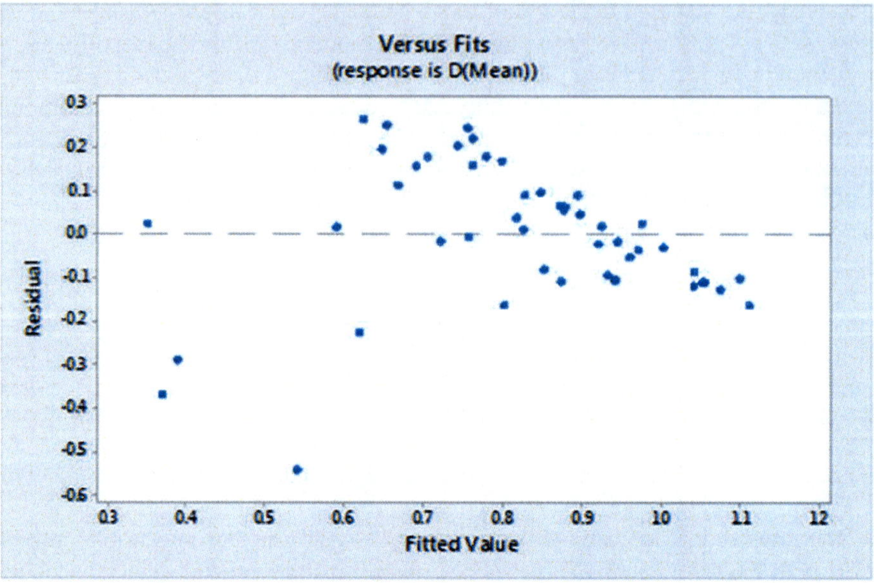

Fig. 4 Plot of residuals versus the predicted response for D (mean)

The plot of residuals versus the predicted response Overall desirability-mean is presented in Fig. 4. The residuals have scattered randomly on the display, suggesting a constant variance of the original observations for all values of the Overall desirability-mean.

A surface plot is used to see how fitted response values relate to two continuous variables based on model equation. A surface plot contains predictors on the x-axis and the y-axis and a continuous surface that represents the fitted response values on the z-axis. Figure 5 shows the surface plot of overall desirability-mean when the software development phase is held at requirements phase. The plot shows that overall desirability-mean D (mean) value can be achieved when severity is trending towards critical level and CR priority between critical and necessary level. Figure 6 depicts the surface plot of overall desirability-mean when the software development phase is held at design level. Between optional and necessary level for CR priority and towards critical level for severity maximum overall desirability-mean can be obtained. In Fig. 7, Overall desirability-mean is maximized when CR priority is at urgent level and severity at critical level while holding the software development phase at implementation level. In Fig. 8, software development phase is held at testing and other phases level. Overall desirability-mean is maximum when CR priority is between necessary and urgent level and severity is at low level.

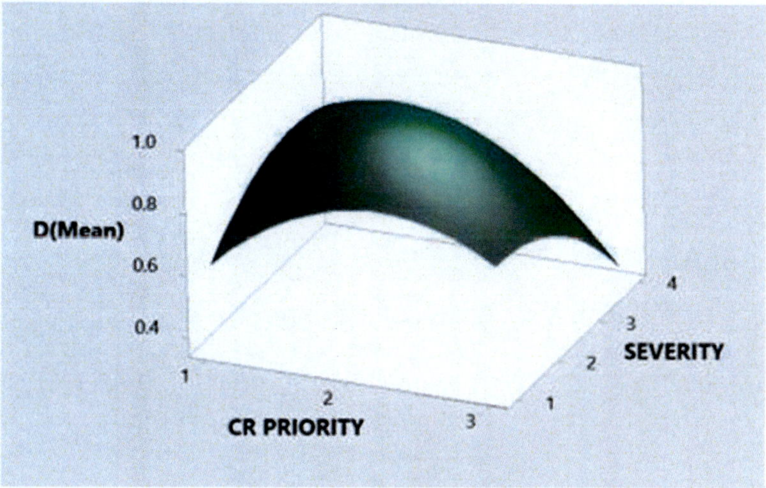

Fig. 5 Surface plot for software development phase-requirements gathering

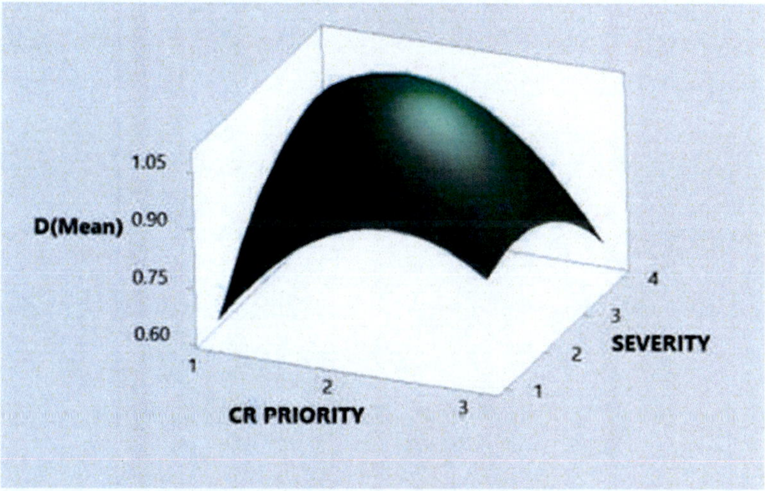

Fig. 6 Surface plot for software development phase-design phase

Table 4 depicts ANOVA on overall Desirability of S/N ratio of the combined response. The P-value of the quadratic term severity is the only significant factor contributing towards variability in the response values. The goodness of fit statistics for the model determines how well the model fits the data. The value of S is found to be 0.19. A low value of S indicates that the model describes the response overall

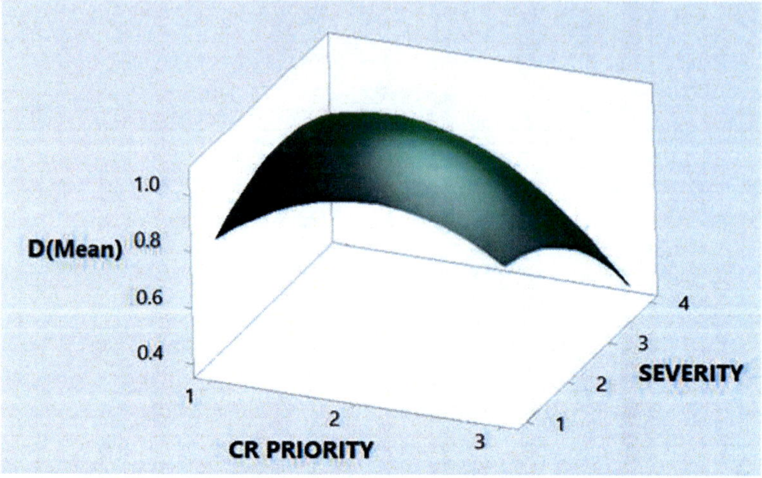

Fig. 7 Surface plot for software development phase-implementation

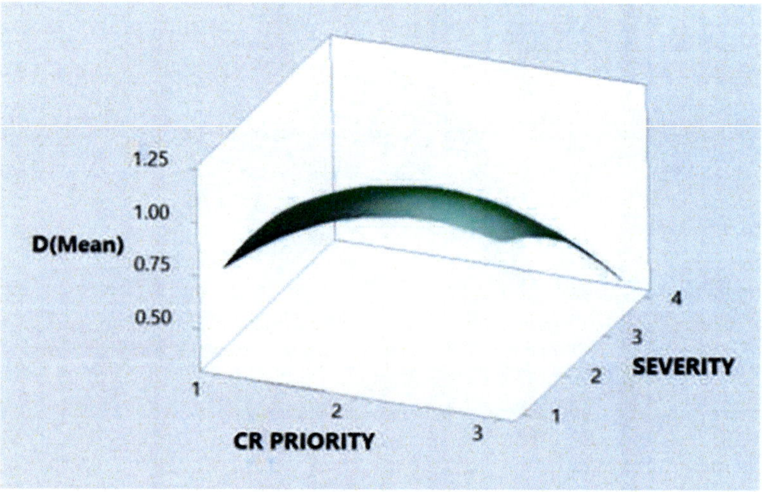

Fig. 8 Surface plot for software development phase-testing and other phases

desirability-S/N ratio pretty well. R^2 value obtained for the model is 28.86%. Higher R^2 value is desirable, as the model fits the data better. R^2 increases when more predictors are added to the model. The regression equation for the overall desirability D for S/ N ratio is given in Eq. 7.

$$\begin{aligned} D(S/N\ Ratio) = & -0.092 + 0.027A + 0.346C \\ & + 0.051B1 - 0.167B2 + 0.185B3 - 0.068B4 \end{aligned}$$

Table 4 Analysis of variance

Source	DF	Adj	Adj	F -value	F -value
Model	14	0.53	0.04	0.96	0.52
Linear	5	0.21	0.04	1.08	0.39
CR priority	1	0.01	0.01	0.24	0.63
Severity	1	0.04	0.04	1.07	0.31
Software development phase	3	0.16	0.05	1.37	0.27
Square	2	0.22	0.11	2.81	0.08
CR priority * CR priority	1	0.00	0.00	0.07	0.79
Severity * Severity	1	0.22	0.22	5.54	0.03
2-Way interaction	7	0.09	0.01	0.34	0.93
CR priority * Severity	1	0.00	0.00	0.08	0.78
CR priority * Software development phase	3	0.04	0.01	0.38	0.77
Severity * Software development phase	3	0.04	0.01	0.38	0.77
Error	33	1.30	0.04		
Total	47	1.82			

$$
\begin{aligned}
&- 0.0165A * A - 0.0674C * C + 0.0087A * C \\
&+ 0.0068A * B1 + 0.0072A * B2 - 0.0587A * B3 \\
&+ 0.0447A * B4 - 0.0425C * B1 + 0.0325C * B2 \\
&- 0.0010C * B3 + 0.0110C * B4 \qquad (7)
\end{aligned}
$$

The Model Adequacy Checking is done using Residual Analysis. The residuals in Fig. 9 plot approximately along a straight line, and so conclusion can be drawn regarding the assumption of normality being satisfied and that there are no issues with normality. Figure 10 shows the residuals scattered at random suggesting a constant variance of the observations for all values of the response overall desirability-S/N ratio.

Surface plot in Fig. 11 holds software development phase at requirements and reveals that overall desirability-S/N ratio is optimum when CR priority values range from optional level to urgent level while severity ranges from medium to high level. Figure 12 reveals optimum overall desirability-S/N ratio when CR priority ranges between optional to urgent whilst severity ranges between medium to high levels holding the third factor software development phase at design level, holding the software development phase at implementation level, Fig. 13 shows optimum values of overall desirability-S/N ratio when CR priority is at optional level and severity ranging from medium to high levels. Figure 14 shows maximum overall desirability-S/N ratio is achieved when CR priority is moving towards urgent level and severity values ranging between medium and high levels.

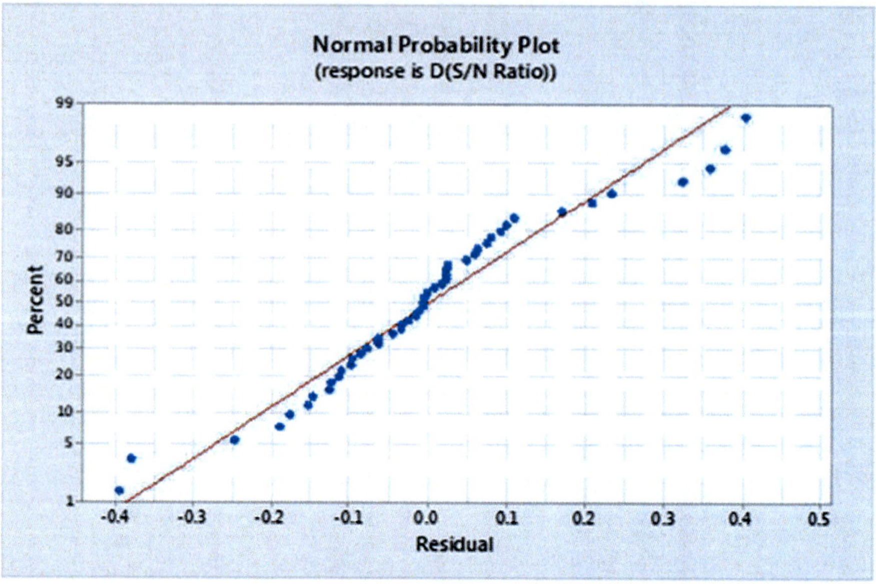

Fig. 9 Normplot of residuals for D(S/N Ratio)

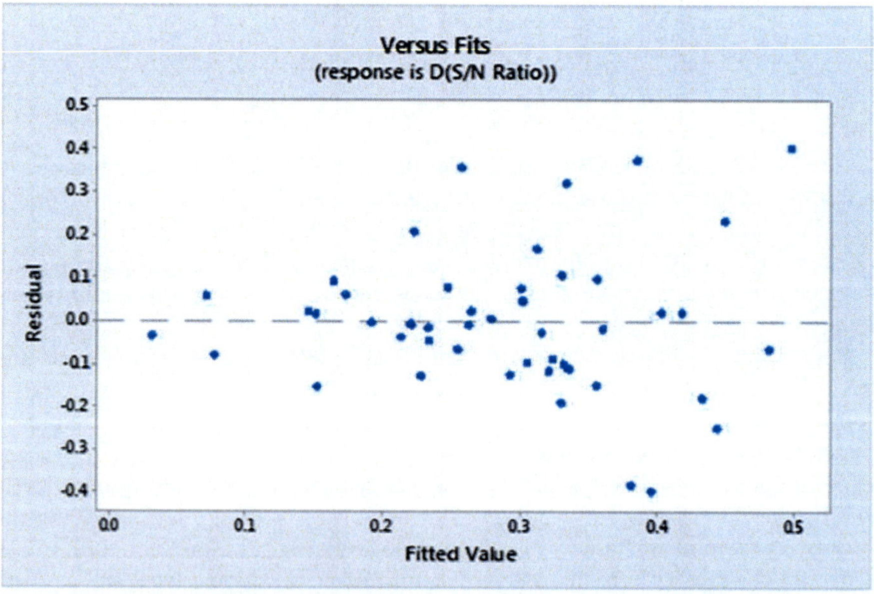

Fig. 10 Residuals versus fits for D(S/N ratio)

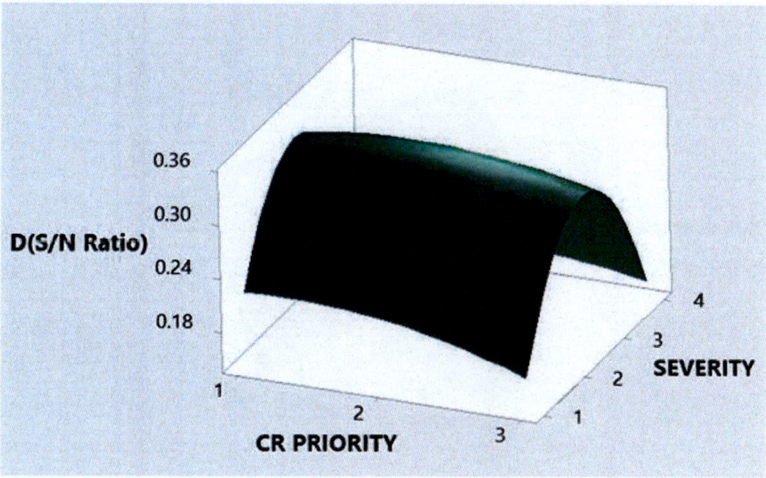

Fig. 11 Surface plot for software development phase-requirements gathering

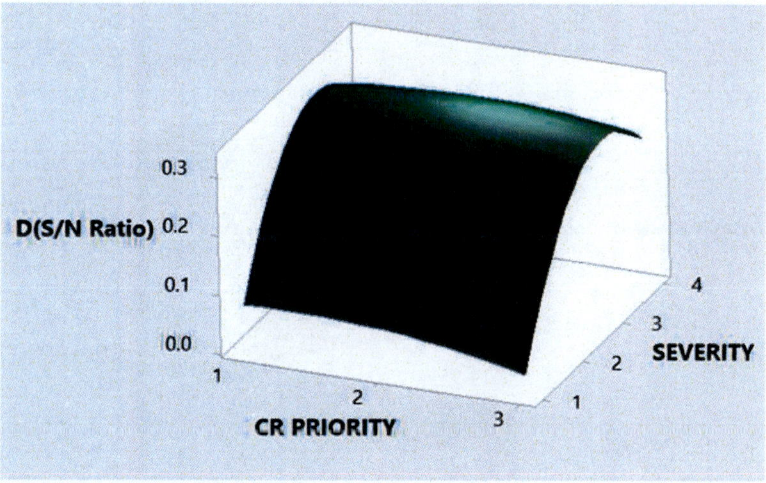

Fig. 12 Surface plot for software development phase-design phase

6 Conclusion

In this chapter, the effect of three input factors CR priority, severity and software development phase on the two responses Time-between-defects and Time-to-eliminate-defects have been evaluated. The two responses were combined using Multi-response desirability function. Interaction between the input factors and their influence on the combined response of overall desirability-mean and overall

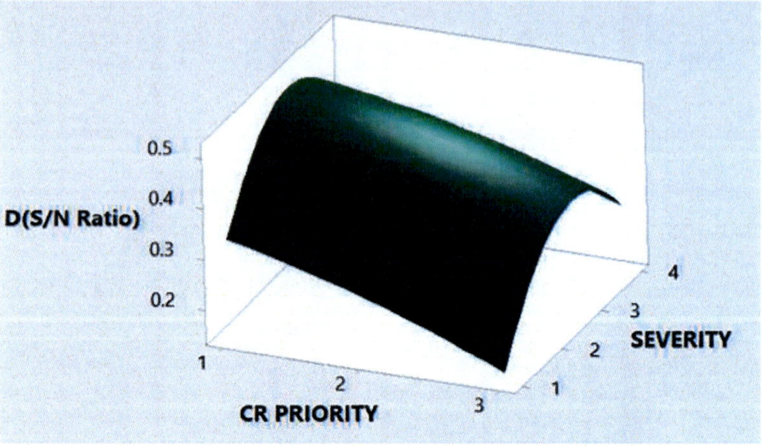

Fig. 13 Surface plot for software development phase-implementation phase

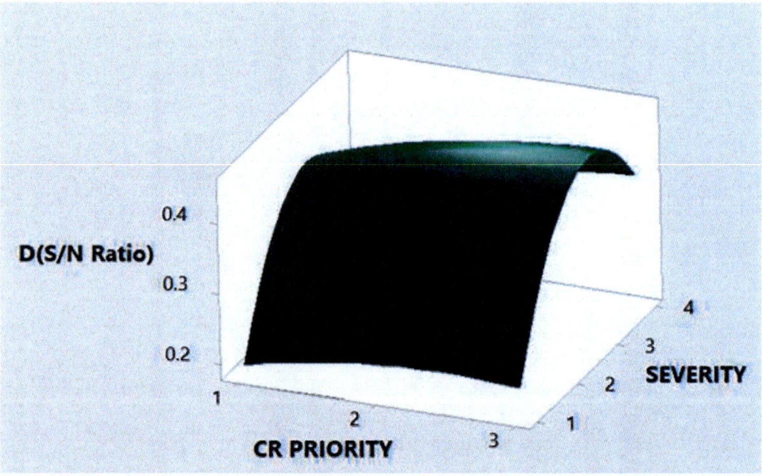

Fig. 14 Surface plot for software development phase-testing and other phases

desirability-S/N ratio has been assessed by the statistical response surface methodology model with full factorial design. A total of 48 runs were conducted in a numerical simulation model as per the design of experiment using Minitab. A multi-objective reliability model has been developed using RSM. The model can be used to predict the desirability values of the mean and the S/N ratios of the combined response for any combination of the input factors. Adequacy of the model was checked by

ANOVA analysis. The optimal levels of the input factors were assessed and a multi-response optimization analysis was conducted for computing the minimum time-between-defects and minimum-time-to-eliminate-defects logged into the system. From this study, results were obtained which are summarized as:

1. The significant factors affecting the combined response of desirability of the mean of time-between-defects and time-to-eliminate-defects are Severity, the quadratic term CR Priority and the interaction term CR Priority and Software Development Phase.
2. A software reliability model relating the input factors to the mean of the overall desirability relating time-between-defects and time-to-eliminate-defects is obtained.
3. Model adequacy checking and the model summary statistics with $S = 0.19$ and $R^2 = 55.98\%$ along with the normal probability plot of residuals and the plot of residuals verses predicted response indicated that the model well fitted the data under study.
4. The quadratic term Severity is the only significant factor contributing towards variability in the combined response value of Overall Desirability-S/N ratio.
5. A reliability model relating the input factors to the Desirability value for S/N Ratio of the combined response of the time-between-defects and time-to-eliminate-defects is obtained.
6. Model Adequacy checking revealed the model fitted the data quite well, though the R-sq value could be improved by adding more predictors in the model.

The thrust in any product development including software is in developing zero defect product. The work in the chapter will guide the development team to understand the effect of the input factors and their optimal levels at which the response of interest produces the optimal result.

References

Arar ÖF, Ayan K (2015) Software defect prediction using cost-sensitive neural network. Appl Soft Comput 33:263–277

Besseris GJ (2010) Taguchi methods in software quality testing. Int J Qual Eng Technol 1(3):339–372

Chong FC, Gwee XF (2018) Ultrasonic extraction of anthocyanin from Clitoria ternatea flowers using response surface methodology. Nat Prod Res 29(15):1485–1487

Derringer G, Suich R (1980) Simultaneous optimization of several response variables. J Qual Technol 12(4):214–219

Durmusoglu ZDU (2018) Assessment of techno-entrepreneurship projects by using analytical hierarchy process (AHP). Technol Soc 54:41–46

Elsayed K, Lacor C (2013) CFD modeling and multi-objective optimization of cyclone geometry using desirability function, artificial neural networks and genetic algorithms. Appl Math Model 37:5680–5704

Erturk E, Sezer EA (2015) A comparison of some soft computing methods for software fault prediction. Expert Syst Appl 42:1872–1879

Gmar S, Helali N, Boubakri A, Sayadi IBS, Tlili M, Amor MB (2017) Electrodialytic desalination of Brackish water: determination of optimal experimental parameters using full factorial design. Appl Water Sci 7(8):4563–4572
Hazir E, Erdinler ES, Koc KH (2018) Optimization of CNC cutting parameters using design of experiment (DOE) and desirability function. J For Res 29(5):1423–1434
Islam MA, Alam MR, Hannan MO (2012) Multiresponse optimization based on statistical response surface methodology and desirability function for the production of particleboard. Compos Part B 43:861–868
Jeong Y, Yoon B (2018) Application of Taguchi model to valuation of information security technology considering security quality failure. Total Qual Manag Bus Excell. https://doi.org/10.1080/14783363.2018.1468246
Kumari A, Karuna B, Satyavathi B (2018) A comprehensive study on equilibrium and kinetics of morpholine extraction from aqueous stream with CA in toluene: experimental evaluation, extraction model and parametric optimization employing desirability function. Chem Eng Res Des 133:243–254
Kuram E, Ozcelik B (2013) Multi-objective optimization using Taguchi based grey relational analysis for micro-milling ofAl7075 material with ball nose end mill. Measurement 46:1849–1864
Lazic L, Kovic SM (2015) Reducing software defects removal cost via design of experiments using Taguchi approach. Softw Qual J 23(2):267–295
Lin H-C, Su C-T, Wang C-C, Chang B-H, Juang R-C (2012) Parameter optimization of continuous sputtering process based on Taguchi methods, neural networks, desirability function, and genetic algorithms. Expert Syst Appl 39:12918–12925
Ma Y, Zhu S, Qin K, Luo G (2014) Combining the requirement information for software defect estimation in design time. Inf Process Lett 114(9):469–474
Moeyersoms J, Fortuny EJ, Dejaeger K, Baesens B, Martens D (2015) Comprehensible software fault and effort prediction: a data mining approach. J Syst Softw 100:80–90
Mohapatra T, Sahoo SS, Padhi BN (2019) Analysis, prediction and multi-response optimization of heat transfer characteristics of a three fluid heat exchanger using response surface methodology and desirability function approach. Appl Therm Eng 151:536–555
Musa JD, Iannino A, Okumoto K (1987) Software reliability: measurement, prediction, application. McGraw-Hill, New York
Myers RH, Montgomery DC (2002) Response surface methodology: process and product optimization using designed experiments, 2nd edn. Wiley, New York
Pai A, Joshi G, Rane S (2019) Integration of agile software development and robust design methodology in optimization of software defect parameters. Int J Syst Assur Eng Manag. https://doi.org/10.1007/s13198-019-00833-6
Phadke MS (1989) Quality engineering using robust design. Prentice Hall, Englewood Cliffs, NJ
Rana R, Staron M, Berger C, Hansson J, Nilsson M, Meding W (2016) Analyzing defect inflow distribution and applying bayesian inference method for software defect prediction in large software projects. J Syst Softw 117:229–244
Ross PJ (1996) Taguchi techniques for quality engineering. McGraw Hill, New York
Saaty TL (1980) The analytic hierarchy process. McGraw-Hill, New York, NY
Samsuri S, Bakri MMM (2017) Optimization of fractional crystallization on crude biodiesel purification via response surface methodology. Sep Sci Technol 53:567–572
Şimşek B, Uygunoğlu T, Korucu H, Kocakerim MM (2018) Analysis of the effects of dioctyl terephthalate obtained from polyethylene terephthalate wastes on concrete mortar: a response surface methodology based desirability function approach application. J Clean Prod 170:437–445
Sivaiah P, Chakradhar D (2018) Analysis and modeling of cryogenic turning operation using response surface methodology. Silicon 10(6):2751–2768
Su C-T (2013) Quality engineering: off-line methods and applications, 1st edn. CRC Press, Boca Raton
Taguchi G (1986) Introduction to quality engineering. Asian Productivity Organization, Tokyo

Multi-criteria Decision Making in Optimal Software Testing-Allocation Problem

Shinji Inoue, Yuka Minamino, and Shigeru Yamada

Abstract We discussed estimating optimal testing-resource allocation for the module testing in software testing phase by applying the notion of the multi-attribute utility theory. Concretely, considering the utilities of software development manager for the reliability, testing-resource and testing-cost, we define testing-resource allocation problems for estimating optimal testing-resource allocation maximizing the utility of the software development manager under the certain testing-strategy. Finally, we show examples of the applications of our approaches by using actual data, and give some consideration on our results and the importance of developing testing management strategy in the software module testing.

1 Introduction

It is well-known that software testing located in the software development process is resource-consuming process. The testing-resource means CPU time, man-month, the number of test cases, and so forth. Generally, software testing consists of the module, integration and system testing. Especially in the module testing, a lot of testing-resource is needed to enhance software reliability of each software module because the software modules are tested independently. Therefore, it is important for software development managers to allocate the testing-resource in the module testing efficiently. The optimal testing-resource allocation problem (Yamada and Ohtera 1990; Nishiwaki et al. 1995; Yamada et al. 1995; Ichimori et al. 2002) is known as one of the interesting problems for software development management.

S. Inoue (✉)
Kansai University, 2-1-1, Ryozenji-cho, Takatsuki-shi, Osaka 569-1095, Japan
e-mail: ino@kansai-u.ac.jp

Y. Minamino · S. Yamada
Tottori University, 4-101, Minami, Koyama-cho, Tottori-shi, Tottori 680-8552, Japan
e-mail: minamino@tottori-u.ac.jp

S. Yamada
e-mail: yamada@tottori-u.ac.jp

A. G. Aggarwal et al. (eds.), *Optimization Models in Software Reliability*,
Springer Series in Reliability Engineering,
https://doi.org/10.1007/978-3-030-78919-0_4

The optimal testing-resource allocation problems discuss how to allocate the testing-resource for each module testing to conduct debugging activities more efficiently. As for most of existing discussions on optimal testing-resource allocation problem, the optimal testing-resource allocation is decided by minimizing the total number of remaining faults in the all software modules under certain constraint of the total testing-resource for conducting the whole module testing.

However, from the point of view of actual software development management, it must be better to consider several evaluation criteria, not only the constraint of the total testing-resource, for deciding the testing-resource allocation in the module testing. As one of the approaches, we discuss optimal testing-resource allocation problems based on the multi-attribute utility theory. The multi-attribute utility theory has been often applied in discussion of software development management. Especially, Kapur et al. (2012) applied the multi-attribute utility theory for developing an optimal software release problem and discussed the estimation procedure for estimating optimal software release time maximizing the utility of the software development manager. We discuss an application of multi-attribute utility theory to the testing-resource allocation problem, which is one of the interesting issues on software development management. Compared with the existing discussion on the optimal testing-resource allocation approach (Nishiwaki et al. 1995; Yamada et al. 1995; Ichimori et al. 2002), our approach gives the another aspects in the software develop management by proposing another optimal problem for testing-resource allocation, such as multi-attribute maximization problem on estimating the optimal testing-resource allocation in the module testing. Further, we show examples of the applications of our approaches by using actual data, in which the multi-attribute utility is formulated from the aspects of the reliability, testing-resource, and testing-cost. And we give some considerations on our results and the importance to develop testing management strategy in the module testing.

2 Section Heading

We introduce an existing approach for optimal testing-resource allocation problem (Nishiwaki et al. 1995; Yamada et al. 1995; Ichimori et al. 2002). In this approach, the software reliability growth process for each module is described by using the testing-effort dependent software reliability growth model (Yamada 2014; Yamada and Ohtera 1990). Let us denote the mean value function of the nonhomogeneous Poisson process (Pham 2000) or the expected number of faults detected up to testing time t by $Z(t)$. The testing-effort dependent software reliability growth model is given as

$$Z(t) = a(1 - \exp[-r \cdot TE(t)]), \tag{1}$$

which is the mean value function of the non-homogeneous Poisson process. In Eq. (1), $TE(t)$ is the total testing-effort expended up to time t, a is the initial faults content, and r is the software fault detection rate per expended testing-effort. Regarding the function $TE(t)$ in Eq. (1), we have

$$TE(t) = \int_0^t te(t)\mathrm{d}t, \tag{2}$$

where $te(t)$ is the testing-effort expenditure expended at testing-time t. And the expected number of remaining faults, $n(t)$, is given as

$$n(t) = a - Z(t) = a \cdot \exp[-r \cdot TE(t)]. \tag{3}$$

The optimal testing-recourse allocation problem is discussed within the following situation: (1) a software system consists of M independent software modules, (2) the number of remaining faults in each module can be estimated by the testing-effort dependent software reliability growth model, (3) the software development managers have to allocate the testing-resource expenditures to each software module testing efficiently, so that the total number of remaining faults in the software system may be minimized.

Let d_i denote the amount of testing-resource expenditure spent for testing software module, i $(i = 1, 2, \cdots, M)$. From Eq. (3), the expected number of remaining faults in the software module i, which is denoted by n_i is given as

$$n_i = a_i \cdot \exp[-r_i d_i], \tag{4}$$

where a_i is the initial fault content for the software module i and r_i represents the fault detection rate per expended testing-resource for the software module i $(0 < r_i < 1)$. Then, the total expected number of remaining faults in the software system is

$$N = \sum_{i=1}^{M} n_i. \tag{5}$$

Then, the software testing-resource allocation problem is formulated as

$$\left.\begin{aligned} &\text{min: } \sum_{i=1}^{M} \delta_i a_i \cdot \exp[-r_i d_i] \\ &\text{subject to } \sum_{i=1}^{M} d_i \le W_P, \quad d_i \ge 0, \end{aligned}\right\}, \tag{6}$$

where W_P is the planned total amount of the testing-resource and δ_i is the weight representing the importance or complexity of the software module i. We should note that the optimal testing-resource allocation problem in Eq. (6) is defined as the minimization problem on the number of remaining faults in the whole software modules under the constraint of the planned total amount of the testing-resource. For solving the Eq. (6), the Lagrange multipliers method is generally applied. Then, we have

$$L = \sum_{i=1}^{M} \delta_i a_i \cdot \exp[-r_i d_i] + \lambda \left(\sum_{i=1}^{M} d_i - W_P \right), \tag{7}$$

where λ is the Lagrange multiplier. The necessary and sufficient conditions for obtaining the optimal solutions are

$$\left. \begin{aligned} &\frac{\partial L}{\partial d_i} = -\delta_i a_i r_i \cdot \exp[-r_i d_i] + \lambda = 0 \\ &\frac{\partial L}{\partial \lambda} = \sum_{i=1}^{M} d_i - W_P \geq 0 \\ &\lambda \left(\sum_{i=1}^{M} d_i - W_P \right) = 0, \qquad \lambda \geq 0 \end{aligned} \right\}. \tag{8}$$

Further, we assume $A_1 \geq A_2 \geq \cdots \geq A_{k-1} \geq \lambda \geq A_k \geq A_{k+1} \geq \cdots \geq A_M$, where $A_i = \delta_i a_i r_i$ $(i = 1, , 2, \cdots, M)$. Then, the optimal testing-resource allocation, d_i^*, is derived as

$$\left. \begin{aligned} &d_i^* = -\frac{1}{r_i}(\ln A_i - \ln \lambda) \qquad (i = 1, 2, \ldots, k-1), \\ &d_i^* = 0 \qquad (i = k, k+1, \ldots, M) \end{aligned} \right\}, \tag{9}$$

where $\ln \lambda$ is given as

$$\ln \lambda = \frac{\sum_{i=1}^{M} \frac{1}{r_i} \ln A_i - W_P}{\sum_{i=1}^{M} \frac{1}{r_i}}. \tag{10}$$

Consequently, the optimal testing-resource allocation is obtained as

$$d_i^* = \max \left\{ 0, -\frac{1}{r_i}(\ln A_i - \ln \lambda) \right\} \qquad (i = 1, 2, \ldots, M). \tag{11}$$

3 Proposed Approach

We propose another approaches for estimating optimal testing-resource allocation with multi-attribute utility of software development manager. Now, we consider the following situation:

(1) A software system consists of M independent software modules.

(2) The number of remaining faults for each software module are estimated by the testing-effort dependent software reliability growth model.
(3) The software development managers allocate the testing-resource to testing for each software module for maximizing their utility, which consists of several attributes being related to the software development management.

We now consider that the following attributes: the testing-resource and the software reliability attributes, respectively. Regarding the testing-resource attribute, we formulate

$$\min : E = \frac{\sum_{i=1}^{M} d_i}{W_P} = \frac{W}{W_P}, \tag{12}$$

since the software managers prefer to spend the testing-resource less than the planned total amount of testing-resource. In Eq. (12), W is the testing-resource expenditure spent up to the end of the module testing. Further, we give the software reliability attribute as

$$\max : R = 1 - \frac{\sum_{i=1}^{M} a_i \cdot \exp[-r_i d_i]}{\sum_{i=1}^{M} a_i} = 1 - \frac{N}{\sum_{i=1}^{M} a_i}. \tag{13}$$

Here we develop the utility function for each attribute based on the following certain situation on the software development management strategy: (1) for the testing-resource attribute, at least 60% of the planned total amount of testing-resources must be consumed, (2) for the software reliability attribute, at least 80% of software faults should be removed, (3) the software development managers take the risk neutral position for each attribute. Then, we set the lowest and highest consumptions for the testing-resource attribute are $E^L = 0.6$ and $E^H = 1.0$, respectively. And the lowest and highest requirements for the software reliability attribute are $R^L = 0.8$ and $R^H = 1.0$, respectively. Then, we obtain the following utility functions for each attribute based on the notion of the risk neutral position:

$$\left.\begin{array}{l} u(E) = 2.5E - 1.5 \\ u(R) = 5R - 4 \end{array}\right\}. \tag{14}$$

From Eq. (14), we define the following optimal testing-resource allocation problem with the additive multi-attribute utility function under the testing management strategy:

$$\left.\begin{array}{ll} \max : u(E, R) & = \delta_R \cdot u(R) - \delta_E \cdot u(E) \\ & = \delta_R \cdot (5R - 4) - \delta_E \cdot (2.5E - 1.5) \\ \text{subject to} & \delta_R + \delta_E = 1, \quad d_i \geq 0 \end{array}\right\}, \tag{15}$$

where δ_R and δ_E are the weight parameters for the attributes R and E, respectively. Consequently, we obtain the optimal testing-resource allocation, d_i^* ($i =$

$1, 2, \ldots, M$), by maximizing the multi-attribute utility function in Eq. (15). From Eq. (15), the optimal testing-resource allocation, d_i^*, is obtained by

$$\left.\begin{array}{l} d_i^* = -\dfrac{1}{r_i} \ln \dfrac{2.5\delta_E \sum_{i=1}^{M} a_i}{5\delta_R a_i r_i W_P} \qquad (i = 1, 2, \cdots, k-1) \\ d_i^* = 0 \qquad (i = k, k+1, \ldots, M) \end{array}\right\}, \tag{16}$$

Then, we have

$$d_i^* = \max\left\{0, -\frac{1}{r_i} \ln \frac{2.5\delta_E \sum_{i=1}^{M} a_i}{5\delta_R a_i r_i W_P}\right\} \qquad (i = 1, 2, \cdots, M). \tag{17}$$

In Eq. (17), we should note that

$$B_1 \geq B_2 \geq \cdots \geq B_{k-1} \geq \frac{2.5\delta_E \sum_{i=1}^{M} a_i}{5\delta_R W_P} \geq B_k \geq B_{k+1} \geq \cdots \geq B_M, \tag{18}$$

where $B_i = a_i r_i$ $(i = 1, 2, \ldots, M)$.

Furthermore, we add the cost attribute for treating more practical situation. For formulating the cost attribute, we set the following cost parameters:

c_1: the debugging cost per fault discovered in the module testing,
c_2: the debugging cost per fault undiscovered during the module testing $(c_1 < c_2)$
c_3: the cost per unit of the testing-resource for the module testing.

Then, the cost function is given as

$$V = c_1 \sum_{i=1}^{M} a_i(1 - \exp[-r_i d_i]) + c_2 \sum_{i=1}^{M} a_i \exp[-r_i d_i] + c_3 \sum_{i=1}^{M} d_i, \tag{19}$$

by following the notion of the testing-effort dependent software reliability growth model. From Eq. (19), the cost attribute is given as

$$\min: \quad C = \frac{V}{C_P}, \tag{20}$$

where C_P represents the planned budget for software testing. And we add the following test management strategy: (4) for the cost attribute, at least 50% of the budget must be consumed in the module testing. The lowest and highest consumptions for the cost attribute are $C^L = 0.5$ and $C^H = 1.0$. Further, we assume the following utility function on the cost attribute: $u(C) = 2C - 1$, which is just one of the examples.

When we consider the three attributes, such as the testing-resource, software reliability, and cost attributes, the optimal testing-resource allocation problem with the three attributes is defined as

$$\left.\begin{aligned}\max: u(E,R,C) &= \delta_R \cdot u(R) - \delta_E \cdot u(E) - \delta_C \cdot u(C) \\ &= \delta_R \cdot (5R-4) - \delta_E \cdot (2.5E-1.5) - \delta_C \cdot (2C-1) \\ \text{subject to} \qquad & \delta_R + \delta_E + \delta_C = 1, \quad d_i \geq 0\end{aligned}\right\}, \tag{21}$$

where δ_C is the weight parameter for the cost attribute. From Eq. (21), the optimal testing-resource allocation, d_i^*, is obtained by

$$\left.\begin{aligned} d_i^* &= -\frac{1}{r_i} \ln \frac{\frac{2.5\delta_E}{W_P} + \frac{2\delta_C c_3}{C_P}}{a_i r_i \left\{\frac{5\delta_R}{\sum_{i=1}^M a_i} + \frac{2(c_2-c_1)\delta_C}{C_P}\right\}} \qquad (i = 1, 2, \ldots, k-1) \\ d_i^* &= 0 \qquad (i = k, k+1, \ldots, M) \end{aligned}\right\}. \tag{22}$$

Then, we have

$$d_i^* = \max\left\{0, -\frac{1}{r_i} \ln \frac{\frac{2.5\delta_E}{W_P} + \frac{2\delta_C c_3}{C_P}}{a_i r_i \left\{\frac{5\delta_R}{\sum_{i=1}^M a_i} + \frac{2(c_2-c_1)\delta_C}{C_P}\right\}}\right\} \qquad (i = 1, 2, \ldots, M). \tag{23}$$

In Eq. (23), we should note

$$C_1 \geq C_2 \geq \cdots \geq C_{k-1} \geq \frac{\frac{2.5\delta_E}{W_P} + \frac{2\delta_C c_3}{C_P}}{\frac{5\delta_R}{\sum_{i=1}^M a_i} + \frac{2(c_2-c_1)\delta_C}{C_P}} \geq C_k \geq C_{k+1} \geq \cdots \geq C_M. \tag{24}$$

4 Numerical Examples

We show numerical examples for our proposed approach by using actual data Yamada and Ohtera (1990). The data consists of 10 modules and 251 faults still remain through the module testing. Table 1 shows the estimated values of a_i and r_i, which have been estimated by following the testing-effort dependent software reliability growth model. "M" means module, then "M1" means the module 1. We set the cost parameters as $c_1 = 1$, $c_2 = 2$, and $c_3 = 5$. And we also set the planned total amount of testing-resource and the budget for the module testing as $W_P = 1.0 \times 10^6$ and $C_P = 1.0 \times 10^6$, respectively.

Tables 2 and 3 show the estimated optimal testing-resource allocation for each module, total amount of optimal testing-resource, and utility along with the several weight patterns for the 2 and 3 attributes, respectively. In Table 2, the optimal total amount of testing-resource increases as the δ_R is increasing and the δ_E is decreasing. And we can say that the utility is ordered as P1 > P5 > P2 > P4 > P3, i.e., the utility is getting higher as the difference between δ_R and δ_E is increased. In Table 3, the optimal total amount of testing-resource is increased as δ_R is increasing and the δ_R and δ_E are decreasing. The order on the utility is P6 > P1 > P5 > P2 > P3 > P4. From Table 3, we can see the differences among the weights must influence the

Table 1 Estimated expected number of remaining faults (2 attributes)

M	a_i	r_i	z_i				
			P1	P2	P3	P4	P5
1	63	5.332×10^{-5}	21.183	5.4920	2.3573	1.0087	0.2615
2	13	2.523×10^{-4}	4.4768	1.1607	0.4974	0.2132	0.0552
3	6	5.262×10^{-4}	2.1465	0.5565	0.2385	0.1022	0.0265
4	51	5.169×10^{-5}	21.851	5.6652	2.4279	1.4050	0.2698
5	15	1.707×10^{-4}	6.6169	1.7155	0.7352	0.3151	0.0817
6	39	5.723×10^{-5}	19.736	5.1168	2.1929	0.9398	0.2437
7	21	9.938×10^{-5}	11.365	2.9466	1.2628	0.5412	0.1403
8	9	1.743×10^{-4}	6.4802	1.6801	0.7200	0.3086	0.0800
9	23	5.057×10^{-5}	22.335	5.7906	2.4817	1.0636	0.2757
10	11	8.782×10^{-5}	11.000	3.3345	1.4291	0.6125	0.1588
N^*			127.19	33.458	14.339	6.1454	1.5933

Table 2 Estimated optimal testing-resource allocation (2 attributes)

	Weight		d_i^*					
	δ_R	δ_E	M1	M2	M3	M4	M5	M6
P1	0.1	0.9	20441	4225.3	1953.5	16397	4794.5	11901
P2	0.3	0.7	45759	9575.8	4518.9	42513	12703	35489
P3	0.5	0.5	61649	12934	6129.1	58905	17666	50294
P4	0.7	0.3	77540	16292	7739.3	75297	22630	65099
P5	0.9	0.1	102858	21643	10305	101412	30538	88687
	Weight		d_i^*					Utility
	δ_R	δ_E	M7	M8	M9	M10	W^*	
P1	0.1	0.9	6177.7	1884.5	579.84	0	68355	1.0428
P2	0.3	0.7	19761	9629.4	27274	13591	220814	0.7636
P3	0.5	0.5	28287	14491	44029	23239	317624	0.7015
P4	0.7	0.3	36813	19352	60784	32888	414434	0.7585
P5	0.9	0.1	50396	27097	87478	48259	568673	0.8793

Table 3 Estimated optimal testing-resource allocation (3 attributes)

	Weight			d_i^*					
	δ_R	δ_E	δC	M1	M2	M3	M4	M5	M6
P1	0.8	0.1	0.1	70464	14797	7022.3	67998	20420	58507
P2	0.6	0.2	0.2	52070	10910	5158.4	49023	14674	41369
P3	0.5	0.25	0.25	44466	9302.5	4387.9	41179	12299	34284
P4	0.4	0.3	0.3	36862	7695.5	3617.4	33335	9923.7	27200
P5	0.2	0.4	0.4	18469	3808.5	1753.6	14363	4178.5	10064
P6	0.1	0.45	0.45	3264.8	595.32	212.98	0	0	0
	Weight			d_i^*					Utility
	δ_R	δ_E	δ_C	M7	M8	M9	M10	$W*$	
P1	0.8	0.1	0.1	33017	17187	53323	28591	371326	0.4430
P2	0.6	0.2	0.2	23147	11560	33928	17423	259262	0.1661
P3	0.5	0.25	0.25	19068	9233.9	25911	12806	212936	0.1024
P4	0.4	0.3	0.3	14988	6907.7	17893	8189.5	166612	0.0966
P5	0.2	0.4	0.4	5119.6	1281.2	0	0	59036.4	0.3537
P6	0.1	0.45	0.45	0	0	0	0	4073.06	0.7266

Table 4 Estimated expected number of remaining faults (3 attributes)

	z_i					
M	P1	P2	P3	P4	P5	P6
1	1.4711	3.9227	5.8840	8.8258	23.532	52.535
2	0.3109	0.8290	1.2435	1.8652	4.9732	11.817
3	0.1491	0.3975	0.5962	0.8943	2.3846	5.3639
4	1.5174	4.0464	6.0695	9.1041	24.275	51
5	0.4595	1.2253	1.8794	2.7568	7.3506	15
6	1.3706	3.6547	5.4820	8.2228	21.925	39
7	0.7893	2.1046	3.1569	4.7353	12.626	21
8	0.4500	1.2000	1.8000	2.6999	7.1988	9
9	1.5511	4.1361	6.2040	9.3058	23	23
10	0.8931	2.3817	3.5724	5.3586	11	11
$N*$	8.9620	23.898	35.846	53.768	138.26	238.49

utility. That is, we can say the testing strategy for the module testing influences on the value of the utility. Further, Tables 1 and 4 show the estimated expected number of remaining faults based on the estimated optimal testing allocation for each module for the cases of 2 and 3 attributes, respectively. Focusing on the column of P5 in Table 1, the number of remaining faults can be reduced to 1.5933 from 251 faults due to the testing-strategy, in which the software development manager set the higher weight for the software reliability attributes. We can obtain the same investigation in Table 4.

5 Concluding Remarks

We discussed approaches for estimating optimal testing-resource allocation based on the multi-attribute utility theory in the module testing of software system. Concretely, applying the software reliability estimated the testing-effort dependent software reliability growth model, testing-resource, and testing-cost attributes for developing the additive linear form multi-attribute function, we formulated a testing-resource allocation problem, which enables us to estimate the optimal testing-resource allocation maximizing the utility of the software development manager under the certain testing-strategy. Further, we showed examples of application of our approaches by using actual data. Then, we discussed the behavior of the utility, optimal testing-resource allocation, and the number of remaining faults, which depend on the weight parameters, such as δ_R, δ_E and δ_C. From the examples of application of our approaches, we investigated that the developing the certain strategy for conducting the differentiated injections to the software reliability, testing-resource, and testing-cost leads to obtaining the higher utility of software developing manager. However, we need to conduct more investigations for our approaches and to figure out the relationship between the testing-strategy and utility by using other actual data and module testing situation.

References

Ichimori T, Tanaka M, Yamada S (2002) An optimal effort allocation problem for both module and integration testing in software development. J Jpn Ind Manag Assoc 53:201–207

Kapur PK, Singh VB, Singh O, Singh JNP (2012) Software release time based on different multi-attribute utility functions. Int J Reliab, Qual Saf Eng 20:1350012

Nishiwaki M, Yamada S, Ichimori T (1995) Testing-resource allocation policies based on an optimal software release problem. J Jpn Ind Manag Assoc 46:182–186

Pham H (2000) Software reliability. Springer, Singapore

Yamada S (2014) Software reliability modeling—fundamentals and applications. Springer Japan, Tokyo/Heidelberg

Yamada S, Ohtera H (1990) Software reliability—theory and practical application, in Japanese. Soft Research Center, Tokyo

Yamada S, Ichimori T, Nishiwaki M (1995) Optimal allocation policies for testing-resource based on a software reliability growth model. Int J Math Comput Model 22:295–301

Release Planning Analysis Through Testing Coverage and Fault Reduction Factor Based Models with Change Point Perspective

Neha, Gurjeet Kaur, and Vinita Jindal

Abstract This study delivers a valuable adjoin to the literature of software reliability growth model (SRGM) and associated software release time problem. In context of SRGM, this study examines the incorporation of testing coverage and Fault Reduction Factor (FRF) simultaneously. Since the fault detection process and coverage rate generally get influenced by several attributes including testing tactics, change in resources etc. therefore the principle of change point analysis was introduced to deal with such changes during the testing phase. Models proposed in the study are developed using Non Homogeneous Poisson Process that can be used to evaluate the reliability of software system quantitatively. We have also discussed goodness-of-fit of the proposed models by statistical estimation on two real life fault datasets. To investigate the optimal time to end the testing of the software, a release planning problem is discussed. The proposed release planning problem takes into consideration the minimization of the development cost subject to the reliability requirement constraint. The developed cost model is modeled with the help of some established cost items such as costs of removing faults in testing phase as well as in operational phase and fixed cost per unit time. The study finally takes into consideration sensitivity analysis in optimal release time problem taking into account the impact of variations in the cost parameters.

Keywords Optimization problem · Release planning · Software reliability growth models · Fault reduction factor · Change point · Testing coverage · Fault reduction factor · Change point · Sensitivity analysis

Neha
Department of Operational Research, University of Delhi, Delhi, India

G. Kaur
Shaheed Sukhdev College of Business Studies, University of Delhi, Delhi, India
e-mail: gurjeetkaur@sscbsdu.ac.in

V. Jindal (✉)
Department of Computer Science, Keshav Mahavidyalaya, Delhi, India

A. G. Aggarwal et al. (eds.), *Optimization Models in Software Reliability*,
Springer Series in Reliability Engineering,
https://doi.org/10.1007/978-3-030-78919-0_5

1 Introduction

Efficient software performance forms a special part in human life. For example in hospitals, railways, national security, defense, online transactions, daily life utility IT instruments etc. all these areas require continuous accessibility of IT tools and programming codes. So, since software is so vital in today era therefore software must be of good quality and hence should be highly reliable. One of the most critical and significant task a development team deals with is building a highly reliable software set-up. Since reliability is a primary consideration of both software developer and users, this has therefore dragged the attention of many academicians and practitioners towards the study of analyzing software related concerns. In recent decades, several approaches have been introduced to measure and enhance the software reliability. Amongst them non homogeneous poison process (NHPP) is recognized to be the most effective and successful analytical method in software reliability engineering. NHPP is a tool that aids in describing the failure phenomenon in the testing period of the software project. The debugging method is considered by these models as a counting process and characterized by mean value function (MVF) and once the MVF is determined reliability can be estimated.

Models illustrating the phenomenon of failure and consequent improvements in reliability due to fault detection/removal are referred to as software reliability growth models (SRGMs). Previously, numerous NHPP based SRGMs have been developed (Goel 1985; Yamada et al. 1984). It has been analyzed in the aforementioned literature references that the faults and testing time are related to each other either exponentially or are S-shaped distributed. These models were blend of both S-shaped and exponential growth curve. The very first SRGM was introduced by Goel and Okumoto (1979), also known as exponential model. The other class of SRGMs introduced later was known as flexible SRGMs (Okumoto and Goel 1979; Yamada et al. 1983).

The model assumed the fault detection rate and initial fault content as constant. Then, later in SRGM development more practical approach catering to real life issues were implemented in SRGMs and hence the reliability was assessed. SRGMs such as testing effort, testing coverage, fault reduction factor (FRF), fault removal efficiency etc. belong to these development (Hsu et al. 2011; Huang et al. 2007; Li and Pham 2017; Zhang et al. 2003). Amongst these considerations, testing coverage and FRF are the two most important challenges that a development team faces and hence are taken in this proposed study to check the influence on reliability of the software.

Testing coverage is observed to be an essential consideration for developers and users equally (Pham and Zhang 2003). From the developer's perspective it helps to assess the software quality and ultimately evaluate the additional endeavour required to improve it. On the other hand, testing coverage helps the users in planning when to buy software system (Pham and Zhang 2003). During testing phase a coverage growth function c(t) that can be stated as the fraction of the code covered up to time t, depicts the coverage growth behavior of software.

To understand the coverage growth behavior, testing coverage based SRGMs have been proposed previously (Li and Pham 2017; Pham 2014) and this incorporation of testing coverage helps in improving the prediction power of the SRGMs. Our proposed study also incorporates this vital factor and we have considered two SRGMs where testing coverage is taken to follow Weibull distribution and Exponentiated Weibull distribution. Due to the robust behavior of Wibull distribution, it has been utilized many times in the field of reliability and survival analysis for example-Geophysics, food science, medical science etc. (Lai et al. 2006). On the other side, Exponentiated Weibull distribution is an extended and advanced version of Weibull distribution as it considers two shape parameters.

The other important metric that needs a concern during testing phase is FRF. During testing phase, detected faults are not always equal to the failure occurred, instead it is a proportion of the failure experienced this specific phenomenon is termed as fault reduction factor (FRF) (Musa 1975). In the generalized basic execution time model, Musa (1975) expressed it as proportionality constant. He assumed that FRF could affect the fault detection and correction framework. Several researchers have emphasized FRF in terms of different ratios such as detectability, associability, fault growth rate and fault exposure ratio (Friedman et al. 1995; Malaiya et al. 1993; Musa 1991; Naixin and Malaiya 1996). Our study illustrates the behavior of both crucial issues namely FRF and testing coverage in NHPP based SRGMs, where FRF is considered as constant factor that influences the detection process in the testing phase. Incorporating these real life issues, we have developed software reliability models that help to accurately and effectively control the reliability growth during testing phase.

SRGMs assume many parameters that represent the various factors or issues. During the implementation of SRGMs in a real test environment for the estimation of reliability, it is considered that SRGM's parameters should remain stable over the testing phase. Although it is not always true, for example, let us assume that after few days of testing, development team observed that the testing needs an additional proficient member and they can make the amendments in the current testing strategies. Due to such kind of changes, the estimated values of some model parameters may change. In the context of SRGM, this point of change in the model is defined as the change point model. Initially, Zhao (1993) proposed change point based SRGM who discussed the change point estimation method in his study. In recent years, various distinctive techniques for change point based SRGMs have been introduced by numerous scholars (Aggarwal et al. 2019; Kaur et al. 2017).

The basic quality characteristics that a user expects from a software system are reliability, planned delivery and development cost. Therefore, the prime focus of a development team is to employ these characteristics so that a long-term profit in the market can be attained from the software development. But software users wish to get a highly reliable software system at lowest prices as soon as possible which conflicts with the developers interest, since their focus is on minimization of the total development cost. Due to this tradeoff between the objectives of the users and development team, management must define the optimal time to terminate the testing and make software available to the users by considering their requirements and satisfying their own objectives as well.

This study discusses such optimization problem that calculates the release time of the software system. In software release time optimization problem, two prime focus of the development team are cost minimization and quality maximization (Yamada and Osaki 1987). Where, cost minimization benefits the development team so that they can provide software at a price desired by the users, there reliability is considered as the most powerful quality measure. Users are more inclined towards the quality attribute of the software, therefore an optimization problem considering both cost model and reliability is considered to identify the optimal release time. We have used cost to model objective function and reliability is used as a constraint with specified lower bound.

Motive of the study

As we have addressed the effect of code coverage and net fault reduction on the reliability of software systems, it is therefore important to integrate these issues into the SRGMs. There are numerous other factors that may change these issues such as expertise of test team, testing methods etc. Therefore, in SRGMs context such changes are modeled with change point analysis. These developed analytical expressions are tested on two real life failure datasets and estimated values of the model parameters are obtained. Further, we have taken into consideration the impact of these factors on the development cost and on the release time of the software system. To determine the optimal values of the incurred development cost and release time we used an optimization problem which is expressed by cost-reliability criteria. We have also discussed the sensitivity analysis that considers variation in the cost parameters so that developers can easily identify the key cost parameter that required more focus during software development.

Organization

The proposed study is systematized as follows,

Section 1—A thorough introduction of the software reliability and its related techniques was given.

Section 2—This accounts for a comprehensive literature view.

Section 3—This section gives the modeling framework of the proposed study.

Section 4—This section of study illustrates the experimental numerical based on the proposed methodology.

Section 5—Theoretical and managerial implications of the study are given in this section.

Section 6—This section concludes the study and provides the future scope of the study.

2 Literature Review

This section presents a categorization of the techniques used and issues encountered in the software development process. It discusses the literature references which lay the groundwork for study of some factors potentially related to software reliability.

2.1 *Software Reliability Growth Models*

Examining the landscape of research in the areas of software reliability engineering-specifically the areas of SRGMs, has evolved since 1979. The term NHPP based SRGM first appeared in scholarly literature by Goel and Okumoto (1979). Authors described the failure detection as Non-Homogeneous Poisson Process (NHPP) considering that hazard rate is a proportion of the leftover faults in the system. Based on the fundamental principle of GO model numerous researches have been conducted that assumed failure and fault removal phenomenon by NHPP (Kapur and Garg 1992; Yamada et al. 1983) and the research is still continuing. Since these models do not assume real life issues a development team faces during testing period of software therefore a number of researches have been done incorporating these issues into the SRGMs namely testing effort, fault efficiency, error generation, fault reduction factor, testing coverage, patching etc. (Chatterjee and Singh 2014; Hsu et al. 2011; Huang and Lyu 2005; Jain et al. 2014b). All these aspects are necessary to be included in tracking the growth of software reliability. Our proposed study considers the fusion of two such critical issues, testing coverage and FRF into SRGMs. Study also assumed that coverage rate is not smooth over the testing time period, there is a change point after that coverage rate gets altered.

2.2 *Testing Coverage*

Previously, a variety of approaches to software reliability evaluation have been proposed by integrating testing coverage into the SRGMs. Chen et al. (1996) discussed the over estimation of reliability due to the observation of similar nature of testing techniques and proposed a technique by addressing both coverage and time to solve this issue. Author applied this technique to SRGMs and investigated the improvement made by them, which resulted in significant reduction in the overestimation of reliability. Further, Malaiya et al. (2002) defined the relation among reliability, time and coverage measures (block, branches etc.) and proposed a logarithmic-exponential model to relate the. Model which was tested over four datasets. Pham and Zhang (2003), suggested an SRGM considering the testing coverage information and further they compared the proposed SRGM with some existing SRGMs. Authors also discuss the release planning problems. Pham (2014) suggested a loglog fault detection rate and coverage based reliability growth models subject to uncertain operating environment and compared the models with existing SRGMs.

Chatterjee and Singh (2014) gave an NHPP based software reliability model with logistic-exponential coverage function under an imperfect debugging environment. They further used the proposed model to determine the optimal time to release the system. Li and Pham (2017) has done a considerable research and proposed a reliability growth model for a uncertain operating environment with imperfect debugging and testing coverage. Tandon et al. (2020) proposed multi-release SRGMs that consider FRF and testing coverage simultaneously to observe reliability growth. Models were validated over two Open Software System datasets. In our proposed study, we have considered these two proposed models and incorporated the change point perspective. Further, we have investigated the termination of testing time for the software system.

2.3 *Fault Reduction Factor*

Fault reduction factor defines a close relationship between the terms faults and failure in testing environment of software development life cycle. It states that faults removed are only a fraction of failure experienced during testing phase (Musa 1975). Several analytical approaches have been proposed that take FRF into consideration to enhance reliability of software. Hsu et al. (2011) proposed FRF based SRGMs that were developed using two cases of FRF. In case one, a constant FRF was taken and the other dealt with the increasing and decreasing behavior of FRF. Jain et al. (2014a) incorporated real life situations and proposed a model that considered imperfect debugging environment of testing phase with multiple change point perspective. Further, authors made an addition to the study and calculated the total expected cost by considering a warranty cost model (Jain et al. 2014b). Chatterjee and Shukla (2016b) proposed a general framework of software reliability model with fault detection and correction process through Weibull-Type FRF to make the model more realistic and flexible under an imperfect debugging environment with change point concept. In 2019 Aggarwal et al. (2019b) introduced an SRGM for multi-release software system by integrating imperfect debugging and time dependent FRF.

2.4 *Change Point Concept*

In practice, the distribution of failure may be influenced by several other parameters such as testing efficiency, resources etc. This phenomenon is termed as change point during software testing phase. Kapur et al. (2007) proposed an SRGM based on severity of the faults using change point. Li et al. (2010) discussed a release policies by including testing effort based SRGM with multiple change points. In 2012, Raju et al. (2012) suggested an SRGM by considering imperfect debugging. Authors also discussed the change point concept and performed parameter estimation to test the model.

Singh et al. (2011) also examined the number of faults removed in the software system using a two dimensional testing coverage based SRGM. The model was developed using Cobb-Douglas production function. Nijhawan and Aggarwal (2015) introduced an SRGM with change point scenario for a multi release software system. In addition to the change point SRGMs, Chatterjee and Shukla (2016a) proposed a test coverage and change point based SRGM. Authors used a time dependent fault detection probability and incorporated into SRGM with s-shaped coverage. Further, Chatterjee and Shukla (2019) introduced a unified methodology to model a testing coverage based reliability growth models by considering change point scenario. Author also discussed the imperfect fault detection process into the proposed SRGM context.

2.5 Release Planning

An emphasis on customer retention ultimately resulted into more attention on releasing the software at scheduled time. Decision of releasing software is primarily focused on the number of removed faults. This scenario results in development of release time optimization problem incorporating SRGM. In past few decades, several researches have been introduced that dealt with different release polices for the software system. Okumoto and Goel (1979) discussed an optimum release strategy based on exponential SRGM. Huang and Lyu (2005) showed the effectiveness of both testing effort and testing efficiency and also discussed the release planning problem.

Li et al. (2010) optimized the release time with the help of testing effort based reliability models. Kapur et al. (2012) introduced a 2-dimensional SRGM for multiple versions of software systems. Authors considered Cobb-Douglas function to derive the model that was further used to formulate an optimization problem for release planning. Kumar et al. (2018) proposed a release time optimization problem using an SRGM incorporating patching. Verma et al. (2020) discussed release planning problem using grey wolf optimizer and results were compared with the other soft computing techniques. In this study we investigate a release planning problem based on SRGMs with a change point perspective. The considered optimization problem consists of cost model as an objective function and software reliability as the constraint. In the next segment, we thoroughly present the proposed methodology.

3 Modelling Framework

This section deals with the notations used, assumptions upon which the SRGMs are proposed and the techniques used for assessing performance of the models. It also discusses the formulation of an optimization problem to calculate the testing termination period and express the optimal release time of software for its operational phase (Fig. 1.).

Notation

$m(t)$	Expected number of faults removed till time t for model.
$a(t)$	Initial fault content at time t
$\varphi(t)$	Fault detection rate
$c(t)$	Time dependent testing coverage function
r	Fault reduction factor
T	Release time of the software
T^*	Optimal release time of the software
τ	Change point
C_1	Cost of removing faults before change point in testing phase
C_2	Cost of removing faults after change point in testing phase
C_3	Cost of removing faults in operational phase
C_4	Cost per unit time
$R(t)$	Reliability at time t
$b_1, s_1(b_2, s_2)$	Scale and shape parameter of Weibull distribution before (after) change point
$k_1, l_1, v_1(k_2, l_2, v_2)$	Shape and scale parameters of Exponentiated Weibull distribution before (after) change point

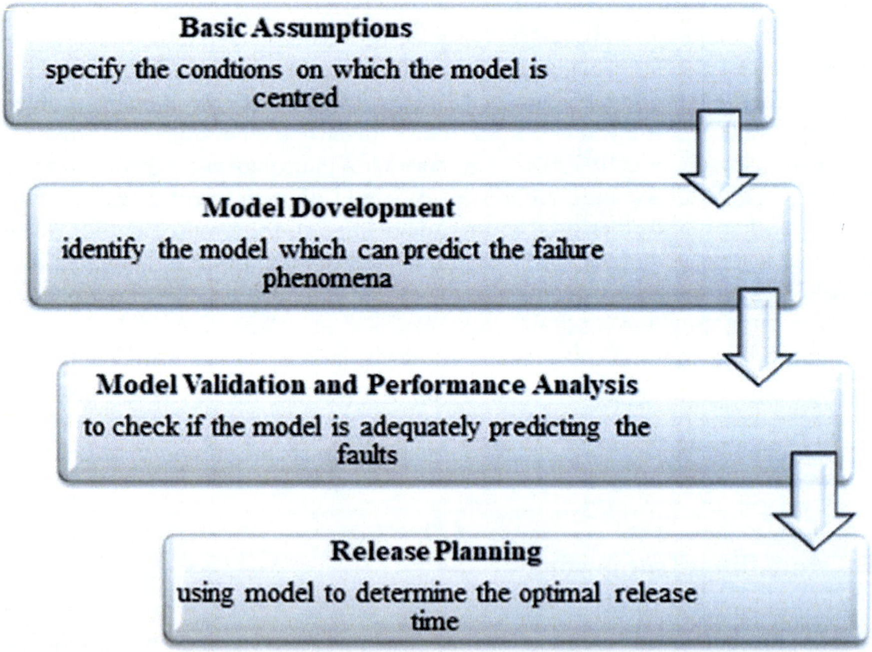

Fig. 1 Illustration of flow of appropriate methodology for modeling and release planning

3.1 Assumptions

In order to develop reliability growth model based on the NHPP, following basic assumptions are considered,

1. Occurrence of the software failure follows NHPP process.
2. Failure rate relies on the fault detection rate and leftover faults in system. Fault detection rate can be considered in terms of the percentage of code examined up to time t.
3. Faults are mutually independent and debugging is perfect.
4. The fault detection rate is a function of constant FRF.
5. Change point perspective is considered.

3.2 Model Development

A general mathematical expression of SRGM can be derived by following differential equation,

$$\frac{dm(t)}{dt} = \varphi(t)(a(t) - m(t)) \tag{1}$$

One can apply different function of $a(t)$ and $\varphi(t)$ to develop an SRGM based on more real life issues of testing phase. A variety of $a(t)$ and $\varphi(t)$ functions reveals numerous assumptions of testing process. In the best possible case, these functions are assumed to be constant, this SRGM was given by Goel and Okumoto (1979). Constant initial fault content $a(t)$ implies perfect debugging process and constant fault detection rate $\varphi(t)$ implies that failure intensity is directly proportional to the left over faults.

The proposed study considers perfect fault debugging i.e., $a(t) = a$. $\varphi(t)$ which is represented in terms of testing coverage function, i.e., $\varphi(t) = \frac{c'(t)}{1-c(t)}$ (Pham and Zhang 2003). Therefore, the above mentioned differential equation can be rewritten as,

$$\frac{dm(t)}{dt} = \frac{c'(t)}{1 - c(t)}(a + m(t)) \tag{2}$$

Also, according to the fourth assumption, this equation becomes

$$\frac{dm(t)}{dt} = r\frac{c'(t)}{1 - c(t)}(a + m(t)) \tag{3}$$

where, r is the FRF. Two cases of testing coverage function $c(t)$ are assumed, in case-1, $c(t)$ is assumed to follow Weibull distribution functions and in case-2, it is following Exponentiated Weibull distribution function. Testing coverage is combined with a constant FRF to incorporate more practicality in the proposed SRGMs. Based on the discussion, mean value function (MVF) for both the SRGMs are as follows, (Tandon et al. 2020).

Model 1

Model 1 is based on 1st case of testing coverage, i.e., it considers $c(t)$ as Weibull distribution function, $c(t) = 1 - e^{-bt^s}$. and constant FRF. The MVF in this is given as follows,

$$m(t) = a\left(1 - e^{-\mathrm{rb}t^s}\right) \tag{4}$$

Model 2

It includes testing coverage as Exponentiated Weibull distribution function which is given by, $c(t) = \left(1 - e^{(-(t/l))^k}\right)^v$. MVF for model 2 is,

$$m(t) = a\left(1 - \left(1 - \left(1 - e^{\left(-\left(\frac{t}{l}\right)^k\right)}\right)^v\right)^{\mathrm{r}}\right) \tag{5}$$

Change point Analysis

As we have discussed that fault detection, correction process and code coverage can be affected by several factors for example testing strategies, variation in test team, resources etc. that implies a discontinuity in the detection, correction process and eventually fault coverage rate. In SRGM context, this discontinuity in model parameters with respect to time is defined as change point analysis. This discontinuity can occur at any moment of time in testing period of software development process. Mathematically, testing coverage using change point perspective can be defined as follows,

$$c(t) = \begin{cases} c_1(t), 0 \le t \le \tau \\ c_2(t), \tau < t \end{cases} \tag{6}$$

where τ represents the change point during testing. We derived the change point SRGM for both cases of testing coverage which are as follows,

Model 1

This case considers testing coverage as Weibull distribution function. Thus solving Eq. 4 using change point concept (Eq. 6) with initial conditions $m(0) = 0$ and $c(0) = 0$, we get the MVF as follows,

$$m(t) = \begin{cases} a\left(1 - e^{-\mathrm{r}b_1 t^{s_1}}\right), 0 \le t \le \tau \\ a\left(1 - e^{-r(b_1\tau^{s_1} + b_2(t^{s_2} - \tau^{s_2}))}\right), \tau < t \end{cases} \tag{7}$$

Table 1 Mean value function of SRGMs

Model	Mean value function ($m(t)$)
Model 1 (without change point)	$a\left(1 - e^{-rb_1 t^{s_1}}\right)$
Model 2 (without change point)	$a\left(1 - \left(1 - \left(1 - e^{\left(-\left(\frac{t}{l_1}\right)^{k_1}\right)}\right)^{v_1}\right)^{r}\right)$
Model 1 (with change point)	$a\left(1 - e^{-r\left(b_1 \tau^{s_1} + b_2\left(t^{s_2} - \tau^{s_2}\right)\right)}\right)$
Model 2 (with change point)	$a\left(1 - \left(\frac{\left(1 - \left(1 - e^{\left(-\left(\frac{\tau}{l_1}\right)^{k_1}\right)}\right)^{v_1}\right)\left(1 - \left(1 - e^{\left(-\left(\frac{t}{l_2}\right)^{k_2}\right)}\right)^{v_2}\right)}{\left(1 - \left(1 - e^{\left(-\left(\frac{\tau}{l_2}\right)^{k_2}\right)}\right)^{v_2}\right)}\right)^{r}\right)$

Model 2

This case assumes testing coverage as Exponentiated Weibull distribution function. Thus, solving Eq. 5 using change point concept (Eq. 6) with initial conditions $m(0) = 0$ and $c(0) = 0$, we get the MVF as follows,

$$m(t) = \begin{cases} a\left(1 - \left(1 - \left(1 - e^{\left(-\left(\frac{t}{l_1}\right)^{k_1}\right)}\right)^{v_1}\right)^{r}\right), 0 \leq t \leq \tau \\ a\left(1 - \left(\frac{\left(1 - \left(1 - e^{\left(-\left(\frac{\tau}{l_1}\right)^{k_1}\right)}\right)^{v_1}\right)\left(1 - \left(1 - e^{\left(-\left(\frac{t}{l_2}\right)^{k_2}\right)}\right)^{v_2}\right)}{\left(1 - \left(1 - e^{\left(-\left(\frac{\tau}{l_2}\right)^{k_2}\right)}\right)^{v_2}\right)}\right)^{r}\right), \tau < t \end{cases} \quad (8)$$

The analytical expressions of the models used in this study are summarized in Table 1.

3.3 Validation and Performance Analysis

Once the model is derived it is then selected for an application on a real life dataset. Its performance can be measured with its capability to meet the experiential data and to forecast the future pattern of the failure in satisfactory manner. Many performance criteria are introduced previously to validate the fitness of models on any actual data. The criteria that are considered in this chapter have explained clearly in Table 2.

Table 2 Performance criteria

Performance criteria	Description	Expression
Mean Square Error (MSE)	Helps to calculate the distance of the expected values from the observed values. Minimum is the distance better the model is. *n* is the number of the observations	$\frac{1}{n}\sum_{i=1}^{n}(m_i(t) - y_i)^2$
Predictive Risk Ratio (PRR)	Helps to assess the estimated values of the experimental values with respect to the model estimates	$\sum_{i=1}^{n}\frac{(m_i(t)-y_i)}{m_i(t)}^2$
Predictive Power (PP)	This metric depends on the distance amid the expected values and the experimental values with respect to the observed values	$\sum_{i=1}^{n}\frac{(m_i(t)-y_i)}{y_i}^2$
Mean Absolute Predictive Error (MAPE)	This measure helps to determine the preciseness of the model's prediction	$\frac{100}{n}\sum_{i=1}^{n}\left\lvert\frac{y_i-m_i(t)}{y_i}\right\rvert$
R^2	It is intended to evaluate the significant affiliation that occurs amid the variables and lies in within 0 and 1. Higher the value better is the fitness	$1-\sum_{i=1}^{n}\frac{(m_i(t)-y_i)}{(m_i(t)-y_i)^2}^2$

3.4 Release Planning of Software System

Once the model is tested, it can also be used to assess the optimum time when software system launches. The problem described in this chapter will help to estimate the time to finish the testing and make the new product accessible for its operating phase. We have used a constraint optimization problem to decide the optimum release time, that reduces the cost of development by considering system's reliability no less than the predefined level of aspiration.

A software system's performance is usually relying on reliability obtained through testing period and it is found that the performance of the software is higher for longer testing. Whereas, extended software testing entails a delay in the planned release time, in addition to increase in the assigned development cost and gradual loss of goodwill. So we use optimization of the cost function and get the optimum cost for the software system to address this contradictory scenario. Okumoto and Goel (1979) suggested a cost model for the software by using three cost parameters namely; debugging cost incurred in testing, debugging cost in operational phase and fixed cost per unit time i.e., C_1, C_3 and C_4 respectively, which is formulated as,

$$C(T) = C_1 m_1(T) + C_3(m_2(T_\infty) - m_2(T)) + C_4 T \tag{9}$$

To incorporate the change point concept Kapur et al. (2009) formulated a release time problem using change point based SRGMs, and the modified cost function includes different cost of fault removal after and before change point, given as,

$$C(T) = C_1 m_1(\tau) + C_2(m_2(T) - m_1(\tau)) + C_3(m_2(T_\infty) - m_2(T)) + C_4 T \quad (10)$$

Here, C_1 is the fault removal cost before change point and C_2 is the fault removal cost after change point and $m_1(t)$ and $m_2(t)$ represent the mean value function for testing process before and after change point. The value of the cost parameters mainly depend on the testing environment, team skills and testing strategies of the team etc.

As reliability is the most crucial factor that highlights the user's requirements, hence it is considered to be a key quality measure of the software system. Therefore, it is important to consider this key quality factor when developing the software. Yamada and Osaki (1987) proposed and constraint optimization problem that minimizes the development cost with respect to reliability as its constraint function and formulated as,

$$\begin{aligned} &\min C(T) \\ \text{Subject to} \quad & \\ &R(T) \geq R_0 \end{aligned} \quad (11)$$

where $T \geq 0$.

We assumed reliability function as $R(T) = \frac{m(T)}{a}$ given by Huang et al. (2002). We use estimated values of the model parameters to perform optimization that ultimately helped calculate the release time and minimize the development cost. The proposed methodology is implemented in the next section to illustrate the numerical example.

4 Numerical Illustration

Here, the estimation of proposed models is put forth and parameters used in the models are estimated. These estimated values are then used to identify the optimum software system release time. Datasets used in this analysis are collected from a bug tracking system Bugzilla (https://bugzilla.gnome.org) and from Firefox (https://bugzilla.mozilla.org). The applications of the models on these collected datasets are given as follows,

First Application (DS-1)

The proposed models are estimated using the non-linear LSE technique. The estimated parameters can be used to get information about the leftover fault in the system and thereby can be used to define the accuracy of the model in predicting the failure. Table 3 depicts the estimation output of model 1 and model 2 for both without and with change point ($\tau = 12$). Further, we have assessed the performance of the proposed models using some fitness criterions namely MSE, PPR, PP, MAPE and R^2 (Table 4).

The values of performance criteria show that the models with change point are more significant in terms of explaining the failure in the system as the R^2 is greater

Table 3 Parameters estimations (DS-1)

Model 1				Model 2			
Parameter	Estimated Values Without change point	Parameter	Estimated Values With change point	Parameter	Estimated values Without change point	Parameter	Estimated Values With change point
a	91.769	a	87.096	a	84.995	a	85.1
r	0.088	r	0.984	r	0.146	r	0.142
b	0.324	b_1	0.041	k	2.062	k_1	1.57
s	1.423	b_2	10.692	l	4.881	k_2	2.509
		s_1	0.177	v	0.303	l_1	3.601
		s_2	0.995			l_2	6.765
						v_1	0.539
						v_2	14.985

Table 4 Goodness of fit criteria (DS-1)

Performance criteria	Model 1		Model 2	
	Without change point	With change point	Without change point	With change point
MSE	5.644	4.3230	3.809	6.2783
PPR	9.475	0.7647	2.084	0.1891
PP	5.832	0.3215	2.108	0.1484
MAPE	7.738	6.2083	4.197	5.2093
R^2	0.992	0.995	0.991	0.994

and hence better than the model where change point is not taken into account. Models fitness is also observed graphically by plotting goodness of fit curve with 95% confidence bound of the estimated faults. From the comparison of visual interpretation of Fig. 2, we can state that models with change point are lying more close to the real values of data thereby indicating better fit of the models.

Second Application (DS-2)

The proposed models are tested on another dataset DS-2. Table 5 depicts the estimation output using model 1 and model 2 for both without and with change point concept ($\tau = 31$). After obtaining the predicted values of the models' parameters, we then evaluate the performance of the models in predicting the data. From Table 6, it is clear that MSE, PRR, PP and MAPE of the model with change point are lowest than the models of without change point perspective and $\boldsymbol{R}^2$ is greater for models with change point. It means change point shows a significant part to enhance the performance of the models.

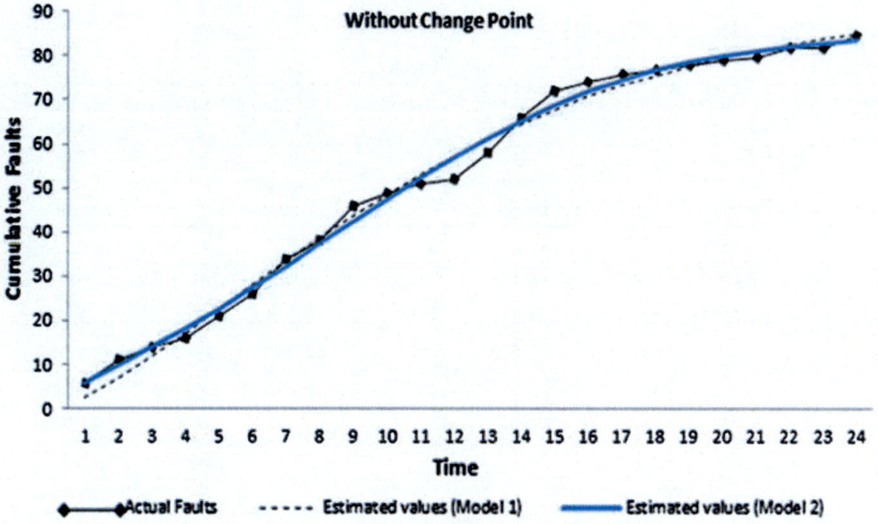

(a)

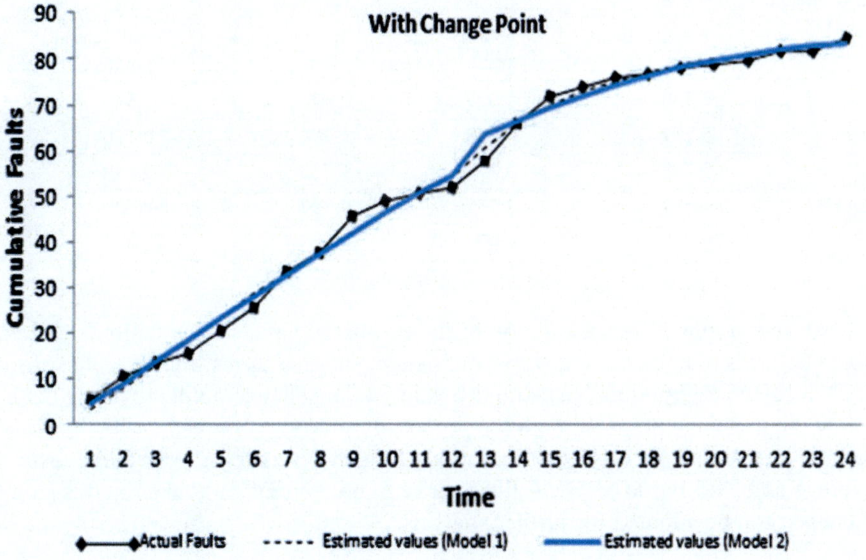

(b)

Fig. 2 Goodness of fit curve for DS-1. **a** Without change point. **b** With change point

Table 5 Parameters estimations DS-2

Model 1				Model 2			
Parameter	Estimated Values Before Change Point	Parameter	Estimated Values After Change Point	Parameter	Estimated Values Before Change Point	Parameter	Estimated Values After Change Point
a	94.332	a	71.246	a	71.327	a	66.225
r	0.17	r	0.156	r	0.949	r	0.123
b	0.717	b_1	1.141	k	4.736	k_1	0.402
s	0.547	b_2	1.533	l	67.310	k_2	3.393
		s_1	0.557	v	0.082	l_1	0.048
		s_2	0.665			l_2	21.904
						v_1	13.328
						v_2	20.635

Table 6 Goodness of fit criteria DS-2

Performance criteria	Model 1		Model 2	
	Without change point	With change point	Without change point	With change point
MSE	3.259	2.029	2.945	1.346
PPR	5.252	0.198	6.274	0.110
PP	5.423	0.299	7.733	0.119
MAPE	4.639	3.585	4.797	2.707
R^2	0.982	0.99	0.983	0.994

Next, the graphical representation of the estimated value of the faults is plotted using goodness of fit curve with 95% confidence bound of the estimated faults. From Fig. 3, it can be interpreted that estimated values using change point are fitting better than without change point for both the proposed models. After testing the fitness of models, we can decipher that incorporation of change point perspective into testing coverage and FRF based SRGMs gives better fit on the real data taken in this study and hence can be adopted for further analysis.

4.1 *Optimal Software Release Time*

To calculate the ideal time for termination of testing, we have used optimization problem given by Eq. 11 with the estimated parameters considered in the proposed SRGMs, we have performed an optimization. While solving the problem, four cost

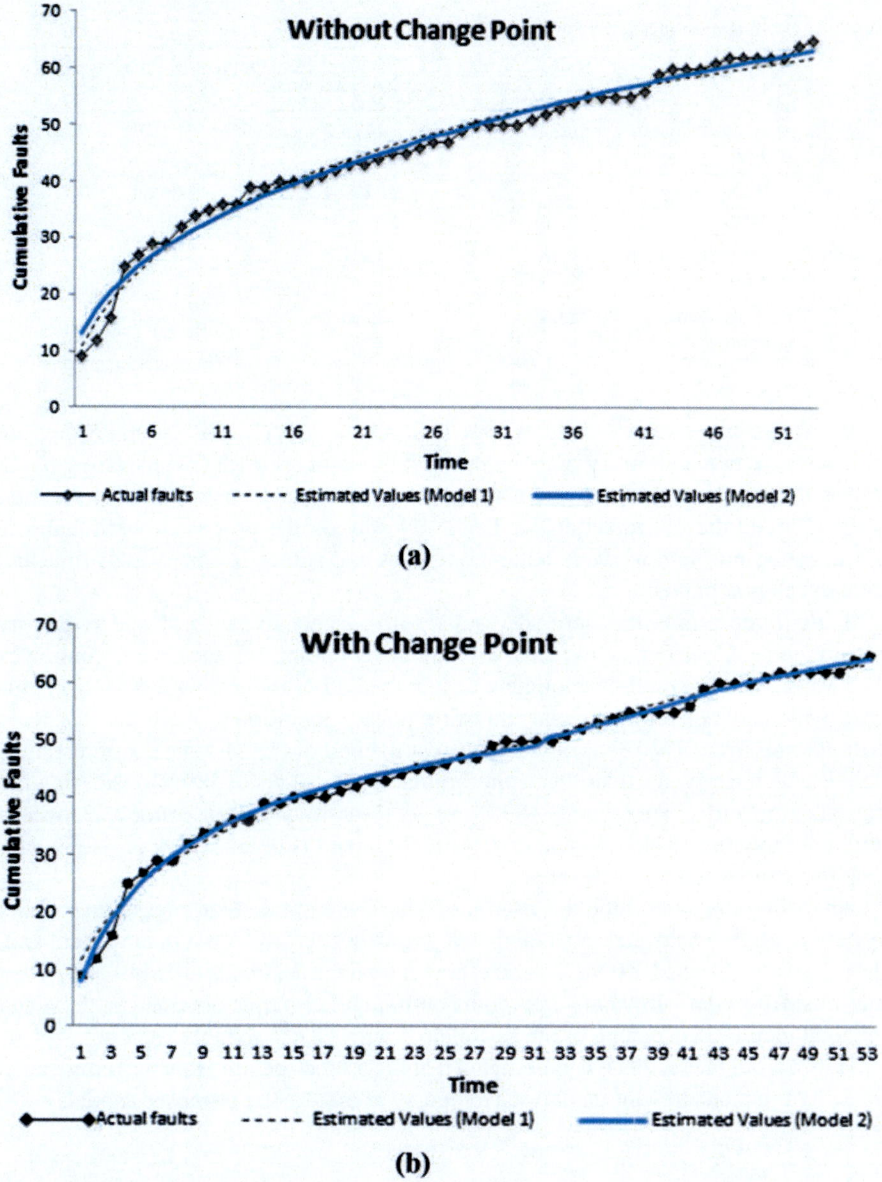

Fig. 3 Goodness of fit curve for DS-2. **a** Without change point. **b** With change point

Table 7 Optimization problem results using DS-1

		Release Time (T^*)	Budget Consumed $C(T^*)$	Reliability $R(T^*)$
Model 1	Without change point	31.5	4416.2	0.979
	With change point	25.7	4451.4	0.964
Model 2	Without change point	24.99	3961.6	0.987
	With change point	25.04	3966.7	0.988

values are assumed to be as follows $C_1 = \$40$, $C_2 = \$50$, $C_3 = \$100$ and $C_4 = \$20$. The costs are taken arbitrary with the fact that the fault removal cost in testing phase is less than the cost of fault removal in operational phase. Using this information we performed the optimization for DS-1. By solving the proposed cost-reliability optimization problem we have obtained the optimal values for the release time and total development cost.

Table 7 represents the optimal results with respect to the cost and reliability requirements. Considering model 1 without change point, release time is coming as 31.5 weeks and the total development cost at the end of the testing is $4416.2. This cost is inclusive of fault removal in testing phase, operational phase and the fixed cost per unit time. The reliability achieved at the end of the 31.5 weeks of testing is 0.979, which is greater than the aspired reliability level $R_0 = 0.9$. Considering the optimization with change point SRGM, we get the software release time as 25 weeks with optimum cost $4451.4. Then reliability obtained is 0.964 which is also greater than the aspired reliability level.

Analyzing proposed model 2 from the Table 7 we obtain that release time of the system is 24.99 weeks and the total development cost is $3961.6. On the other hand, the reliability obtained is 0.987 which is higher than the desired level. Similarly, when we consider model with change point the optimal release time obtained is 25.04 and the total incurred cost came out to be $3966.7 with 98.8% reliable software.

Hence from the results it can be stated that the all the specification and constraints made by the management team or developers are met by the proposed models.

4.2 Variations in Cost Parameters

Sensitivity analysis is the study to examine the consequence of variations in parameters of mathematical model or system on the outputs or performance of the system. It is observed that cost parameters are usually dependent on the capacity of the efficiency of the research team and can be modified. Using the sensitivity analysis, we are analyzing the economic effect of cost adjustment on the total cost usage and optimal

Table 8 Sensitivity analysis of variations with no change point

		Changed cost parameters	T^*	$C(T^*)$	$R(T^*)$
Model 1	10% increased	$C_1 = 44$	31.119	4775.4	0.977
		$C_3 = 110$	32.496	4433.7	0.982
		$C_4 = 22$	30.950	4478.7	0.976
	10% decreased	$C_1 = 36$	31.947	4056.5	0.981
		$C_3 = 90$	30.407	4395.2	0.975
		$C_4 = 18$	32.198	4352.5	0.981
Model 2	10% increased	$C_1 = 44$	24.986	4297.5	0.988
		$C_3 = 110$	25.1	3972.0	0.988
		$C_4 = 22$	24.924	4011.6	0.988
	10% decreased	$C_1 = 36$	24.989	3625.8	0.988
		$C_3 = 90$	24.492	3951.0	0.987
		$C_4 = 18$	24.989	3911.7	0.988

release time. The main advantage of this methodology is that it allows focusing on recognition of key or prime parameters. Our proposed study carried out the analysis using the relative change in the cost parameters. Mathematically, relative change is given as,

$$relative\ Change = \frac{new\ value - old\ value}{old\ value}$$

We consider the case wherein we evaluate the change in total cost and release time by increasing (and decreasing) the cost parameters by 10%. Tables 8 and 9 depict the output using the DS-1 and a graphical representation of relative changes are also plotted in the Figs. 4, 5, 6 and 7.

Table 8 shows the obtained values release time, cost and reliability for the different values of the parameters when SRGM do not have change point incorporated. It clearly depicts that for the increments in the cost parameters C_1 and C_4 of model 1 and model 2, optimal time and reliability both decreases while on contrary increase in C_3 increases the testing time and reliability as well. On the other hand, if we decrease the cost parameters by 10% the release time and optimal reliability decrease with change in C_1 and C_4, whereas fault removal cost during operational phase gives early release of the software system with less reliability.

Now, considering models with change point we examined the variations in cost parameters and Table 9 depicts the optimal values obtained for release time, cost and reliability. These results clearly show that for the increment in cost of fault removal of before and after change point in testing phase, optimal time decreases for both the models. When the cost C_1 increased and decreased by 10%, the reliability remains unchanged for model 1. And when C_2 increased and decreased by 10%, reliability for model 2 remains unchanged. This implies that cost fluctuations before the change

Table 9 Sensitivity analysis of variations with change point models

		Changed cost parameters	T^*	$C(T^*)$	$R(T^*)$
Model 1	10% increased	$C_1 = 44$	25.703	4680.5	0.964
		$C_2 = 55$	25.055	4584.2	0.961
		$C_3 = 110$	26.859	4480.2	0.969
		$C_4 = 22$	25.116	4502.2	0.962
	10% decreased	$C_1 = 36$	25.71	4222.3	0.964
		$C_2 = 45$	26.302	4317.3	0.967
		$C_3 = 90$	24.347	4417.3	0.957
		$C_4 = 18$	26.366	4417.5	0.967
Model 2	10% increased	$C_1 = 44$	24.482	4304.5	0.985
		$C_2 = 55$	25.036	3967.0	0.988
		$C_3 = 110$	25.511	3974.6	0.990
		$C_4 = 22$	25.012	4016.7	0.980
	10% decreased	$C_1 = 36$	24.152	3637.0	0.984
		$C_2 = 45$	23.271	3984.5	0.978
		$C_3 = 90$	24.123	3958.3	0.983
		$C_4 = 18$	25.373	3915.4	0.990

point do not alter the reliability of the system by model 1 and change in cost after change point for model 2 do not alter the reliability of the system.

Based on the performed numerical experiment, we can plot the graph for the relative change in the release time, reliability and total cost for variations in each cost parameters Figs. 4, 5, 6 and 7. Following conclusions can be drawn from the above experimental analysis,

1. The increment in fault removal cost in testing phase and fault removal cost per unit time result in early release of the software system for both the models (with and without change point concept). While decrease in these cost parameter results in delay in the planned delivery for all the proposed models.
2. For the increment in fault removal cost in operational phase a delay is observed in release time for all the proposed models.
3. Amongst all the cost parameters change in fault removal cost in operational phase depicts a large variation in software release time. However, from the budget perspective cost of eliminating fault in testing processe shows a wide difference in overall budget consumption.

5 Theoretical and Managerial Implications

A combined analysis of testing coverage and FRF with change point perspective during the testing process provides an insight of what needs to be addressed in the area

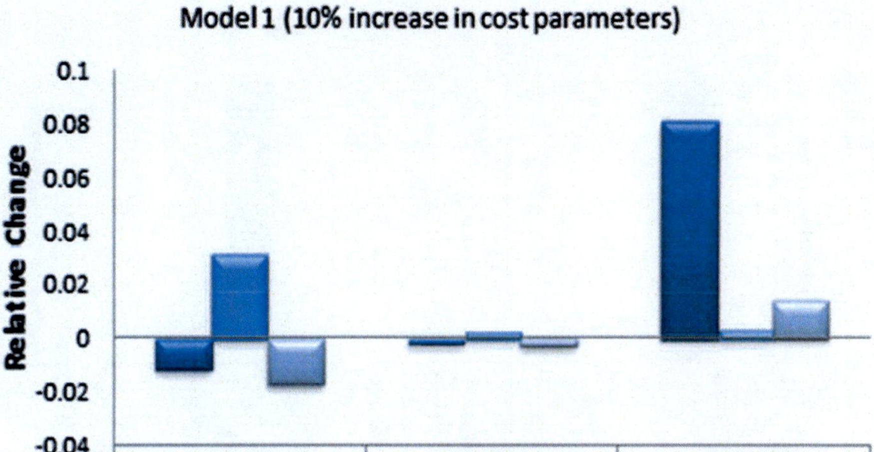

	Time	Reliability	Budget Consumed
C1	-0.012095238	-0.002042901	0.081336896
C3	0.031619048	0.003064351	0.003962683
C4	-0.017460317	-0.003064351	0.014152439

(a)

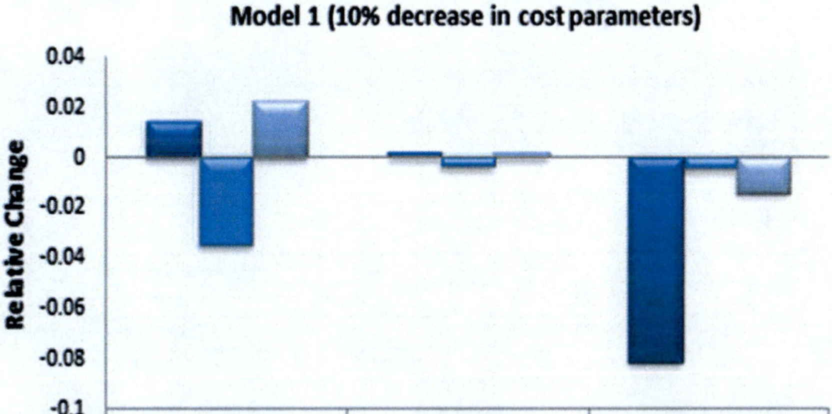

	Time	Reliability	Budget Consumed
C1	0.014190476	0.002042901	-0.081450115
C3	-0.034698413	-0.004085802	-0.004755219
C4	0.02215873	0.002042901	-0.014424166

(b)

Fig. 4 Relative change using model 1 (without change point)

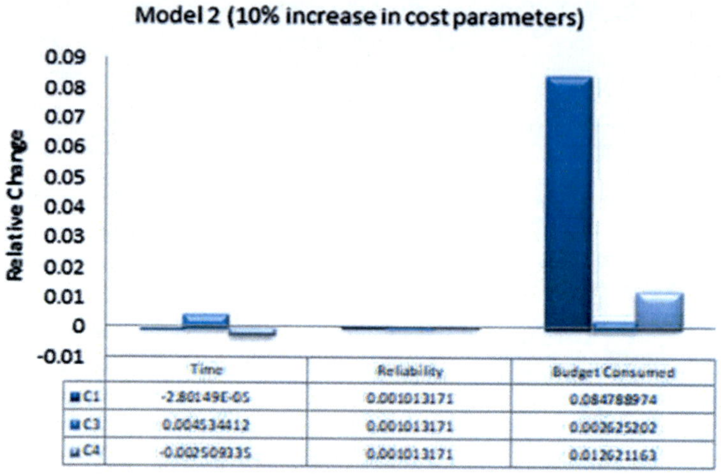

(a)

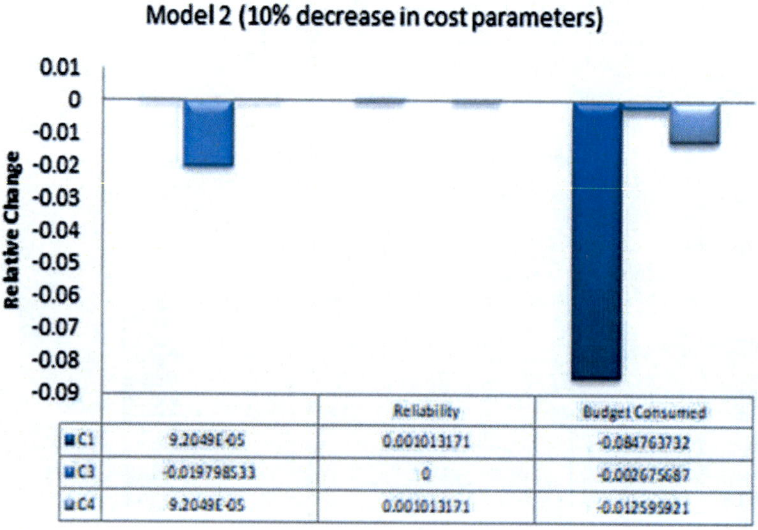

(b)

Fig. 5 Relative change using model 2 (without change point)

of software reliability. It offers a reference framework by explaining how software reliability increases by integrating these real life and vital factors to the SRGM with change point. Estimated values clearly indicate that models are accurately predicting the fault content. Similar concepts were kept in past as well (Aggarwal et al. 2019; Kapur et al. 2009). After estimation models are then used to evaluate the optimal release time and expected software system. Change is observed in total development

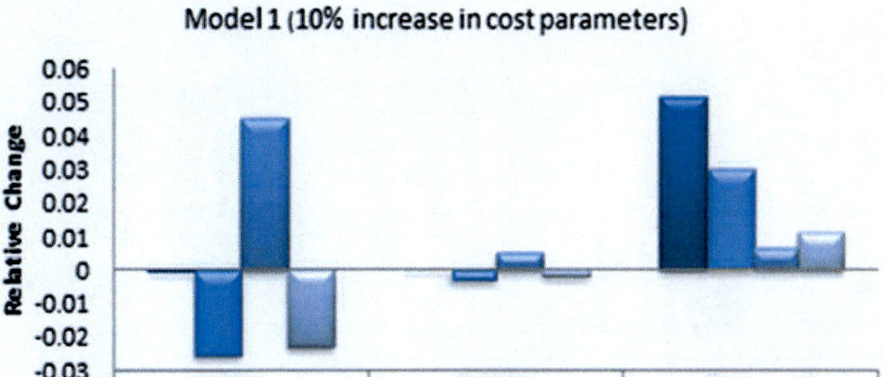

	Time	Reliability	Budget Consumed
C1	-3.89045E-05	0	0.051466954
C2	-0.025248988	-0.003112033	0.029833311
C3	0.044934641	0.005186722	0.006469875
C4	-0.022875817	-0.002074689	0.01141214

(a)

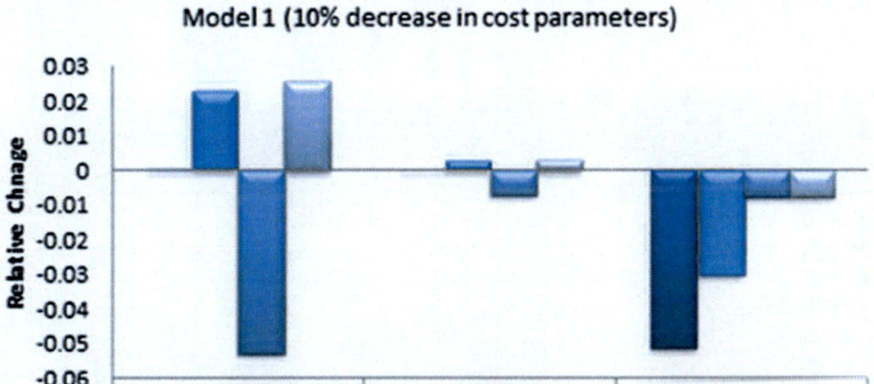

	Time	Reliability	Budget Consumed
C1	0.000233427	0	-0.051466954
C2	0.023264862	0.003112033	-0.030125354
C3	-0.05279334	-0.007261411	-0.007660511
C4	0.025754746	0.003112033	-0.007615582

(b)

Fig. 6 Relative change using model 1 (with change point)

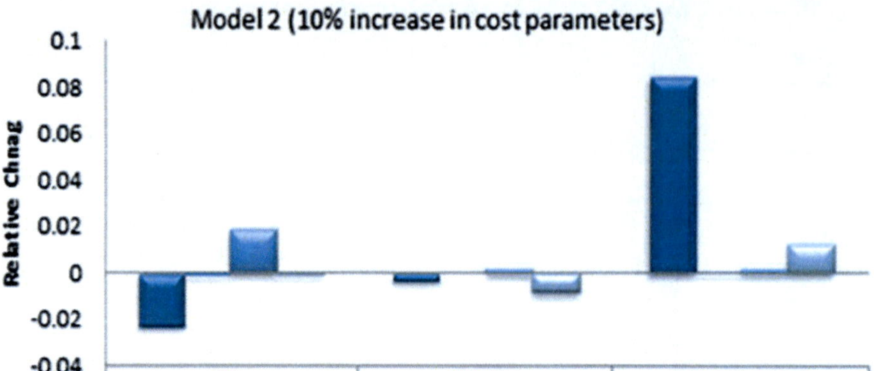

	Time	Reliability	Budget Consumed
C1	-0.022381951	-0.003036437	0.085158948
C2	-0.000259559	0	7.56296E-05
C3	0.018708196	0.002024291	0.00199158
C4	-0.00119397	-0.008097166	0.012604936

(a)

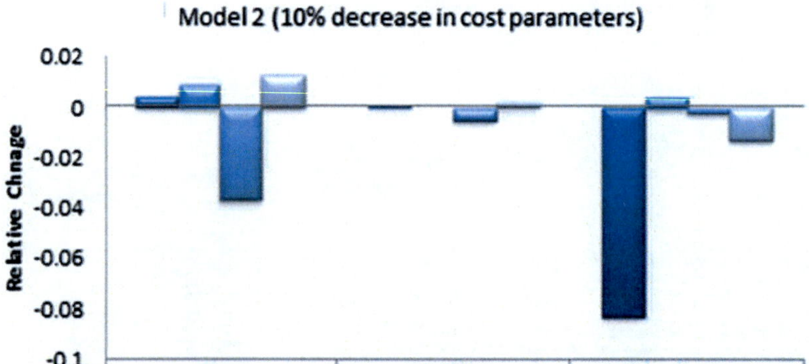

	Time	Reliability	Budget Consumed
C1	0.004372567	0.001012146	-0.083116949
C2	0.009124488	0	0.004487357
C3	-0.03671758	-0.005060729	-0.002117629
C4	0.013197564	0.002024291	-0.012932664

(b)

Fig. 7 Relative change using model 2 (with change point) cost with respect to different testing time. Similar results hold true in past research as well (Chatterjee and Shukla 2017; Jain et al. 2014a, b).

Beside theoretical implications, this article documents the managerial implication too. In the ever growing competitive environment of software systems, software development team needs to focus on building trust and enhancing user's growth rate. Therefore, using testing coverage and FRF together into the SRGMs with change point concept increases the reliability growth of the software which eventually helps development team to gain user's trust. This study also offers means to identify the software release time by taking care of user's requirements.

6 Conclusion, Limitations and Directions for Future Research

In this chapter, we obtained the SRGMs and examined the optimal release time by reducing the overall development cost, relative to the software reliability not less than a given aspiration level. The main focus of our study is to develop SRGMs from the point of view of change point and provide optimum time to make the product available for the operational phase. Proposed SRGMs include real life problems such as testing coverage and FRF, where FRF is believed to be constant and two distributions are considered for testing coverage, the first being Weibull distribution and the second being Exponentiated Weibull distribution. It is observed that coverage rate is itself influenced by various factors such as testing strategies, change in resources etc. In SRGM, this phenomenon of change in detection and consequently in coverage rate is considered as change point analysis and models are derived using change point concept. These models are tested using two datasets and comparison is addressed by integrating different performance measurements. The numerical example demonstrates that the proposed models give significant outcomes with improved fitness and predictive capabilities.

Estimated values of proposed MVF parameters are used to compute the release time of the software system with the help of optimization technique. Optimization is performed by reducing the development cost subject to constraint on reliability. Since cost is most often based on the development team's testing efficiency and expertise, therefore variations in cost parameters may influence the overall cost and also the release time. To understand this, we have discussed sensitivity analysis based on the variation in cost parameters by 10%. As results, we obtained that fault removal cost in operational phase affects the release time more as compared to other cost parameters. On the other hand, fault removal cost in testing phase has more influence on the total budget consumption than the other cost parameters.

Limitations and Future Directions

Following are the limitations observed from the proposed study,

1. The analysis concentrates on the single release of the software system. However, software usually comes with multiple releases. The proposed research can be extended in future by implementing the multi-release principle of software system.
2. The suggested model assumes the perspective of a single change point in the coverage rate function. Whereas multiple change points can be incorporated into the model in the future.
3. To measure the release time, we used a non-linear optimization problem that can be further calculated through meta-heuristic techniques and a comparison can be examined.
4. The cost model includes three cost parameters, such as fault removal costs during testing process, fault removal cost in operational phase and cost of removing faults per unit time. However, several costs influence the total development cost such as penalty cost, opportunity cost, warranty cost etc., hence for future perspective cost model can be integrated using these costs components.

References

Aggarwal AG, Dhaka V, Nijhawan N, Tandon A (2019a) Reliability growth analysis for multi-release open source software systems with change point. In: System performance and management analytics. Springer, pp 125–137

Aggarwal AG, Gandhi N, Verma V, Tandon A (2019b) Multi-release software reliability growth assessment: an approach incorporating fault reduction factor and imperfect debugging. Int J Math Operat Res 15:446–463

Chatterjee S, Shukla A (2016a) Effect of test coverage and change point on software reliability growth based on time variable fault detection probability. JSW 11:110–117

Chatterjee S, Shukla A (2016b) Modeling and analysis of software fault detection and correction process through Weibull-type fault reduction factor, change point and imperfect debugging. Arab J Sci Eng 41:5009–5025

Chatterjee S, Shukla A (2017) An ideal software release policy for an improved software reliability growth model incorporating imperfect debugging with fault removal efficiency and change point. Asia-Pac J Oper Res 34:1740017

Chatterjee S, Shukla A (2019) A unified approach of testing coverage-based software reliability growth modelling with fault detection probability, imperfect debugging, and change point. J Softw Evolut Process 31:e2150

Chatterjee S, Singh J (2014) A NHPP based software reliability model and optimal release policy with logistic–exponential test coverage under imperfect debugging. Int J Syst Assur Eng Manage 5:399–406

Chen M-H, Lyu MR, Wong WE (1996) An empirical study of the correlation between code coverage and reliability estimation. In: Proceedings of the 3rd international software metrics symposium, Berlin, Germany. IEEE, pp 133–141

Friedman MA, Tran PY, Goddard PI (1995) Reliability of software intensive systems. William Andrew, USA

Goel AL (1985) Software reliability models: assumptions, limitations, and applicability. IEEE Trans Softw Eng 1411–1423

Goel AL, Okumoto K (1979) Time-dependent error-detection rate model for software reliability and other performance measures. IEEE Trans Reliab 28:206–211

Hsu C-J, Huang C-Y, Chang J-R (2011) Enhancing software reliability modeling and prediction through the introduction of time-variable fault reduction factor. Appl Math Model 35:506–521

Huang C-Y, Lyu MR (2005) Optimal release time for software systems considering cost, testing-effort, and test efficiency. IEEE Trans Reliab 54:583–591

Huang C-Y, Lo J-H, Kuo S-Y, Lyu MR (2002) Optimal allocation of testing resources for modular software systems. In: Proceedings of 13th international symposium on software reliability engineering, 2002, Annapolis, USA. IEEE, pp 129–138

Huang C-Y, Kuo S-Y, Lyu MR (2007) An assessment of testing-effort dependent software reliability growth models. IEEE Trans Reliab 56:198–211

Jain M, Manjula T, Gulati T (2014a) Imperfect debugging study of SRGM with fault reduction factor and multiple change point. Int J Math Oper Res 6:155–175

Jain M, Manjula T, Gulati T (2014b) Prediction of reliability growth and warranty cost of software with fault reduction factor, imperfect debugging and multiple change point. Int J Oper Res 21:201–220

Kapur P, Garg R (1992) A software reliability growth model for an error-removal phenomenon. Softw Eng J 7:291–294

Kapur P, Kumar A, Yadav K, Khatri SK (2007) Software reliability growth modelling for errors of different severity using change point. Int J Reliab Quality Safety Eng 14:311–326

Kapur P, Garg R, Aggarwal AG, Tandon A (2009) General framework for change point problem in software reliability and related release time problem. Int J Reliab Quality Safety Eng 16:567–579

Kapur P, Pham H, Aggarwal AG, Kaur G (2012) Two dimensional multi-release software reliability modeling and optimal release planning. IEEE Trans Reliab 61:758–768

Kaur G, Aggarwal AG, Kedia A (2017) A study of optimal testing resource allocation problem for modular software with change point. Ann Comput Sci Inf Syst 14:77–84. https://doi.org/10.15439/2018KM11

Kumar V, Singh V, Dhamija A, Srivastav S (2018) Cost-reliability-optimal release time of software with patching considered. Int J Reliab Quality Safety Eng 25:1850018. https://doi.org/10.1142/S0218539318500183

Lai C-D, Murthy D, Xie M (2006) Weibull distributions and their applications 63–78. https://doi.org/10.1007/978-1-84628-288-1

Li Q, Pham H (2017) A testing-coverage software reliability model considering fault removal efficiency and error generation. PloS One 12:e0181524

Li X, Xie M, Ng SH (2010) Sensitivity analysis of release time of software reliability models incorporating testing effort with multiple change-points. Appl Math Model 34:3560–3570

Malaiya YK, Von Mayrhauser A, Srimani PK (1993) An examination of fault exposure ratio. IEEE Trans Software Eng 19:1087–1094

Malaiya YK, Li MN, Bieman JM, Karcich R (2002) Software reliability growth with test coverage. IEEE Trans Reliab 51:420–426

Musa JD (1975) A theory of software reliability and its application. IEEE Trans Softw Eng 1:312–327

Musa JD (1991) Rationale for fault exposure ratio K. ACM SIGSOFT Softw Eng Notes 16:79

Naixin L, Malaiya YK (1996) Fault exposure ratio estimation and applications. In: Proceedings of ISSRE'96: 7th international symposium on software reliability engineering. IEEE, pp 372–381

Nijhawan N, Aggarwal AG (2015) On development of change point based generalized SRGM for software with multiple releases. In: 4th international conference on reliability, Infocom Technologies and Optimization (ICRITO) (Trends and Future Directions), Noida, India. IEEE, pp 1–6

Okumoto K, Goel AL (1979) Optimum release time for software systems based on reliability and cost criteria. J Syst Softw 1:315–318

Pham H (2014) Loglog fault-detection rate and testing coverage software reliability models subject to random environments Vietnam. J Comput Sci 1:39–45

Pham H, Zhang X (2003) NHPP software reliability and cost models with testing coverage. Eur J Oper Res 145:443–454

Raju ON, Rakesh D, SubbaReddy K (2012) SRGM with imperfect debugging using capability analysis of log-logistic model. Int J Comput Technol 2:30–33

Singh O, Garmabaki AH, Kapur PK (2011) Unified framework for developing two dimensional software reliability growth models with change point. In: 2011 IEEE international conference on quality and reliability, Bangkok, Thailand. IEEE, pp 570–574

Tandon A, Neha, Aggarwal AG (2020) Testing coverage based reliability modeling for multi-release open-source software incorporating fault reduction factor. Life Cycle Reliabil Safety Eng 1–11. https://doi.org/10.1007/s41872-020-00148-7

Verma V, Neha N, Aggarwal AG (2020) Software release planning using grey wolf optimizer. In: Soft computing methods for system dependability. IGI Global, M'Hamed Bougara University, Algeria

Yamada S, Osaki S (1987) Optimal software release policies with simultaneous cost and reliability requirements. Eur J Oper Res 31:46–51

Yamada S, Ohba M, Osaki S (1983) S-shaped reliability growth modeling for software error detection. IEEE Trans Reliabil 32:475–484

Yamada S, Ohba M, Osaki S (1984) S-shaped software reliability growth models and their applications. IEEE Trans Reliab 33:289–292

Zhang X, Teng X, Pham H (2003) Considering fault removal efficiency in software reliability assessment. IEEE Trans Syst Man Cybernet-Part A Syst Humans 33:114–120

Zhao M (1993) Change-point problems in software and hardware reliability. Commun Stat-Theory Methods 22:757–768

Understanding Interactions Among Software Development Attributes and Release Planning Problem Through ISM and MAUT

Vibha Verma, Anu G. Aggarwal, and Hoang Pham

Abstract Satisfaction expected from the software product is determined by role of individual attributes in the development process and release decisions. This chapter proposes a study to analyse the relationship among various attributes of the software development process and its importance form customer's point of view. The attributes were carefully chosen based on the industry practices, theory and literature. Further the attributes were compared to see which attributes is affected/influenced by which attribute. A hierarchical model that highlights the importance of attributes at various levels along with their interrelationships was developed through Interpretive Structural Modelling (ISM). The results show that budget, reliability, release time and post-release support are highly affected attributes. Then an optimal release problem with multiple objectives is developed incorporating the most affected attributes and solved using Multi-Attribute Utility Theory (MAUT). Here developer aims to optimize the development cost i.e. budget and the reliability under fixed availability of budget considering that the developer provides warranty benefits to the customers. Through this chapter we are able to identify the relations among software attributes and role of most influenced attributes in release planning. The detailed methodology of the proposed study has been demonstrated through an empirical example.

Keywords ISM · MICMAC analysis · MAUT · Software development attributes · Release time · Warranty time · Customer satisfaction · Casual interrelationships

---

V. Verma · A. G. Aggarwal (✉)
Department of Operational Research, University of Delhi, New Delhi, India

H. Pham
School of Engineering, Rutgers University, New Brunswick, NJ, USA
e-mail: hopham@soe.rutgers.edu

A. G. Aggarwal et al. (eds.), *Optimization Models in Software Reliability*, Springer Series in Reliability Engineering,
https://doi.org/10.1007/978-3-030-78919-0_6

1 Introduction

A software firm's performance can be measured by the visibility of its product in the market and from customer feedback about the functioning of the product. To achieve a high level of satisfaction among customers it becomes necessary for organizations to look towards software from management, development and maintenance perspective (Yamada and Tamura 2016; Pham 2007). Successful management of software projects can be gained by carefully monitoring all the attributes of the development process such as budget, test cases, release time, post-release support, reliability etc. It can be well supported by a structural model that clearly specifies the importance of attributes and investigates possible interactions among them (Galin 2018).

To establish the hierarchical structure of the attributes based on their importance in the development process and interactions among them, Interpretive Structural Modelling (ISM) has been used. ISM is a methodology that systematizes the elements under study by interpreting the relationship among them and presents it with a hierarchical or multilevel structure. This imperative technique helps firms and researchers by converting a complex web of elements into a readable and understandable structure or model (Warfield 1974). ISM has been applied in various disciplines such as vendor selection (Mandal and Deshmukh 1994), education quality (Mahajan et al. 2014), software engineering project (Samantra et al. 2016), supply chain (Agi and Nishant 2017), aircraft industry (Pitchaimuthu et al. 2019), knowledge management (Maheshwarkar and Sohani 2019), etc. But it has not been practically or theoretically applied to the attributes of the software development process that define performance of software products.

The performance of the software product is as perceived by users based on their expectations. The expectations tend to change with change in technology. Software products become obsolete due to change in requirements which is result of fast technological changes. This problem can be dealt with the timely release of the software product or timely upgradations. Hence it is very important to determine the optimal release time (Pham 1996; Zhang and Pham 1998). We refer to optimal time because very large as well as the small duration for testing has a negative effect on firms' performance (Song et al. 2018). Testing software for longer time periods increases reliability but it incurs a lot of costs and disturbs the budget (Pham and Zhang 1999a, b). Also, testing for longer duration delays the introduction of the software product into the market when its demand is high and thus leads to loss of opportunity (Huang and Lyu 2005).

On the other hand release of a premature product (spending less amount of time on testing) leads to customer dissatisfaction due to failures experienced by users during the execution period (Pham and Zhang 1999a). So, optimal release time helps in maintaining the product's image into the market by minimizing the possibility of failure during the operation or implementation phase (Chatterjee and Shukla 2017). Even after testing the software for optimal period a developer can never be sure of no failures during execution. To make a software bug free it has to be tested infinitely which is a costly affair. Hence, some developers differentiate their product by providing back up support for their product after its release. This is termed as warranty.

Warranty is a kind of written contract or obligation between developers and customers to highlight the responsibilities of developers as well as customers during the product usage period (Park and Pham 2010, 2012; Bai and Pham 2006). It can also be defined as the confidence of developers in their product that it will perform satisfactorily and will not fail up to a given time if used as per the guidelines. This ensures that any failure in the operational phase will be attended by developers free of cost under specified terms and conditions (Li and Pham 2017; Sgarbossa and Pham 2010). But this service cannot be provided for whole lifetime. An unlimited warranty will cause heavy losses to software firms. Hence it is also very crucial to decide about the optimal time for which warranty should be provided (Blischke 2019). So to overcome these difficulties, authors have used Multi-Attribute Utility Theory (MAUT) for determination of optimal release time and optimal warranty time with objective of minimizing the budget usage and maximizing the reliability based on failure phenomenon characterized by imperfect debugging (Verma et al. 2019) and delayed S-shaped Fault Reduction Factor (FRF) (Musa 1975).

In the present study, the aim is to integrate the attribute evaluation process using ISM with MAUT for software release planning. This helps in aggregating experts' opinion in rating the influence of attributes on each other and thus based on the results manage release time problem. So, in particular, the research gap identified is the analysis of most influential and important attributes along with their interrelationships integrated with MAUT to determine optimal release and warranty time under the fixed budget by achieving reliability goal.

To bridge the gap the chapter fulfills following research objectives:

(1) To identify software development attributes that affect customer satisfaction through a literature review.
(2) To analyze and interpret the significance and interrelationship among the attributes.
(3) To develop a hierarchical structure of the identified attributes to highlight importance level using ISM.
(4) To effectively and systematically analyze and classify the attributes based on their driving and dependence power (MICMAC Analysis).
(5) Optimal Release and Warranty Planning based on ISM results using MAUT such that it establishes trade-off with reliability and development cost.

The rest of the chapter is organized as; Next section discusses the detailed literature review. Section 3 presents the proposed methodology followed by its implementation through numerical and result discussion in the next section. Section 5 presents the theoretical and managerial implications drawn from the study. Finally Sect. 6 wraps the chapter with conclusions, limitations and future scope.

2 Research Background

In this section, we discuss the motivation behind the study through an extensive literature review of the concepts used in this chapter.

2.1 *Software Attributes*

The software attributes considered in the study have been considered after a thorough study of literature and practices followed in the industries. These attributes represent the viewpoint of the developer and the factors that may influence customer satisfaction covering the management, development and maintenance aspect of the software products. The overall satisfaction received from the software product is based on individual attributes of the development process (Zhang and Pham 2000). The attributes have been collected together through extensive literature reviews and practices followed in industries. These attributes along with its references have listed in Table 1.

2.2 *ISM*

For the purpose of analyzing software development attributes that affect customer satisfaction, we have used ISM. ISM was proposed by Warfield (1974) and it helps in mapping the possible relations among attributes (Malone 1975; Watson 1978). This is an interpretive method where a group of experts decides whether and how certain elements are related. As a result, ISM presents a hierarchical structure (Model) highlighting the importance of attributes. Warfield and Cárdenas (1994) defined it as a method that clarifies the interrelations and structure of elements of the system under study. ISM helps to present a complex system in a simple manner by identifying the structure. The well-defined models developed from unclear models of system helps to build theory by answering questions such as what and how.

ISM has been applied in various disciplines such as vendor selection (Mandal and Deshmukh 1994), education quality (Mahajan et al. 2014; Sahney et al. 2010), supply chain (Agi and Nishant 2017), aircraft industry (Pitchaimuthu et al. 2019), knowledge management (Maheshwarkar and Sohani 2019), barrier analysis for product-service system (Kuo et al. 2010), total quality management (Talib et al. 2011), sustainable manufacturing (Dubey et al. 2015), governance (Grover et al. 2007; Lal and Haleem 2009), etc. In the field of software, it has been used to identify and model risk factors of software engineering projects (Samantra et al. 2016), on the attributes of software quality and assessment of environmental factors affecting environmental factors (Capiluppi et al. 2020; Galin 2018).

Table 1 List of software attributes

Attribute	Description	References
Budget (A_1)	Project budget allotted by management for software development	Verma et al. (2020), Zmud (1980), Nan and Harter (2009), Tam et al. (2020)
Requirement analysis (A_2)	Documentation of functional and non-functional requirements	Chakraborty et al. (2012), Tam et al. (2020), Zhang and Pham (2000)
Quality/reliability (A_3)	The probability that software will not fail up to given time if used under specified conditions	AL-Badareen et al. (2011), Kapur et al. (2011), Li and Pham (2017), Yamada and Tamura (2016)
Release time (A_4)	Time set for introducing the product into the market based on targets set by management in terms of cost, reliability etc	Chatterjee and Shukla (2017), Song et al. (2018), Huang and Lyu (2005), Sgarbossa and Pham (2010)
Application type (A_5)	Software systems are of different types based on their significance	Glass and Vessey (1995), Pham (2003), Kapur et al. (2011)
Test cases (A_6)	It helps developers to identify more fault-prone areas and accordingly depute the resources	Yamada and Tamura (2016), Kapur et al. (2011)
Dedicated resources (A_7)	It helps to deploy resources cost-efficiently	Yamada and Tamura (2016), Kapur et al. (2011)
Fault tolerance (A_8)	This is a technique that helps prevent system failure after the release of the product which may occur due to some dormant faults	Yamada and Tamura (2016), Kapur et al. (2011), Kovalev et al. (2020)
Post release support/back up support (A_9)	The support provided by developers to the customer for any failure after the release of the product	Kimura et al. (1999), Kumar et al. (2014), Bai and Pham (2006), Park and Pham (2012)

2.3 MAUT for Release Planning

MAUT has been extensively used for release management of software projects. This technique helps to maximize the overall utility of more than one objective. Each objective is given through single-attribute utility functions (SAUF) which are further combined to form multi-attribute utility functions (MAUF). Release planning problems can be solved using MAUT based on the analysis of attributes. Minamino et al. (2015) used MAUT for release planning based on failure phenomenon considering change point.

Some other works done using MAUT are by Garmabaki et al. (2012), Pachauri et al. (2014) and many more. All these studies are based on some failure model characterized by some features that define the failure behavior. In this chapter, we have considered the failure model with delayed S-shaped (Yamada et al. 1984) FRF and error generation proposed by Aggarwal et al. (2019).

3 Proposed Methodology

Existing research has outlined the potential role of ISM in defining Interrelationships among attributes but it has not been used in the field of software to identify hierarchical structure and relation among attributes of the software development process. Based on our review of literature related to release time management by software developers we identified there is a lack of study about the attributes that affect the whole development process and thus satisfaction. The presence of a large number of related variables makes the structure of the system more complex which may not be expressed in a clear fashion. ISM is an aid to simplify this structure. On the other hand, MAUT has been used in literature to determine optimal release time by achieving multiple objectives at a time. The steps followed in the chapter for fulfilling research objectives have been summarized in Fig. 1. All these steps have been explained in detail in further subsections.

3.1 ISM Method to Study the Relationship of Software Attributes

The steps of ISM are as followed (Talib et al. 2011; Attri et al. 2013):

Fig. 1 Proposed methodology

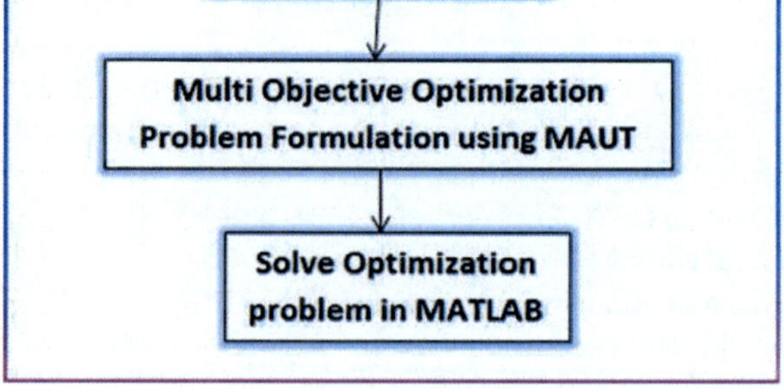

(i) Establish objectives to be fulfilled through ISM.
(ii) Define attributes based on a literature survey and industry practices. The 9 attributes derived from the literature have already been discussed in Sect. 2.
(iii) Determine contextual or relative relations among attributes through a survey from experts. The wordings like "is influencing" or "influenced by" are used to relate the attributes.
(iv) Form Structural Self-Interaction Matrix (SSIM) based on expert opinion collected through surveys. Based on the responses received from survey relations among attributes are established. The expert assigns a symbol for a direct relationship between two attributes. Let i and j be any two attributes. The association among attributes is represented using V (symbolizing i influences j), A (symbolizing j influences i), X (symbolizing i and j influence each other) and O (symbolizing i and j are not associated). These symbols form the SSIM matrix.
(v) Formulate Initial and then Final Reachability Matrix (FRM) from SSIM by including transitive relations. The SSIM matrix is converted to the Initial reachability matrix with '0' and '1' by converting the four symbols of SSIM according to the specified rules. The rules are

- For V replace (i, j)th entry by '1' and (j, i)th by '0'.
- For A replace (i, j)th entry by '0' and (j, i)th by '1'.
- For X replace (i, j)th entry by '1' and (j, i)th also by '1'.
- For O replace (i, j)th and (j, i)th entry by '0'.
  The FRM is built after checking for any transitive links in the initial one. A relation is transitive if i influences j and j influences k then i should influence k.

(vi) Carry out level partitioning of the final reachability matrix.
Partitioning of attributes into different levels is done based on three types of sets namely reachability, antecedent, and intersection set. These sets are derived based on the influencing behaviour of attributes. Reachability set consists of attributes that influence other attributes including itself. Antecedent sets are attributes are influenced by other attributes including it also while the intersection of attributes in these two sets is categorized under intersection set. If the reachability set is the same as the intersection set than that attribute is levelled as level 'I'. Then all the attributes in level 'I' are extracted and the iterative process is continued till all attributes have achieved their level. These levels help to develop a hierarchical level of these attributes.
(vii) Develop the ISM model by removing the transitive links from FRM to give a pictorial structure.
Based on the FRM and level partitioning a model is developed by removing the transitive links for simplicity. The top-level of the model consists of attributes extracted at first level, similarly, the levelling goes on.
(viii) Analyze the Model established.

3.2 MICMAC Analysis

MICMAC Analysis (Attri et al. 2013) is used to determine the power of attributes in either influencing other attributes or getting influenced by other attributes. This procedure analyses attribute for indirect classification based on the driving and dependence power of attributes. Here driving means how influential is the attribute and dependence refer to influenced attribute. The values of driving power and dependence power are calculated from the FRM. Adding vertical values gives driving power while horizontal gives dependence power of the corresponding attribute. Plotting the values divide attributes into four segments (quadrant) namely autonomous, Linkage, Dependent and Independent. These quadrants can be explained as follows:

- Quadrant I (Autonomous): Low driving and dependence power i.e. they have weak links and act as detached with other attributes but have a strong connection with few strong attributes.
- Quadrant II (Dependent): Low driving and high dependence power i.e. these attributes are easily influenced by other attributes.
- Quadrant III (Linkage): High driving and dependence power i.e. these attributes are very influential as well as get influenced if there is any change in these attributes.
- Quadrant IV (Independent): High driving and low dependence power i.e. they hardly ever get influenced by other attributes.

3.3 MAUT to Determine Optimal Release and Warranty Time

This technique has a strong theoretical and practical foundation based on expected utility theory. Also, it provides feasibility to consider various attributes. Li et al. (2011) was first to manage release time problem using MAUT in software. The critical decision is the determination of optimal release and warranty time to create a trade-off between the development costs, reliability level without losing any opportunity in the market. So budget and reliability are considered as two attributes for MAUT. Multiple objectives are expressed by SAUF which are categorized by the worst and best case. For example here for our problem best achievable reliability is 0.99 and worst would be below 0.60. Similarly, for cost best would be to develop the software by utilizing 50% of the budget and the worst case would be to utilize the whole budget.

The changes in reliability performance can be measured through the Software Reliability Growth Model (SRGM) that quantitatively evaluates the reliability by analyzing the failure behavior of the software system. Here, we have considered an SRGM with delayed S-shaped Fault Reduction Factor and imperfect debugging i.e. the model considers that the number of faults experienced may not be the same as the number of faults removed and error constant increases at a constant rate. $m(t)$ is the Mean Value Function (MVF) for the cumulative number of faults removed which has been derived by solving differential Eq. (1) (Aggarwal et al. 2019).

$$\frac{dm(t)}{dt} = r(t)(a(t) - m(t)) \tag{1}$$

where

$$(t) = r * b(t),\ b(t) = \frac{b^2 t}{1 + bt},\ \text{and } a(t) = a + \alpha m(t) \tag{2}$$

Solving the above differential equation (Eq. 1) using (2) and applying initial condition $m(0) = 0$, we get

$$m(t) = \frac{a}{1 - \alpha}\left(1 - (1 + bt)^{r(1-\alpha)} e^{-btr(1-\alpha)}\right) \tag{3}$$

where, $a(t)$ represents initial fault content,

$r(t)$ is a fault detection rate,

$b(t)$ is FRF.

and r is the proportionality constant.

Based on our problem the firm has to maximize reliability and minimize development cost. So,

$$\text{Max } R(t) = e^{m(t+x) - m(t)} \tag{4}$$

$$\text{Min} \frac{C(t)}{C_B} \tag{5}$$

where, C_B is the Budget for the software project and $C(t)$ the expected software development cost is given by,

Expected Development Cost = Testing Cost + Warranty Cost + Operational Phase Cost

$$C(T, W) = C_1 T + C_2 m(T) + C_3 W + C_4(m(T + W) - m(T)) + C_5(m(\infty) - m(T + W)) \tag{6}$$

where,

C_1 = Fixed Testing Cost

C_2 = Fault removal cost during Testing

C_3 = Fixed Warranty Cost

C_4 = Fault removal cost during Warranty

C_5 = Fault removal cost during the Operational Phase.

Before developing the MAUF in the additive linear form, SAUF is developed for each attribute of MAUT as shown in Eq. (7). The two SAUF are for cost and reliability respectively.

$$U(R) = \frac{R(t) - R^{\min}}{R^{\max} - R^{\min}} \text{ and } U(C) = \frac{C^{\min} - C(t)}{C^{\min} - C^{\max}} \tag{7}$$

$U(R)$ represents SAUF for reliability attribute where $R^{\min} \leq R(t) \leq R^{\max}$ and $U(C)$ represents cost SAUF where $C^{\max} \leq C(t) \leq C^{\min}$.

The two objectives of minimizing development cost and maximizing reliability have been combined by multiplying cost-utility with '−' sign (Li et al. 2011). By maximizing MAUF the optimal release time and warranty time are obtained. MAUF is given by Eq. (8).

$$\text{Max } U(R, C) = w_r U(R) - w_c U(C) \tag{8}$$

where, $w_r + w_c = 1$, w_r is reliability weight and w_c is the weight assigned to the cost attribute.

Overall MAUT can be summarized into the following four steps (Dyer 2005; Von Winterfeldt and Fischer 1975):

(i) Select Attributes that have to be either maximized or minimized.
(ii) SAUF for each attribute is developed (Eq. 7).
(iii) Weights assigned to attributes are determined. It can either be done through the lottery system or by experts.
(iv) Apply the additive approach to form MAUF (Eq. 8)
(v) Solve MAUT using a optimization technique.

4 Empirical Illustration and Result Discussion

The objective of the study is to develop a hierarchal model of software development attributes that highlights the interrelationships. Using the inferences further a multi optimization problem is formulated using MAUT for determination of optimal release and optimal warranty time.

4.1 ISM

As discussed in Sect. 3.1, nine attributes of the software development process were identified and used for ISM. Five experts from IT firms with more than 10 years of experience were considered for finalizing the SSIM matrix. If 3 or more experts gave a similar response then it was finalized but if responses were more scattered it

Table 2 SSIM matrix

Attribute	A_1	A_2	A_3	A_4	A_5	A_6	A_7	A_8	A_9
A_1	1	A	X	A	A	A	A	A	A
A_2		1	O	O	A	O	O	O	O
A_3			1	X	O	A	A	A	V
A_4				1	A	O	A	O	X
A_5					1	V	V	O	V
A_6						1	O	O	O
A_7							1	O	V
A_8								1	V
A_9									1

was given 'O' relation. The SSIM matrix finalized for further computation is given in Table 2. The meaning of the symbols is as discussed in Sect. 3.1 whereas A_i represents the ith attribute.

Table 3 shows the initial reachability matrix obtained from SSIM (Table 2). This shows the relation among elements in binary form. The procedure for replacing the values has been explained in Sect. 3.1. Then final reachability matrix (Table 4) is obtained by incorporating transitive relations of attributes. The asterisk in the Table 4 represents the values that were initially not present in the Initial reachability matrix and were added later based on the methodology discussed in Sect. 3. The table also shows the driving power and dependence power of each attribute which will be used for dividing attributes into different segments (MICMAC analysis).

After getting the FRM next step is to partition the attributes into different levels. The three iterations of the process have been shown in Tables 5, 6 and 7 respectively. From Table 5 we see that attribute A_1, A_3, A_4 and A_9 i.e. Budget, Quality/Reliability, Release time and Post back up support are categorized into level 1. These attributes

Table 3 Initial reachability matrix

Attribute	A_1	A_2	A_3	A_4	A_5	A_6	A_7	A_8	A_9
A_1	1	0	1	0	0	0	0	0	0
A_2	1	1	0	0	0	0	0	0	0
A_3	1	0	1	1	0	0	0	0	1
A_4	1	0	1	1	0	0	0	0	1
A_5	1	1	0	1	1	1	1	0	1
A_6	1	0	1	0	0	1	0	0	0
A_7	1	0	1	1	0	0	1	0	1
A_8	1	0	1	0	0	0	0	1	1
A_9	1	0	0	1	0	0	0	0	1

Table 4 Final reachability matrix

Attribute	A_1	A_2	A_3	A_4	A_5	A_6	A_7	A_8	A_9	Driving power
A_1	1	0	1	1*	0	0	0	0	1*	4
A_2	1	1	0	0	0	0	0	0	0	2
A_3	1	0	1	1	0	0	0	0	1	4
A_4	1	0	1	1	0	0	0	0	1	4
A_5	1	1	1*	1	1	1	1	0	1	8
A_6	1	0	1	0	0	1	0	0	0	3
A_7	1	0	1	1	0	0	1	0	1	5
A_8	1	0	1	0	0	0	0	1	1	4
A_9	1	0	1*	1	0	0	0	0	1	4
Dependence	9	2	8	6	1	2	2	1	7	

Table 5 Iteration 1

Attribute	Reachability set	Antecedent set	Interaction set	Level
A_1	A_1,A_3,A_4,A_9	$A_1,A_2,A_3,A_4,A_5,A_6,A_7,A_8, A_9$	A_1,A_3,A_4,A_9	I
A_2	A_1,A_2	A_2,A_5	A_2	
A_3	A_1,A_3,A_4,A_9	$A_1,A_3,A_4,A_5,A_6,A_7,A_8, A_9$	A_1,A_3,A_4,A_9	I
A_4	A_1,A_3,A_4,A_9	A_1,A_3,A_4,A_5,A_7,A_9	A_1,A_3,A_4,A_9	I
A_5	$A_1,A_2,A_3,A_4,A_5,A_6,A_7,A_9$	A_5	A_5	
A_6	A_1,A_3,A_6	A_5,A_6	A_6	
A_7	A_1,A_3,A_4,A_7,A_9	A_5,A_7	A_7	
A_8	A_1,A_3,A_8,A_9	A_8	A_8	
A_9	A_1,A_3, A_4, A_9	$A_1,A_3,A_4,A_5,A_7,A_8, A_9$	A_1,A_3,A_4,A_9	I

Table 6 Iteration 2

Attribute	Reachability set	Antecedent set	Interaction set	Level
A_2	A_2	A_2,A_5	A_2	II
A_5	A_2,A_5,A_6,A_7	A_5	A_5	
A_6	A_6	A_5,A_6	A_6	II
A_7	A_7	A_5,A_7	A_7	II
A_8	A_8	A_8	A_8	II

Table 7 Iteration 3

Attribute	Reachability set	Antecedent set	Interaction set	Level
A_5	A_5	A_5	A_5	III

are very much influenced by all other attributes and come under the category of dependent variables (Fig. 3). Also, they form the top level of the ISM hierarchy.

Table 6 shows that attribute A_2, A_6, A_7 and A_8 i.e. requirement analysis, test cases, dedicated resources and fault tolerance lie at the second level. This forms the middle level of the ISM hierarchy. These variables are neither very much influenced by other attributes or influence others. Similarly, Table 7 shows A_5 i.e. application type is at the lowest level and thus is most important in determining other attributes. Table 8 summarizes levels for all the attributes.

Figure 2 represents the ISM model developed from partitioned levels and the final

Table 8 Summary of level partitioning

Attribute	Reachability set	Antecedent set	Interaction set	Level
A_1	A_1,A_3,A_4,A_9	$A_1,A_2,A_3,A_4,A_5,A_6,A_7,A_8, A_9$	A_1,A_3,A_4,A_9	I
A_2	A_1,A_2	A_2,A_5	A_2	II
A_3	A_1,A_3,A_4,A_9	$A_1,A_3,A_4,A_5,A_6,A_7,A_8, A_9$	A_1,A_3,A_4,A_9	I
A_4	A_1,A_3,A_4,A_9	A_1,A_3,A_4,A_5,A_7,A_9	A_1,A_3,A_4,A_9	I
A_5	$A_1,A_2,A_3,A_4,A_5,A_6,A_7,A_9$	A_5	A_5	III
A_6	A_1,A_3,A_6	A_5,A_6	A_6	II
A_7	A_1,A_3,A_4,A_7,A_9	A_5,A_7	A_7	II
A_8	A_1,A_3,A_8,A_9	A_8	A_8	II
A_9	A_1,A_3, A_4, A_9	$A_1,A_3,A_4,A_5,A_7,A_8, A_9$	A_1,A_3,A_4,A_9	I

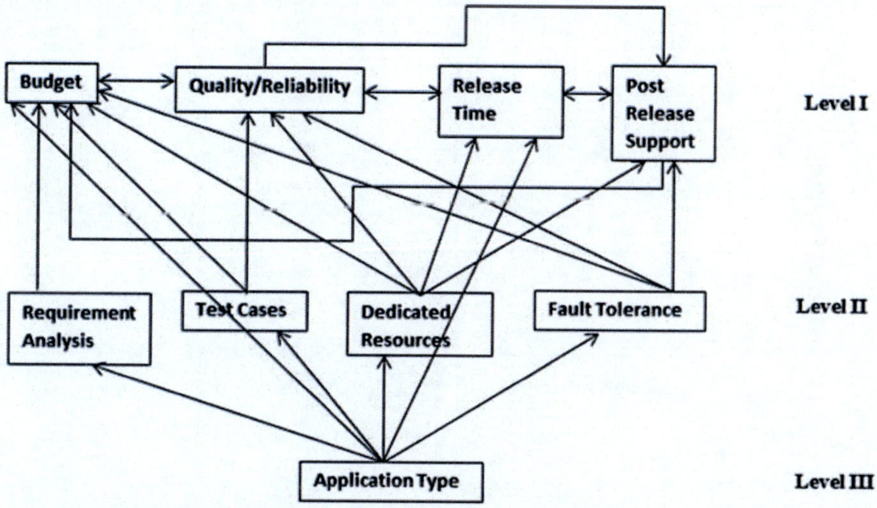

Fig. 2 ISM based hierarchical model of software development attributes

reachability matrix after removing transitive links. This model shows the interrelationships and hierarchy of attributes. The arrow in the model implies 'influences'. The 9 attributes were divided into three levels indicating that attribute at the third level (Application type) is the largest influencer. This is so because all the development decisions are taken after considering the type of application to be developed. For example, the strategy and resource requirements for aircraft are different from those of the traffic signals. So A_5 influences the middle and top levels. The budget is influenced if developers want to improve the quality or reliability of the software system. It is also influenced if the developer decides to test the software for a longer duration. Further, it can be seen that back support also affects the budget. All these relations can clearly be identified through ISM.

4.2 *MICMAC Analysis*

Figure 3 presents the classification of attributes based on driving power and dependence power (Table 4). This helps in the indirect classification of attributes. Two attributes application type and dedicated resources are classified as independent variables. There is no linkage attribute i.e. all the attributes are stable. Three attributes namely fault tolerance; test cases and requirement analysis are autonomous variables. These attributes have both weak driving and dependence power and do not have much influence on the system. Then variables with high dependence power are placed into the fourth cluster. These are dependent on each of the attributes of the system.

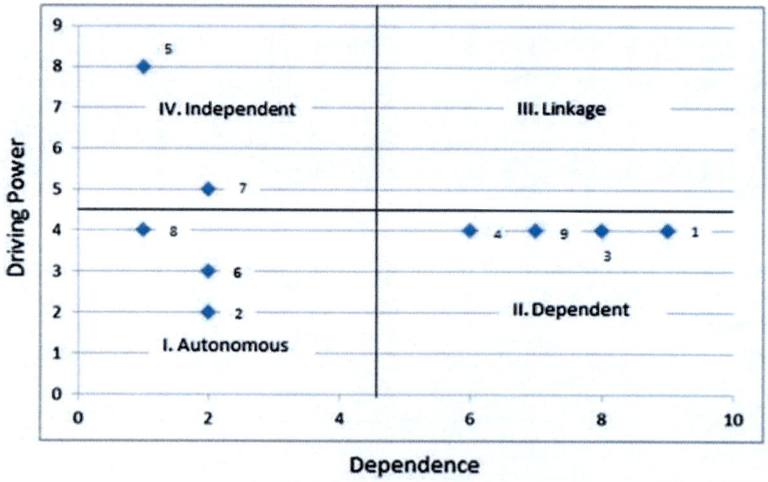

Fig. 3 Driving power and dependence graph (MICMAC analysis)

The hierarchy (Fig. 2) suggests that Budget, Quality/Reliability, Release time and post-release support (back up support) provided to customers by developers is highly influenced by other attributes of the development process. Hence it is very crucial for developers to monitor the resources (in monetary terms) consumed in achieving an acceptable level of reliability by performing testing for the optimal duration. These two, in turn, define the optimal duration of back support provided after the sale of the product. By optimal, we mean that the time is set such that it neither makes developers suffer nor it makes customers suffer due to a bad product.

4.3 MAUT

Next, we formulate an optimization problem to determine optimal release time and warranty time with the objective of minimizing development cost and maximizing reliability using MAUT (Eq. 8). After incorporating management restrictions into Eq. (7) for cost and reliability attributes we get Eq. 9. Management seeks a minimum reliability of 60% and maximum to be 99% when they are supposed to spend at least 50% of the allotted budget. So, replacing $R^{\min} = 0.6$, $R^{\max} = 0.99$, $C^{\min} = 0.5$ and $C^{\max} = 1$ in Eq. (7). We get,

$$U(R) = \frac{R(t) - 0.6}{0.99 - 0.6} \text{ and } U(C) = \frac{0.5 - C(t)}{0.5 - 1} \tag{9}$$

Solving Eq. (9) We get,
$U(R) = \frac{100R(t)}{39}$ and $U(C) = 2C(t) - 1$

where $R(t)$ and $C(t)$ is given by Eqs. (4), (5) and (6) respectively.

Further, the weights for both the objective is taken as 0.5 i.e. $w_r = 0.5$ and $w_c = 0.5$ because we have only two attributes here. So the MAUF is given as;

$$\text{Max } U(R, C) = 0.5 \times U(R) - 0.5 \times U(C) \tag{10}$$

The $m(t)$ considered for the study is given by Eq. (3). The estimated parameter values of $m(t)$ on fault dataset of Tandem Computers (Wood 1996) are $a = 108.554$, $b = 0.654$, $\alpha = 0.011$ and $r = 0.263$. These values are replaced in MAUF (Eq. 10) to solve the optimization problem. The cost values assumed are as follows (Eq. 6):

$$C_1 = 50, C_2 = 130, C_3 = 70, C_4 = 230, C_5 = 3000$$

Solving this problem in MATLAB gives optimal release time to be of 39 and 32 weeks of warranty with the reliability of 82.75. The process was conducted using 1.8498×10^4 units of cost. The results of optimization using MAUT have been further exemplified using the Cost curve (Fig. 4), the Reliability curve (Fig. 5) and the Utility function curve (Fig. 6). These figures suggest that an increase in testing time improves reliability and vice versa but results in more cost.

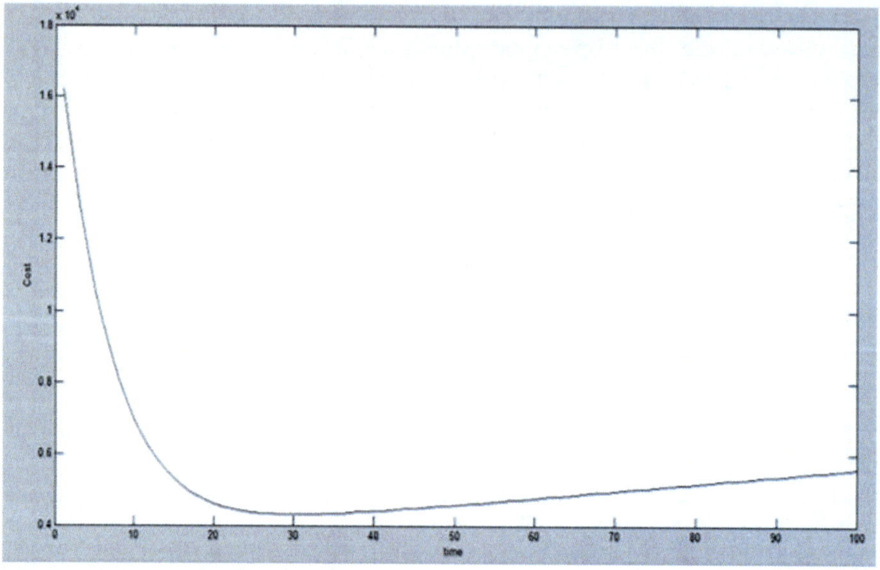

Fig. 4 Behaviour of cost function (Eq. 6)

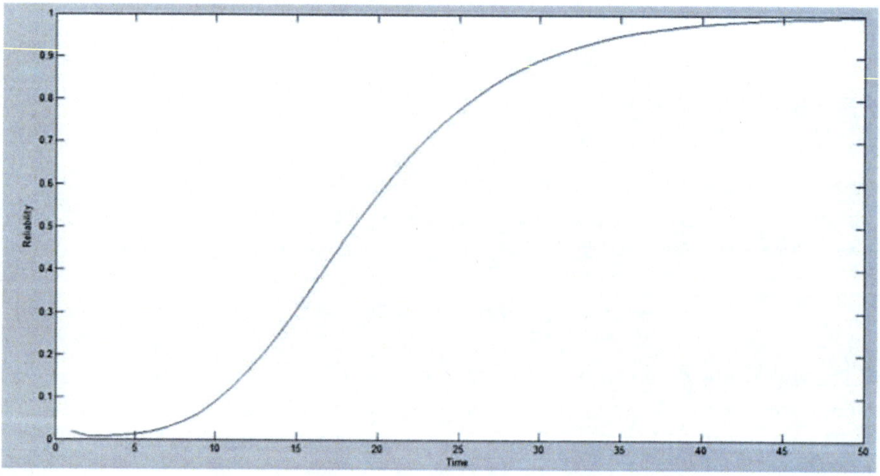

Fig. 5 Behaviour of reliability function (Eq. 4)

The reliability of the software system tends to improve with time. If the developers decide to spend more on testing then the reliability of the product will increase but it will delay its introduction in the market. Therefore developers set target reliability to achieve within an optimal testing period that helps them avoid any kind of losses. These losses could be in monetary terms or brand image.

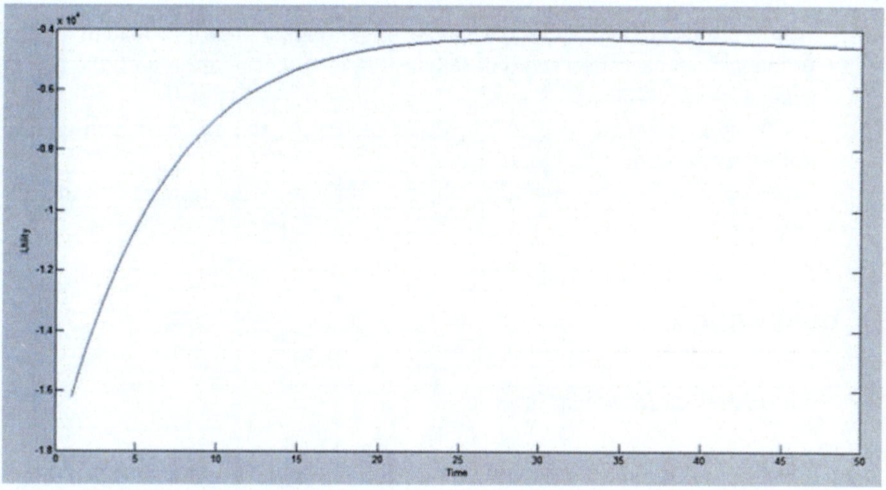

Fig. 6 Utility function behaviour (Eq. 10)

In this example problem we have assumed equal importance for the objectives. But this may not be the case in reality so, we check for the sensitivity of the weights assigned to the two objectives of minimizing cost and maximizing reliability. The sensitivity results are shown in Table 9.

Following observations can be made from Table 9:

(i) The utility value helps to represent the preference relationship among cost and reliability.
(ii) As the testing time increases the numerical value of utility decreases.
(iii) Testing time is also visibly effecting the optimal warranty time. Software tested for longer durations is released with less warranty time period.

Table 9 Sensitivity analysis results

w_r	w_c	T (Weeks)	W (Weeks)	$R(t)$	Development Cost ($\times 10^4$)	$U(R, C)$
0.1	0.9	37.002	33.92	76.93	1.8545	−7.817
0.2	0.8	37.86	33.10	79.61	1.8525	−6.9343
0.3	0.7	38.052	32.42	80.17	1.8486	−6.0688
0.4	0.6	38.87	32.23	82.41	1.8508	−5.1881
0.5	0.5	39	32	82.75	1.8498	−4.3222
0.6	0.4	39.29	31.99	83.47	1.8503	−3.4543
0.7	0.3	40.012	31.87	85.17	1.8533	−2.5856
0.8	0.2	40.89	31.77	87.01	1.8566	−1.7203
0.9	0.1	41	31.67	87.23	1.8564	−0.85860

(iv) Above point can be supported because longer testing durations lead to reliable product. It can be observed that reliability has direct relation with the testing time.
(v) As weight assigned to cost objective increases, the optimal testing time duration decreases and vice-versa.
(vi) Increase in weight of reliability objective improves the reliability and vice-versa.

5 Implications

5.1 Theoretical Implications

Many theoretical aspects can be highlighted from the study. The hierarchical model developed in the study establishes the importance and relations among software attributes and factors that affect customer satisfaction. These attributes have been derived from past studies and industrial practices (Capiluppi et al. 2020). The attributes (Table 1) such as budget allocated for software development process, requirement analysis, quality/reliability, type of application being developed, test cases designed to highlight more fault prone areas, resources deployed for various activities, fault tolerance of the system and warranty support provided by the developer has been used to develop hierarchical model based on their relations and MICMAC Analysis (Glass and Vessey 1995; Verma et al. 2020; Yamada and Tamura 2016). These relations were concluded after obtaining data form some IT experts. The results are in line with the past researches. ISM tool can be backed up for its use in various studies because it helps to break complex structure of elements involved in a process into layers. These layers depict connections among attributes and their importance. The hierarchical model developed using ISM clearly shows that all decisions are based on type of application being developed.

Further, MICMAC analysis determined the attribute power in influencing and getting influenced by other attributes. The critical decisions regarding software release and warranty duration by creating trade-off between the development cost and reliability level without losing any market opportunities is the ultimate objective of software development firms. Hence MAUT helps to maximize the software utility based on single utility functions of reliability and cost (Pachauri et al. 2014). The results are in correspondence to various studies related to determination of release time using MAUT (Garmabaki et al. 2012; Minamino et al. 2015).

Sensitivity analysis on the weights assigned to SAUF of the objective function shows how prioritising reliability objective or cost objective impacts the results. Firms can analyse these results for making optimistic decisions of software development. This is so because if more priority is given to reliability then developers can improve reliability but it will lead to more cost and if cost is given more priority then developers may deliver a less quality software product. So, weights should be assigned consciously based on type of application being developed and clients' requirements.

5.2 Managerial Implications

The study conducted in this chapter provides deep insights to software developers and managers of software development firms. Understanding relation among attributes of software development process that affects customer satisfaction is very crucial for management team to plan for future strategies that will help gain monetary benefits. The hierarchical structure derived using ISM represents interrelationships of software development attributes (Agi and Nishant 2017; Attri et al. 2013; Maheshwarkar and Sohani 2019; Pitchaimuthu et al. 2019). Understanding of attribute relations helps the management to decide how much time, effort and cost has to be utilised or should be utilised for various tasks during the software development process. These relations has been further clarified and supported by MICMAC analysis which explains their dependence. It shows which attributes are more influencing and which are getting influenced by other attributes.

The application of these two tools ISM and MICMAC analysis on the data obtained from experts suggests that all the attributes are influenced by the type of application being developed. All the decisions taken for development is based on application type. Same results can be observed from driving power and dependence power graph of attributes. This graph was result of the MICMAC analysis. Release and Warranty planning, Budget and Reliability are fully dependent on other attributes and are influenced by all aspects of the software development process whereas attributes such as requirement analysis, test cases and fault tolerance are autonomous variables. These variables affect other variables in the model are also affected by application type.

Further in the study utility of software has been determined based on reliability and development cost attributes. Optimal release time of the software was determined to be 39 weeks with 32 weeks of optimal warranty period. Reliability of the software system after testing was of 83.19. The whole process consumed 1.8498×10^4 units of cost. The various curves plotted namely utility, reliability and cost curves suggest that increase in testing time improves reliability but at same time leads to increase in cost. Also this decision delays the introduction of the product into the market which may make it obsolete and tend to heavy losses for the firm. Sensitivity analysis on the utility weights of cost and reliability shows how preference relationship among cost and reliability affects testing time, warranty time and software utility (Table 9).

Developers need to prevent software development process from any kind of crisis. A crisis can be explained as developers' incompetence to develop complete and quality software. This can occur if development projects are running over budget or over time, and if there is lack of resources and if software is not able to handle complex instructions. Hence it is crucial for developers to maximize software utility and determine optimal release time, warranty time and development cost.

6 Conclusions, Limitations and Future Scope

Understanding the behaviour of software attributes is important to plan future achievements because software products tend to become obsolete with the change in technology and change in customer requirements. Hence it is crucial for software firms to understand the behavior of development attributes. This helps in designing strategies so that firms have monetary benefits along with the satisfaction of their customers. Hence, ISM has been introduced as a method to describe the interrelationships between the attributes of software product and divide these attributes under various levels that represents the level of importance. The attributes that were observed to be most influenced/affected by other attributes were further taken into consideration to formulate the release problem. This aids the software engineers to take decisions regarding the attributes that needs to be focussed more than others for business benefits. The model helps management understand these attributes and make wise release decisions and hence control the customer satisfaction.

The main contributions of the chapter are:

(i) The study identified 9 attributes such as Budget, requirement analysis, Release time, application type, etc. which after the development process of software product hence impact the customer perception about the product.

(ii) The ISM model depicts a hierarchy that highlights all relations among the attributes.

(iii) These attributes were classified into different levels of importance where application type resulted to be the most influential attribute.

(iv) The top-level of ISM hierarchy implies that it is influenced by the other two levels and it can be determined by monitoring the attributes at those levels.

(v) MICMAC analysis helped to categorize attributes into different quadrants based on their driving and dependence power. This helped to understand the role of each attribute in the development process.

(vi) The firm has to determine a release and warranty time under a fixed budget of project development by achieving a considerable amount of reliability. Warranty for a product is dependent on its reliability. A more reliable product is provided with back support for a small duration.

(vii) MAUT was applied to form the multi-objective optimization problem and then solved in MATLAB.

(viii) The optimal release time resulted to be of 39 weeks and optimal warranty time came out to be of 32 weeks. For example if we consider that a computer software is being developed then developers need to decide the release time and warranty they need to provide. In this case firm decide to test the software for 39 weeks and then release it in to the market with 32 weeks of post release support. In this period developers take the responsibility of correcting the faults free of cost.

(ix) The product was able to achieve the reliability of 82.75 using 1.8498×10^4 units of budget.

(x) Sensitivity Analysis results reveal that there is significant effect of weight assigned to a particular objective.

Overall it can be said that this chapter explores the most influential software attributes and their interrelationship using ISM and MAUT helps in the release as well as warranty management for a software product.

Although ISM has been extensively applied in many disciplines, there are some limitations to using this technique. According to Watson (1978), ISM is inflexible in nature because attributes cannot be added, deleted or redefined during the process. To accommodate certain changes it can be repeated for individual's which will involve more cost and time or the process has to be started again after modifications.

There are some problems which can be tackled by extending the present work. This approach does not assign weights to the attributes. It can be done using any multi-criteria decisions making approach. The ISM model obtained can be validated using Structural Equation Modelling. The MAUT considers only two attributes namely cost and reliability. More attributes can be added after studying the literature. Also, the weights assigned to these attributes in MAUT can vary and may not be equal.

References

Aggarwal AG, Gandhi N, Verma V, Tandon A (2019) Multi-release software reliability growth assessment: an approach incorporating fault reduction factor and imperfect debugging. Int J Math Oper Res 15(4):446–463. https://doi.org/10.1504/IJMOR.2019.10016194

Agi MA, Nishant R (2017) Understanding influential factors on implementing green supply chain management practices: an interpretive structural modelling analysis. J Environ Manag 188:351–363

AL-Badareen AB, Selamat MH, Jabar MA, Din J, Turaev S, Malaysia S (2011) Users' perspective of software quality. In: The 10th WSEAS international conference on software engineering, parallel and distributed systems (SEPADS 2011). pp 84–89

Attri R, Dev N, Sharma V (2013) Interpretive structural modelling (ISM) approach: an overview. Res J Manag Sci 2319:1171

Bai J, Pham H (2006) Cost analysis on renewable full-service warranties for multi-component systems. Eur J Oper Res 168(2):492–508

Blischke W (2019) Warranty cost analysis. CRC Press

Capiluppi A, Ajienka N, Counsell S (2020) The effect of multiple developers on structural attributes: a study based on java software. J Syst Softw 110593

Chakraborty A, Baowaly MK, Arefin A, Bahar AN (2012) The role of requirement engineering in software development life cycle. J Emerg Trends Comput Inf Sci 3(5):723–729

Chatterjee S, Shukla A (2017) An ideal software release policy for an improved software reliability growth model incorporating imperfect debugging with fault removal efficiency and change point. Asia-Pacific J Oper Res 34(03):1740017

Dubey R, Gunasekaran A, Sushil ST (2015) Building theory of sustainable manufacturing using total interpretive structural modelling. Int J Syst Sci Oper Logist 2(4):231–247

Dyer JS (2005) MAUT—multiattribute utility theory. In: Multiple criteria decision analysis: state of the art surveys. Springer, pp 265–292

Galin D (2018) Software quality factors (attributes)

Garmabaki AH, Aggarwal AG, Kapur P, Yadavali V (2012) Modeling two-dimensional software multi-upgradation and related release problem (a multi-attribute utility approach). Int J Reliab Qual Saf Eng 19(03):1250012

Glass RL, Vessey I (1995) Contemporary application-domain taxonomies. IEEE Softw 12(4):63–76

Grover D, Shankar R, Khurana A (2007) An interpretive structural model of corporate governance. Int J Bus Gov Ethics 3(4):446–460

Huang C-Y, Lyu MR (2005) Optimal release time for software systems considering cost, testing-effort, and test efficiency. IEEE Trans Reliab 54(4):583–591

Kapur P, Pham H, Gupta A, Jha P (2011) Software reliability assessment with OR applications. Springer

Kimura M, Toyota T, Yamada S (1999) Economic analysis of software release problems with warranty cost and reliability requirement. Reliab Eng Syst Saf 66(1):49–55

Kovalev I, Kovalev D, Chefonov V, Testoedvov N, Koltyshev A, Krivogornitsyn A (2020) The development and reliability analysis environment of fault-tolerance multiversion software. In: IOP conference series: materials science and engineering, 2020, vol 1. IOP Publishing, p 012033

Kumar V, Kapur P, Shrivastava A, Sharma R (2014) Optimal strategies for price-warranty decision model of software product with dynamic production cost. In: 3rd International conference on reliability, infocom technologies and optimization (ICRITO) (Trends and future directions). IEEE, pp 1–6

Kuo TC, Ma H-Y, Huang SH, Hu AH, Huang CS (2010) Barrier analysis for product service system using interpretive structural model. Int J Adv Manuf Technol 49(1–4):407–417

Lal R, Haleem A (2009) A structural modelling for e-governance service delivery in rural India. Int J Electron Gov 2(1):3–21

Li Q, Pham H (2017) NHPP software reliability model considering the uncertainty of operating environments with imperfect debugging and testing coverage. Appl Math Model 51:68–85

Li X, Li YF, Xie M, Ng SH (2011) Reliability analysis and optimal version-updating for open source software. Inf Softw Technol 53(9):929–936

Mahajan R, Agrawal R, Sharma V, Nangia V (2014) Factors affecting quality of management education in India. Int J Educ Manag

Maheshwarkar M, Sohani N (2019) Knowledge management evaluation criteria for industries: identification and interpretive structural modelling. Int J Knowl Manag Stud 10(3):227–250

Malone DW (1975) An introduction to the application of interpretive structural modeling. Proc IEEE 63(3):397–404

Mandal A, Deshmukh S (1994) Vendor selection using interpretive structural modelling (ISM). Int J Oper Prod Manag

Minamino Y, Inoue S, Yamada S (2015) Multi-attribute utility theory for estimation of optimal release time and change-point. Int J Reliab Qual Saf Eng 22(04):1550019

Musa JD (1975) A theory of software reliability and its application. IEEE Trans Softw Eng 3:312–327

Nan N, Harter DE (2009) Impact of budget and schedule pressure on software development cycle time and effort. IEEE Trans Softw Eng 35(5):624–637

Pachauri B, Kumar A, Dhar J (2014) Software reliability growth modeling with dynamic faults and release time optimization using GA and MAUT. Appl Math Comput 242:500–509

Park M, Pham H (2010) Warranty cost analyses using quasi-renewal processes for multicomponent systems. IEEE Trans Syst Man Cyber Part A Syst Humans 40(6):1329–1340

Park M, Pham H (2012) A new warranty policy with failure times and warranty servicing times. IEEE Trans Reliab 61(3):822–831

Pham H (1996) A software cost model with imperfect debugging, random life cycle and penalty cost. Int J Syst Sci 27(5):455–463

Pham H (2003) Software reliability and cost models: perspectives, comparison, and practice. Eur J Oper Res 149(3):475–489

Pham H (2007) System software reliability. Springer Science & Business Media

Pham H, Zhang X (1999) Software release policies with gain in reliability justifying the costs. Ann Softw Eng 8(1–4):147–166
Pham H, Zhang X (1999) A software cost model with warranty and risk costs. IEEE Trans Comput 48(1):71–75
Pitchaimuthu S, Thakkar JJ, Gopal P (2019) Modelling of risk factors for defence aircraft industry using interpretive structural modelling, interpretive ranking process and system dynamics. Meas Bus Excel
Sahney S, Banwet D, Karunes S (2010) Quality framework in education through application of interpretive structural modeling. TQM J
Samantra C, Datta S, Mahapatra SS, Debata BR (2016) Interpretive structural modelling of critical risk factors in software engineering project. Benchmarking Int J
Sgarbossa F, Pham H (2010) A cost analysis of systems subject to random field environments and reliability. IEEE Trans Syst Man Cybern Part C (Appl Rev) 40(4):429–437
Song KY, Chang IH, Pham H (2018) Optimal release time and sensitivity analysis using a new NHPP software reliability model with probability of fault removal subject to operating environments. Appl Sci 8(5):714
Talib F, Rahman Z, Qureshi M (2011) An interpretive structural modelling (ISM) approach for modelling the practices of total quality management in service sector. Int J Model Oper Manag 1(3):223–250
Tam C, da Costa Moura EJ, Oliveira T, Varajão J (2020) The factors influencing the success of on-going agile software development projects. Int J Project Manag 38(3):165–176
Verma V, Anand S, Aggarwal AG (2019) Software warranty cost optimization under imperfect debugging. Int J Qual Reliab Manag
Verma V, Neha N, Aggarwal AG (2020) Software release planning using grey wolf optimizer. In: Soft computing methods for system dependability. IGI Global, pp 1–44
Von Winterfeldt D, Fischer GW (1975) Multi-attribute utility theory: models and assessment procedures. In: Utility, probability, and human decision making. Springer, pp 47–85
Warfield JN (1974) Developing interconnection matrices in structural modeling. IEEE Trans Syst Man Cybern 1:81–87
Warfield JN, Cárdenas AR (1994) A handbook of interactive management. Iowa State University Press Ames
Watson RH (1978) Interpretive structural modeling—a useful tool for technology assessment? Technol Forecast Soc Chang 11(2):165–185
Wood A (1996) Predicting software reliability. Computer 29(11):69–77
Yamada S, Tamura Y (2016) Software reliability. In: OSS reliability measurement and assessment. Springer, Switzerland
Yamada S, Ohba M, Osaki S (1984) S-shaped software reliability growth models and their applications. IEEE Trans Reliab 33(4):289–292
Zhang X, Pham H (1998) A software cost model with error removal times and risk costs. Int J Syst Sci 29(4):435–442
Zhang X, Pham H (2000) An analysis of factors affecting software reliability. J Syst Softw 50(1):43–56
Zmud RW (1980) Management of large software development efforts. MISQ 45–55

أ

Software Reliability Modeling and Assessment Integrating Time Dependent Fault Reduction Factor in Random Environment

Nidhi Nijhawan and Vikas Dhaka

Abstract Growing demand of software in all application domains have led to the rising expectations and requirement for more reliable software systems from user end. Paradoxically, while achieving the reliability goals, the complexity of software turn to be very high and consequently it becomes critical to have influential approaches which evaluate reliability measures accurately. Based on distinct set of assumptions, a very large number of software reliability growth models (SRGMs) have already been developed over past few decades and still ongoing to evaluate various reliability metrics. In this chapter, we derive a software reliability model with the key consideration that the operating environment of software is unalike from the controlled testing environment and is accountable to affect software execution and its reliability significantly. To deal with randomness of operating environment and variations of fault detection rate subject to time we consider time dependent fault reduction factor in random environment. In addition, to suggest release time of the software, cost and reliability criteria are discussed and illustrated with numerical example. To conduct the comprehensive evaluation of goodness of fit, we worked out several selection criteria and comparative analysis with the existing models and it is worth noting that results offered by the proposed model are dependable and highly consistent with the observations procured from the real life data sets.

Keywords Software reliability · SRGM · Fault detection and correction process · Fault reduction factor · Random environment · Release time

---

N. Nijhawan (✉)
Shaheed Rajguru College of Applied Sciences for Women, University of Delhi, Delhi, India
e-mail: nidhi.nijhawan@rajguru.du.ac.in

V. Dhaka
Daulat Ram College, University of Delhi, Delhi, India
e-mail: vikasdhaka@dr.du.ac.in

A. G. Aggarwal et al. (eds.), *Optimization Models in Software Reliability*, Springer Series in Reliability Engineering,
https://doi.org/10.1007/978-3-030-78919-0_7

1 Introduction

The chapter describes software reliability model formulation considering time dependent fault reduction factor in random environment which help in reliability assessment. The chapter presents parameter estimation of the formulated model based on actual data sets followed by the discussion of the findings. Further, the performance analysis on the basis of number of model selection criteria and comparison with existing work is presented in the chapter. The chapter also discusses optimal release time schedule for the software and is illustrated with numerical example.

With the surge in the requirement of software in every direction, software firms work rigorously to improve quality and to provide reliable products to their users for their retention in the competitive market but in practice the likelihood of software failures cannot be avoided and ignored. Software glitches or bugs may be reported when a system fails to perform as anticipated which may end up with complains for unreliable software and in turn with the loss of efforts, money and reputation of the organization. Therefore, reliability has always been the major concern for its developers, users and researchers. For many years software reliability modeling has proved to be an efficient approach for evaluating key metrics such as number of left over faults which may cause failures, rate with which faults are detected or failures are occurred, cost, release time and reliability of software.

2 Literature Survey

In literature, abundance of models have been proposed and among those the Non Homogeneous Poisson Process (NHPP) models are widely used and have contributed immensely in quantifying reliability improvement (Goel and Okumoto 1979; Ohba 1984; Xie 1991; Musa 2004; Pham 2006; Kapur et al. 2011). Some authors have derived NHPP based SRGMs which are efficient to address realistic concept such as time delay between fault detection and its correction while some have incorporated fault reduction factor in their studies to state the relationship between number of number of faults detected and the number of faults corrected (Xie et al. 2007; Lin 2011; Peng et al. 2014). It is commonly noted that SRGMs are worked out on testing data and then are deployed to predict failures and reliability in the field assuming environment in the field and environment during the development of the software are of similar kind. Nevertheless, few researchers have considered uncertainty of the operating environment in their work to consider issues which may affect software failure rate and the reliability significantly. In recent past, an SRGM is proposed which outlined on uncertainty of operating environments with testing coverage (Pham 2013). Further, few NHPP models have been derived subject to random environments while considering distinct distributions for fault detection rate within the modeling process (Pham 2014, 2016). Of late, some authors have extended the study to present models which account for uncertainty of operating environments and discussed optimal release time and predictive analysis assuming various fault detection rate function (Song et al. 2017a, b, 2019).

In the present study, we propose an NHPP based new modeling approach which uniquely integrates time dependent FRF with random environments to predict reliability growth and other dynamic measures which are subject to time and explicitly affected by uncertainty inherent in operating environment. The framework of proposed model is attempted to incorporate both the realistic subjects simultaneously-time variant FRF and random field environment which have not been captured together in any particular model so far. The present work is intended to investigate repercussions emanated from the variant concerns which may be enhanced with time and environmental factors and their influence on reliability growth.

3 Modeling Framework

In this section we discuss the basic concepts and ground work which are required in building the models followed by the model development.

3.1 Background Work

Here, we present the brief details regarding the fault reduction factor and random environment which are key aspects of the present work.

- FRF: Fault Reduction Factor

FRF may be regarded as one of the most dominant parameter in appraising software reliability growth and is defined as cumulative number of faults corrected in proportion to that of the failures occurred by time t (Musa et al. 1987). Here it is denoted as $B(t)$ and expressed mathematically as:

$$B(t) = \frac{m_c(t)}{m_d(t)} \tag{1}$$

where $m_c(t)$ denotes the cumulative number of faults corrected while $m_d(t)$ denotes the cumulative number of faults detected. Alternatively Eq. (1) may be written as:

$$m_d(t) = \frac{m_c(t)}{B(t)} \tag{2}$$

Over recent years few authors have explored different patterns an FRF may take up and considered them in their modeling scheme to characterize the impact of some crucial factors namely time lag, defect density, test case coverage, imperfect debugging, dynamic resources allocation strategy, fault dependencies etc. on software debugging process. Following that, some authors used constant FRF in the model formulation under imperfect debugging (Jain et al. 2014), some have considered an S-shaped FRF in their modeling approach with perfect and imperfect debugging (Pachauri et al. 2015) while some of them have discussed different trends for FRF such as constant, increasing or decreasing as per the influential degrees of affecting factors and obtained distinct forms of failure intensity function (Hsu et al. 2011).

In practice, curves depicting FRF may not necessarily follow overall constant, increasing or decreasing trend but the amalgamation of different trends may be seen oftentimes. In some cases we may come across with many ups and downs in the same curve whereas in other cases curve goes down first and then rises up or vice versa. To demonstrate such conflicting combination of trends for FRF, we plotted few curves using randomly chosen real data sets available in literature (Aggarwal et al. 2017) which are shown in Fig. 1a–f.

Thus, in an effort to cater realistic trends of FRF we consider Exponentiated Weibull distribution in the present work, which was primarily introduced by Mudholkar and Srivastava (1993). It is worth noting that EW distribution has proved to be the finest choice for FRF as it is fairly adequate and flexible to track all possible fluctuations which are subject to time or operating conditions. To this end, we give EW density function, $B(t)$ which is used here to characterize FRF as follows:

$$B(t) = \frac{Nkv}{l^k} t^{k-1}[1 - \exp(-(t/l)^k]^{v-1} \exp(-(t/l)^k), t \geq 0, k > 0, l > 0, v > 0 \tag{3}$$

The corresponding cumulative FRF, $B^C(t)$ is given as below:

$$B^C(t) = N[1 - \exp(-(t/l)^k]^v \tag{4}$$

where (k, v) denote the shape parameters, l denotes the scale parameter and N is a constant.

Figure 2a, b representing the density function and the corresponding distribution function for distinct values of parameters shows the flexible nature of the underlying EW probability distribution and are also shown in Aggarwal et al. (2017).

- Random Environment

In recent past years some research efforts have been made to establish models considering random environment including (Yang and Xie 2000; Zhang et al. 2002; Chang et al. 2014; Pham 2016; Zhu and Pham 2018) so as to incorporate impact of factors or issues affecting software failure rate and other reliability measures. These issues may evolve during software execution in the field due to vagaries and uncertainty involved in the operating environment. In the present study we intend to model fault removal process considering factor representing randomness of operating environment and address it as a random variable, η which is assumed to be gamma distributed

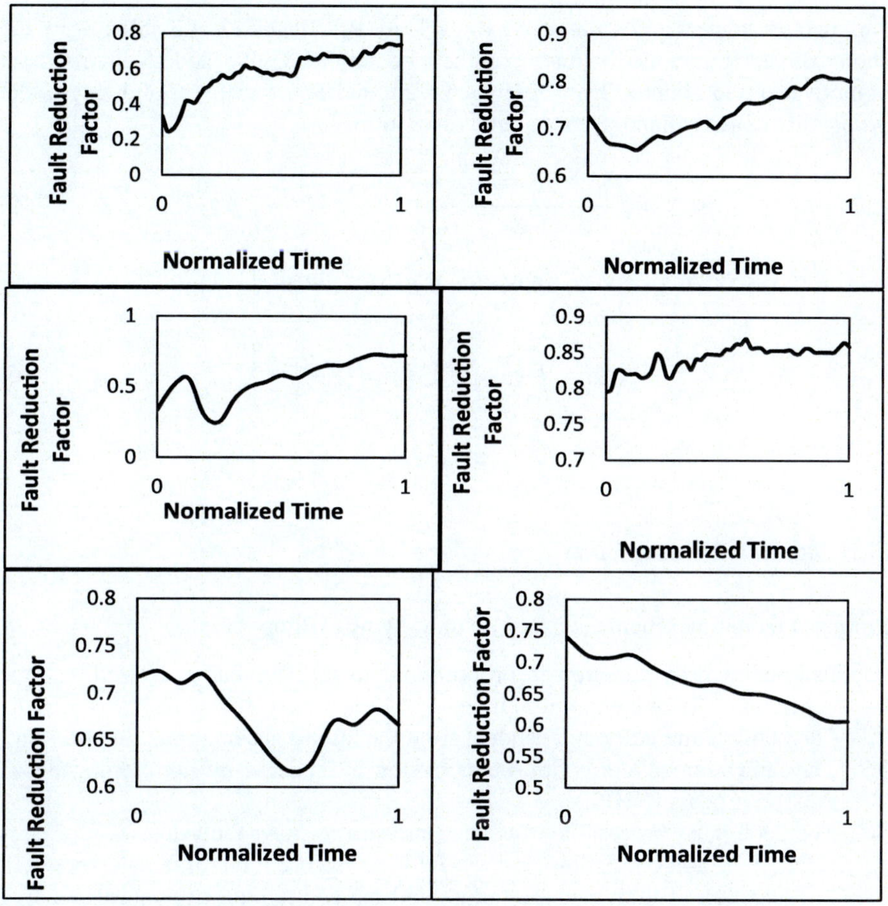

Fig. 1 **a–f** Distinct possible variations in FRF curves

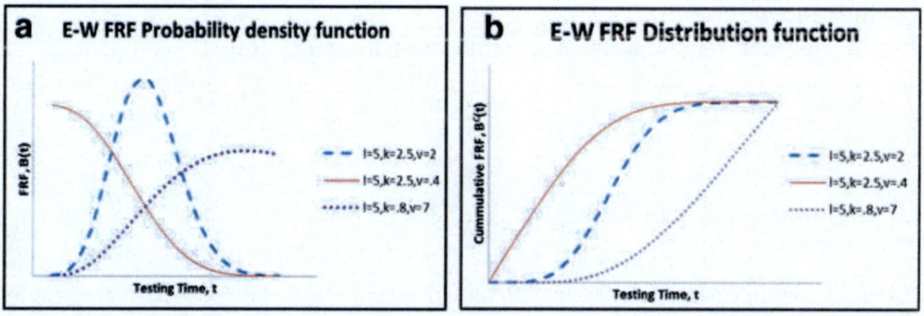

Fig. 2 **a**, **b**Density function and corresponding distribution function of EW-FRF for distinct parameter values

with two parameters and is denoted by $\eta \sim \Gamma(\theta_1, \theta_2)$. Besides its flexible nature, the choice of taking gamma distribution in the modeling to signify random environment is motivated and adopted from reference (Zhu and Pham 2018). Following is the probability density function for gamma distribution:

$$g(\eta) = \frac{(\theta_1)^{\theta_2}.(\eta)^{\theta_2 - 1}.e^{-\theta_1.\eta}}{\Gamma(\theta_2)}, \quad \theta_1, \theta_2 > 0; \ \eta \geq 0 \tag{5}$$

The corresponding Laplace transform of $g(\eta)$ is given as below:

$$G^*(\tau) = \int_0^\infty e^{-\eta.\tau} g(\eta) d\eta = \left[\frac{\theta_1}{\theta_1 + \tau}\right]^{\theta_2} \tag{6}$$

3.2 *Model Assumptions*

Proposed modeling scheme is based on underlying assumptions:

1. Failure intensity function is proportional to leftover faults as well as time dependent FRF at any point in time.
2. Each fault is mutually independent from the failure observation point of view.
3. The fault removal/correction phenomenon is modeled by non-homogeneous Poisson process (NHPP).
4. The debugging process is a two-step process namely; fault detection process (FDP) and fault correction process (FCP) to represent the time delay between the detection of fault causing failure and the correction of the corresponding fault.
5. During the fault removal process, new faults may introduce in the software with introduction rate $\beta(t)$.
6. To consider randomness of operating environment into the modeling, we multiply time independent REF (η) with proportionality which is a function of FRF.

3.3 *Model Formulation*

Based on assumptions stated above, mean value function (MVF) for corrected faults incorporating REF and fault introduction phenomena can be obtained on solving following differential equations:

$$\frac{dm_c(t)}{d(t)} = \eta.b.B(t)(a(t) - m_c(t)); \tag{7}$$

$$\frac{da(t)}{d(t)} = \beta(t)\frac{dm_c(t)}{d(t)} \tag{8}$$

where $a(t)$ denotes total fault content in the software at time t.

Solving above Eqs. (7) and (8) with conditions $a\,(t\,=\,0) = a,\ m_c(t=0) = 0,\ \beta(t) = \beta\ \forall\ t$, we obtain:

$$m_{c(\eta)}(t) = \frac{a}{(1-\beta)}[1 - e^{-\eta .b.(1-\beta).B^C(t)}] \tag{9}$$

Here a represents initial number of faults present in the software before testing starts.

Incorporating REF factor η, the equation describing MVF is given as:

$$m_c(t) = \int_0^\infty m_{c(\eta)}(t)\, g(\eta) d\eta \tag{10}$$

$$= \int_0^\infty \frac{a}{(1-\beta)}(1 - e^{-\eta .b.(1-\beta).B^C(t)})\, g(\eta) d\eta \tag{11}$$

Using Eq. (6) to apply Laplace transformation, the above equation becomes:

$$= \frac{a}{(1-\beta)}\left[1 - G^*\left(b\,(1-\beta)B^C(t)\right)\right] \tag{12}$$

Thus mean number of faults corrected considering REF is achieved as:

$$m_c(t) = \frac{a}{(1-\beta)}\left[1 - \left(\frac{\theta_1}{\theta_1 + b\,(1-\beta)B^C(t)}\right)^{\theta_2}\right] \tag{13}$$

To this end, software reliability can be assessed using MVF of proposed SRGM (given in Eq. (13)) and using the definition of reliability which states that it is the probability that a software does not experience failure in the interval $(t, t+\delta t)$ given that last failure occurred at time t (Xie 1991) and is expressed mathematically as follows:

$$R(\delta t|t) = \exp(-(m(t+\delta t) - m(t)));\quad \delta t > 0,\ t \geq 0 \tag{14}$$

4 Estimations and Analysis

In this section, we provide data sets used to validate the proposed model, discussed various model selection and validation criteria and presented goodness of fit curves followed by tables for parameter estimates that ended with detailed discussion of the findings of the present work.

4.1 *Real Data Sets*

To validate our study, we have used three real data sets in all: DS-I, DS-II and DS-III which present records of cumulative number of faults detected and faults corrected. DS-I is dataset for 87 test days and has been collected from literature (Liu et al. 2016) and is tabulated in Table 1 whereas data sets DS-II and DS-III for 17 test weeks each have been published by Musa et al. (1987), Wu et al. (2007) and is shown in Table 2 (Table 3).

4.2 *Model Selection and Validation Criteria*

This section briefly discusses eight selection criteria which are frequently used to validate the accuracy and predictability of a model.

- Coefficient of determination (R^2)

R^2 measures the proportion of the total variations in a data set that is explained by the model. It must be a value between 0 and 1 and cannot assume negative values.

$$R^2 = 1 - \frac{\sum_{i=1}^{n} (\hat{y}_i - y_i)^2}{\sum_{i=1}^{n} (y_i - \bar{y}_i)^2}; \quad \bar{y}_i = \frac{\sum_{i=1}^{n} y_i}{n}$$

The closer the value of R^2 to 1, the better the model explains the variation in data (Tjur 2009).

- Mean squared error (MSE)

The difference between the observed and the estimated values is regarded as residual or error and it may turn out to be positive, negative or zero. Thus it is suggested to minimize the squares of the residuals. The measure MSE is calculated by taking the mean of the square of the error terms, ε_i ($i = 1, 2, \ldots, n$). Squaring the error terms augments the errors and thus MSE desirable in situations demanding low tolerance for error.

Table 1 DS-I

Time (weeks)	Faults detected	Faults corrected	Time (weeks)	Faults detected	Faults corrected
1	23	6	30	869	660
2	96	15	31	875	689
3	163	17	32	892	712
4	250	20	33	895	744
5	282	23	34	898	773
6	343	35	35	898	794
7	375	44	36	904	811
8	398	44	37	916	811
9	424	44	38	916	837
10	445	55	39	916	849
11	459	84	40	916	855
12	465	105	41	916	860
13	541	128	42	919	863
14	593	163	43	924	869
15	637	203	44	939	884
16	637	302	45	948	927
17	640	483	46	956	942
18	640	523	52	971	942
19	640	529	61	983	942
20	672	529	63	988	942
21	703	529	64	991	942
22	706	529	65	994	945
23	730	532	68	994	959
24	747	535	69	994	962
25	756	541	70	994	965
26	785	581	71	994	968
27	797	608	86	997	968
28	828	622	87	1000	968
29	866	631			

$$MSE = \frac{1}{n}\sum_{i=1}^{n}\varepsilon_i^2$$

Lower value of MSE implies the better fitting (Kapur et al. 2011). In context for models using time lag between detection and correction process, we use MSE which is calculated by taking average of MSE values obtained independently for faults detected and for faults corrected. Thus,

Table 2 DS-II

Time (weeks)	Faults detected	Faults corrected
1	12	3
2	23	3
3	43	12
4	64	32
5	84	53
6	97	78
7	109	89
8	111	98
9	112	107
10	114	109
11	116	113
12	123	120
13	126	125
14	128	127
15	132	127
16	141	135
17	144	143

Table 3 DS III

Time (weeks)	Faults detected	Faults corrected
1	1	0
2	2	2
3	4	3
4	5	5
5	13	12
6	22	18
7	28	25
8	35	33
9	39	36
10	42	36
11	42	39
12	46	42
13	47	46
14	47	47
15	49	48
16	51	50
17	54	54

$$MSE_{Avg} = \frac{1}{2}(MSE_{\text{d}} + MSE_c)$$

- Mean absolute percentage error (MAPE)

Another way to show the deviation is in relative terms rather than absolute terms. To compute the measure, each error term, ε_i; $(i = 1, 2, \ldots, n)$ in absolute is expressed as a percentage of the actual value, y_i.

$$MAPE = \left(\frac{1}{n}\sum_{i=1}^{n}\frac{|\varepsilon_i|}{y_i}\right)100$$

The lower value of MAPE indicates better goodness of fit (Kim and Kim 2016).

- Predictive power (PP)

It measures the distance of model estimates from the actual data and is obtained by summing the squares of the proportion of error terms in the response of actual values, y_i; $i = 1, 2, \ldots, n$. It is given as:

$$PP = \sum_{i=1}^{n}\left(\frac{\hat{y}_i - y_i}{y_i}\right)^2$$

Low values of PP shows successful fit (Pham 2006).

- Bias

It is usually defined as average of prediction error which is calculated as difference between estimated values and actual values. It is calculated as:

$$Bias = \frac{1}{n}\sum_{i=1}^{n}(\hat{y}_i - y_i)$$

Lower values of Bias show good fitting to the data (Pillai and Nair 1997).

- Variation

It is commonly known as standard deviation of prediction error and calculated as:

$$Variation = \sqrt{\frac{\sum_{i=1}^{n}(y_i - \hat{y}_i - Bias)^2}{n-1}}$$

Lower the variation better is the goodness of fit (Pillai and Nair 1997).

- Root mean square prediction error (RMSPE)

Another useful measure for evaluating the performance of a model is the RMSPE which is a measure of closeness with which a model predicts the observations. It is computed as:

$$RMSPE = \sqrt{(Bias)^2 + (Variation)^2}$$

The lower the RMSPE, the better the goodness of fit (Pillai and Nair 1997).

- Akaike's information criterion (AIC)

It is widely used to investigate goodness of fit of stochastic models. It estimates the relative amount of information lost by a given model which is related to degrees of freedom carried by a model. It is defined using likelihood function, L as follows:

$$AIC = -2 \log L + 2k$$

where $L = \prod_{i=1}^{n} \frac{(\hat{y}_i - \hat{y}_{i-1})^{y_i - y_{i-1}}}{(y_i - y_{i-1})!} e^{-(\hat{y}_i - \hat{y}_{i-1})}$ and k is the number of parameters estimated in the model and n is the number of actual observations. Lower AIC indicates less amount of information lost and thus better quality and fitting of a model (Akaike 1974).

4.3 Parameter Estimates

We obtain the estimates of the proposed model by non-linear regression using SPSS-23 software. Estimation of parameters is carried out in two phases. In Phase 1, we estimate parameters of the fault reduction factors then carried their estimated value to Phase 2 to determine the parameters involved in proposed model. The estimation methodology and their results are discussed in detail in following subsections:

- Phase I Estimation

Using definition of FRF given in Eq. (1), actual values of FRF have been computed for all three datasets: DS-I, II, and III and are plotted in Fig. 3a–c. As aforementioned, we have used EW distribution to represent FRF, the parameter estimates of which are summarized in Table 4 for all data sets.

- Phase II Estimation

Using values of estimates for FRF obtained from Phase-I, we have carried out estimation of parameters for proposed model and further for the purpose of model validation, few well known existing models (which have not considered RFE) namely Goel-Okumoto (GO) model (Goel and Okumoto 1979), Delayed S shaped (DS)

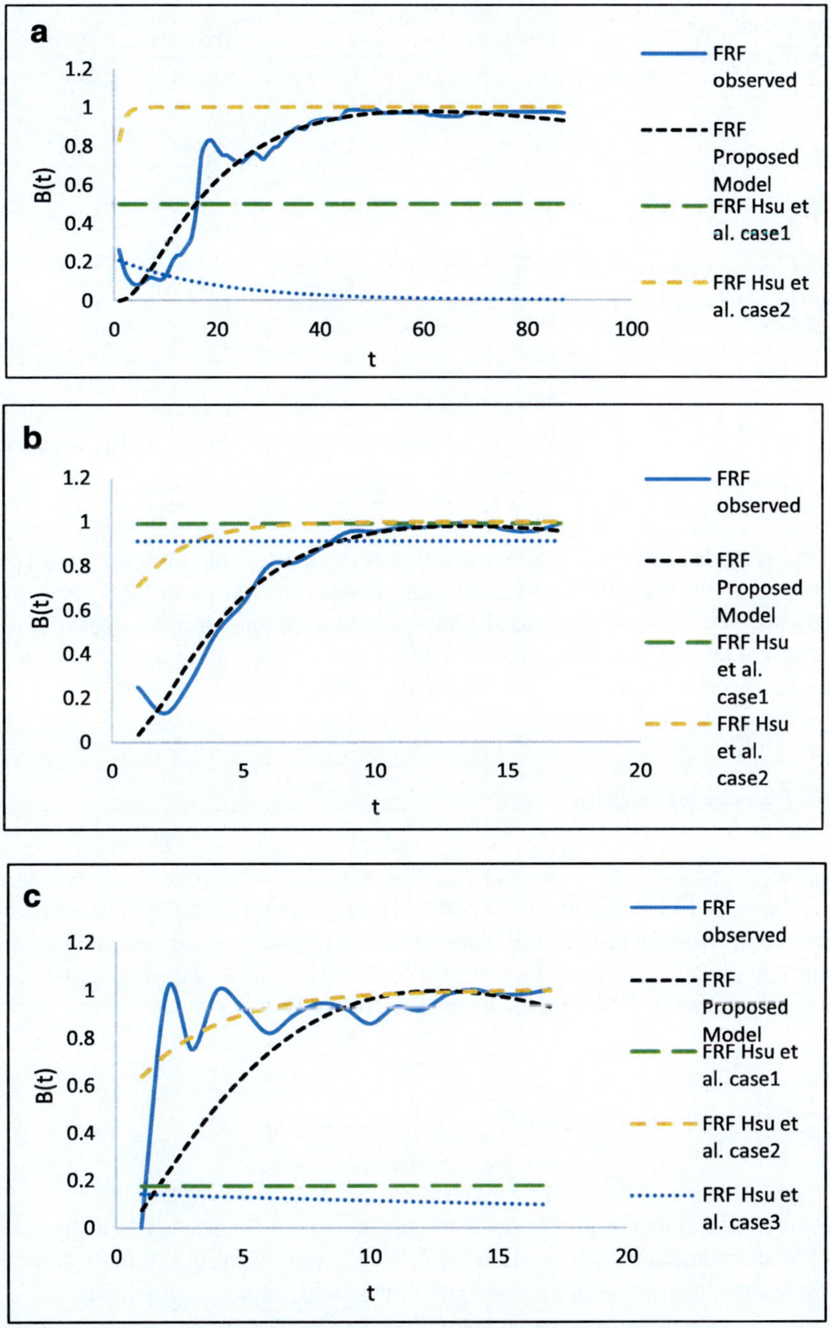

Fig. 3 **a** Statistical goodness of fit curves of FRF for DS-I. **b** Statistical goodness of fit curves of FRF for DS-II. **c** Statistical goodness of fit curves of FRF for DS-III

Table 4 Estimates of EW-FRF parameters

Parameters	DS-I	DS-II	DS-III
N	410.569	84.044	24.421
l	0.502	0.320	11.294
k	0.214	0.244	0.815
v	36.040	25.146	5.200

Table 5 Description of existing models used for comparison

Model	Parameters	MVF, $m(t)$
GO	a, b	$a(1-e^{-bt})$
Delayed S shaped	a, b	$a(1-(1+bt)e^{-bt})$
Inflexion S shaped	a, b, c	$\frac{a(1-e^{-bt})}{1+ce^{-bt}}$
Yamada imperfect debugging	a, b, c	$a(1-e^{-bt})(1-\frac{c}{b})+act$

model (Yamada and Osaki 1985), Inflection S shaped (IS) model (Ohba 1984) and Yamada imperfect debugging model (YID) model (Yamada et al. 1992) (described briefly in Table 5) have been used to compare model parameters and values of several comparison criteria (discussed in Sect. 4.2) and provided the results in Tables 6, 7 and 8 for respective datasets DS-I, II and III.

4.4 Curves of Fitting

Visual representation of the results by means of graphs and curves has always been a prominent way which illustrates the performance of the proposed work and findings. This section includes graphs which are created to figure out the goodness of fit of the present modeling scheme. Figure 4a–c captures the curves showing actual versus estimated number of faults corrected for DS-I, II and III.

4.5 Discussion of Findings

Here we outline some important observations of the present work.

Figure 3a–c showing goodness of fit curves of FRF for DS-I, II and III exhibit that EW distribution used for estimating FRF is proved to be successful in capturing the variations incurred with time and fits considerably close to actual data.

i. On comparing estimated model parameters and values computed using comparison criteria namely R^2, MSE, MAPE, PP, Bias, Variation, RMSPE and AIC as presented in Tables 6, 7 and 8, it is established that proposed model provides

Table 6 Parameter estimates of proposed model for DS-I

Model	Parameters	R^2	MSE	PP	Bias	Variation	MAPE	RMS-PE	AIC
Proposed model	$a = 969.647$, $b = 0.629$, $\theta_1 = 54.381$, $\theta_2 = 9.070$, $\beta = 0.009$	0.987	1419.38	6.94	0.94	75.79	12.78	75.79	1188.0
GO model	$a = 1139.409$, $b = 0.028$	0.930	7620.83	210.25	15.27	175.61	64.78	176.27	1542.7
Delayed S shaped model	$a = 1003.468$, $b = 0.078$	0.977	2520.61	29.68	7.46	101.11	24.49	101.39	1251.2
Inflexion S shaped model	$a = 960.537$, $b = 0.120$, $c = 11.893$	0.982	1929.51	17.89	3.76	88.36	19.72	88.44	1202.4
Yamada imperfect debugging model	$a = 1131.367$, $b = 0.029$, $c = 0.0001$	0.930	7647.88	210.57	15.24	175.92	64.85	176.58	1546.9

Table 7 Parameter estimates of proposed model for DS-II

Model	Parameters	R^2	MSE	PP	Bias	Variation	MAPE	RMS-PE	AIC
Proposed model	$a = 157.450$, $b = 0.994$, $\theta_1 = 8.400$, $\theta_2 = 2.166$, $\beta = 0.002$	0.99	24.83	5.20	0.77	10.27	24.15	10.30	121.1
GO model	$a = 231.624$, $b = 0.058$	0.94	130.89	72.52	2.28	23.59	84.46	23.69	141.4
Delayed S shaped model	$a = 147.514$, $b = 0.263$	0.979	46.31	16.59	1.28	14.03	37.59	14.09	126.2
Inflexion S shaped model	$a = 129.919$, $b = 0.507$, $c = 17.304$	0.982	39.64	8.73	0.74	12.98	28.05	13.00	169.5
Yamada imperfect debugging model	$a = 231.626$, $b = 0.058$, $c = 0.0001$	0.94	130.88	72.52	2.28	23.59	84.45	23.69	143.4

Table 8 Parameter estimates of proposed model for DS-III

Model	Parameters	R^2	MSE	PP	Bias	Variation	MAPE	RMS-PE	AIC
Proposed model	$a = 60.535$, $b = 0.986$, $\theta_1 = 0.530$, $\theta_2 = 0.662$, $\beta = 0.060$	0.993	2.46	0.61	-0.045	3.14	9.25	3.14	70.7
GO model	$a = 789.917$, $b = 0.004$	0.947	18.79	14.85	0.92	8.67	46.19	8.72	76.5
Delayed S shaped model	$a = 65.715$, $b = 0.182$	0.980	7.13	3.66	0.53	5.34	23.27	5.37	72.6
Inflexion S shaped model	$a = 121.855$, $b = 0.57$, $c = 0.999$	0.948	18.27	15.91	1.07	8.55	47.18	8.61	77.7
Yamada imperfect debugging model	$a = 104.520$, $b = 0.032$, $c = 0.029$	0.947	18.85	14.50	0.89	8.68	45.79	8.73	78.6

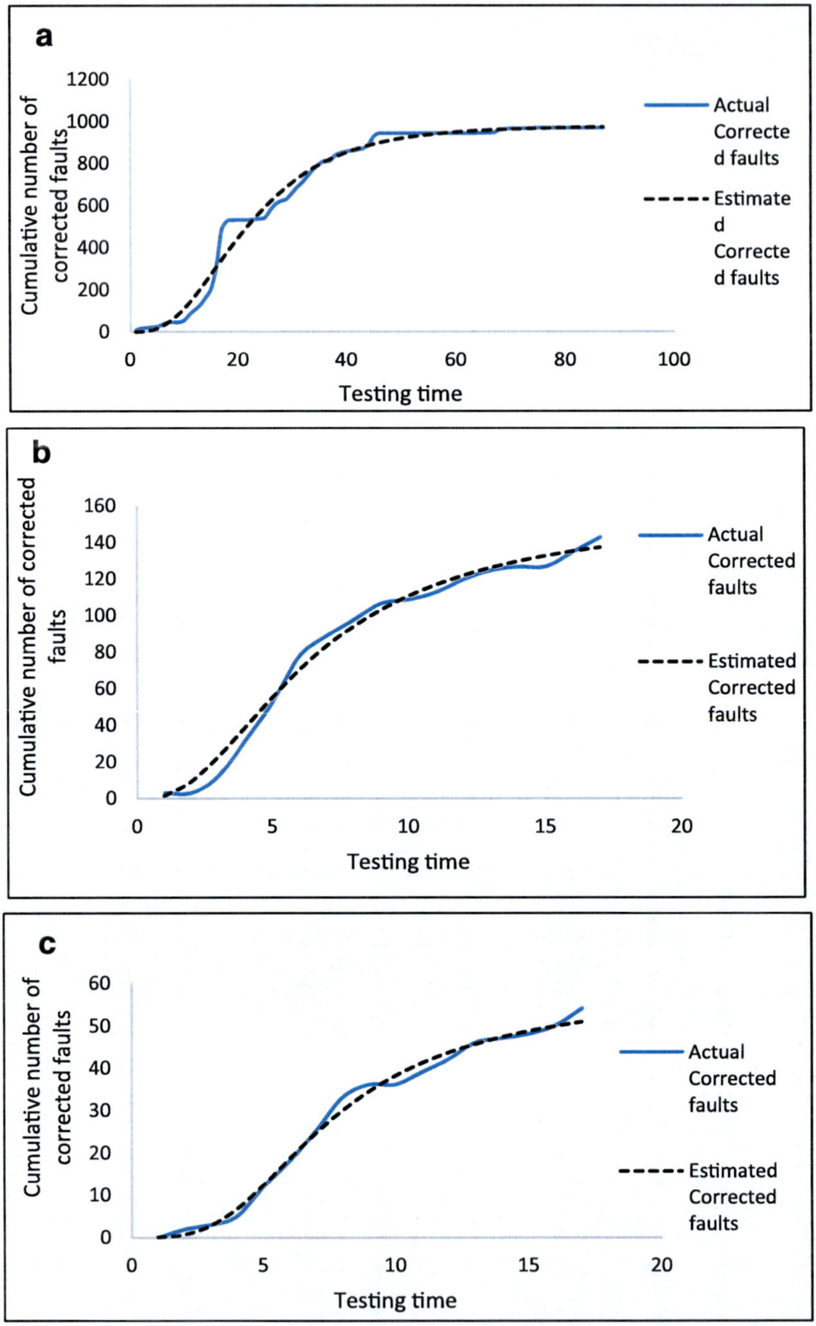

Fig. 4 **a** Statistical goodness of fit curves of fault correction process for DS-I. **b** Statistical goodness of fit curves of fault correction process for DS-II. **c** Statistical goodness of fit curves of fault correction process for DS-III

significantly better results and descriptive power than well-known existing models for all datasets considered in the present study.

ii. We acknowledge that use of two phase estimation method in the proposed work is proved to be reasonable since estimates of FRF obtained in Phase I when used to compute mean number of faults corrected give rise to the values which are quite close to actual number of corrected faults. However, estimates for number of faults detected (computed using FRF and Eq. (1)) also shows outstanding results in virtue of close fitting to corresponding actual values. On the other hand using conventional approach to estimate parameters involved in the mean value function for fault detection (correction) process may not necessarily show good fit to FRF (as shown in Fig. 3a–c) and consequently may not offer acceptable values for fault correction (detection) process. Specifically, from Table 9, it can be observed that values of average MSE, MSE_{Avg} (average of MSE for detected and MSE for corrected faults) for the proposed model for DS-I, II and III came out to be 1293.18, 69.3, 5.42 which are adequately smaller than those calculated for three distinct models (based on three pattern for FRF used) proposed in published work (Hsu et al. 2011). Results summarized in Table 9 and graphs plotted in Fig. 3a–c establish the success and relevance of using two phase parameter estimation approach in the present study.

5 Release Time of Software

The decision problem of utmost concern for management of software firms is to determine when to end testing and release the product into the market to make it available for its users. This decision strongly depends on the criteria used by the management. In practice, there is always a trade-off between cost and the reliability a consequently release time may vary from minimizing total cost incurred to maximizing reliability (Huang 2005; Kapur et al. 2011; Song et al. 2018). In this section, with the help of numerical illustration, we attempt to suggest release time of software while considering both the important criteria and the resultant values of proposed SRGM. We begin by creating a cost function using distinct cost parameters viz. per unit cost to remove faults during testing phase (before release), C_1; per unit cost to remove faults during operational phase (after release), C_2 and testing cost per unit testing time, C_3. Using cost parameters and MVF of Eq. (13), associated total cost incurred can be presented as:

$$C(T) = C_1 m_c(T) + C_2[m_c(T_l) - m_c(T)] + C_3(T) \tag{15}$$

where T denotes release time and T_l denotes software life cycle (>T). We may differentiate Eq. (15) with respect time to find time point that minimizes the cost function, $C(T)$. Let $T = T_c$ denotes the release time that corresponds to minimum cost. Besides cost, reliability plays significant role in decision making. Generally, an appropriate level of reliability, R_0 is targeted to achieve at the time of release.

Table 9 Comparison of proposed SRGM with existing FRF based models (Hsu et al. 2011)

Comparison criterion	Data set	Proposed model	Hsu et al. (2011) case-1	Hsu et al. (2011) case-2	Hsu et al. (2011) case-3
MSE_c	DS-I	1419.376	136,064.5	26,614.75	584,484.2
	DS-II	24.830	306.387	223.043	245.214
	DS-III	2.462	790.150	11.006	938.161
MSE_d	DS-I	1166.373	306.366	306.358	341.591
	DS-II	113.801	48.809	33.674	48.814
	DS-III	8.393	23.742	18.120	31.58
MSE	DS-I	1293.186	68,185.44	13,460.55	292,412.9
	DS-II	69.316	177.598	128.358	147.014
	DS-III	5.426	406.946	14.563	484.870
Estimated parameters	DS-I	Provided in Table 6	$a = 1008.179, B_0 = 0.498,$ $r = 0.120$	$a = 1008.168, B_0 = 0.995,$ $r = 0.060,\ k = 3.645$	$a = 3215.028, B_0 = 0.217,$ $r = 0.092,\ k = 052$
	DS-II	Provided in Table 7	$a = 154.21, B_0 = 0.993,$ $r = 0.101$	$a = 144.308, B_0 = 0.501,$ $r = 0.189,\ k = .543$	$a = 154.390, B_0 = 0.914,$ $r = 0.153,\ k = 2.5E - 5$
	DS-III	Provided in Table 8	$a = 167.440, B_0 = 0.178,$ $r = 0.138$	$a = 90.437, B_0 = 0.5,$ $r = 0.062,\ k = .317$	$a = 167.440, B_0 = 0.146,$ $r = 0.193,\ k = .024$

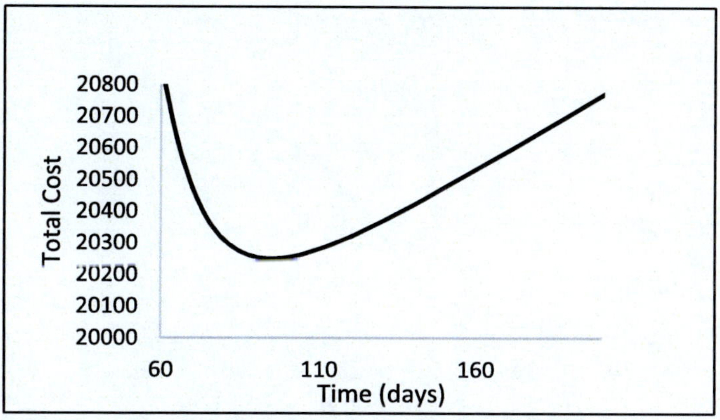

Fig. 5 Curve satisfying cost criterion

Suppose $T = T_R$ represents the release time when reliability requirement, R_0 is satisfied. Thus the optimal release time may be identified as $T^* = \max(T_c, T_R)$.

5.1 *Numerical Illustration*

To illustrate we use estimates of FRF and other parameters of proposed SRGM for DS-I (listed in Tables 4 and 6) and assume $C_1 = \$20$, $C_2 = \$50$ and $C_3 = \$6$ to figure out optimal release time of software. To minimize cost, we use Maple software and obtained minimum cost $C(T) = \$20249.56$ at $T = T_c = 93.8 \approx 94$ *days*. The corresponding cost curve is drawn in Fig. 5. Using definition of reliability (given in Eq. (15)) with $\delta t = 1$ and substituting parameter estimates of MVF (given in Eq. (13)) we plot reliability curve which is shown in Fig. 6. It is noted that reliability achieved at $T = T_c = 93.8 \approx 94$ *days* is 0.825. If $R_0 = 0.85$, it can be satisfied at $T = T_R = 97$ *days* with associated cost \$20,251.24 which is clearly seen from reliability and cost curves. Thus we conclude with decision to release the software at $T^* = \max(T_c, T_R) = \max(94, 97) = 97$*days*. If management raises the reliability requirement from $R_0 = 0.85$ to $R_0 = 0.90$ then $T_R = 105$ *days* which indicates testing needs to be continued to attain the desired level of reliability which leads to relatively higher cost of \$20,267.35. Accordingly, $T^* = \max(T_c, T_R) = \max(94, 105) = 105$ *days*.

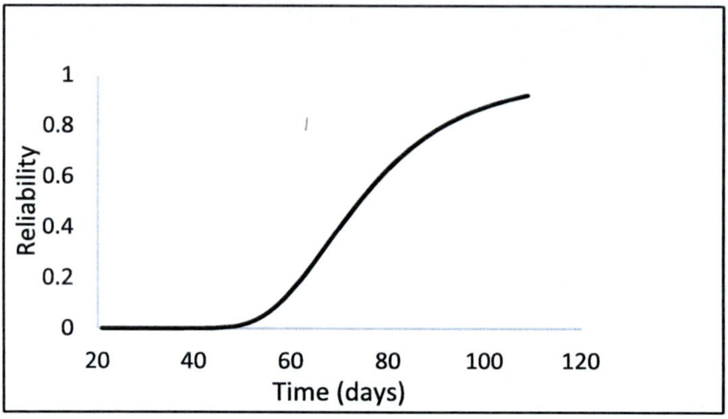

Fig. 6 Curve satisfying reliability criterion

6 Conclusion

In the present work, we propose an SRGM to model fault correction process and to investigate reliability of a software system which is a primary field of interest for software developers and many research studies. The present approach makes use of two important dynamic aspects-time dependent FRF and REF to represent real life phenomenon and elucidate time delayed fault correction process incorporating effects of random environmental factors. The proposed model is applied to three fault data sets of real software projects. Based on various selection and comparison criteria, we analyzed the performance of the proposed work. The investigation was illustrated by the number of related goodness of fit curves which indicates that the proposed model fit considerably close to actual data and better than the other well established existing models in the literature. We also address cost and reliability criteria which aids in making decision regarding release time of software and worked out a numerical example to suggest release time of software based on both the criteria. We conclude that proposed SRGM shows remarkable performance and is sufficiently capable to be used as an effective tool for making reliability assessment and taking decisions on release time.

7 Future Scope

The present study may be extended to derive multi-release software reliability growth models and to study reliability growth process of open source software systems which are offered with timely upgradations. Further, incorporating the concept of change point may aid in augmenting the present modeling scheme.

References

Aggarwal AG, Dhaka V, Nijhawan N (2017) Reliability analysis for multi-release open-source software systems with change point and exponentiated Weibull fault reduction factor. Life Cycle Reliab Saf Eng 6(1):3–14

Akaike H (1974) A new look at statistical model identification. IEEE Trans Autom Control 19:716–719

Chang IH, Pham H, Lee SW, Song KY (2014) A testing-coverage software reliability model with the uncertainty of operation environments. Int J Syst Sci Oper Logist 1:220–227

Goel AL, Okumoto K (1979) Time dependent error detection rate model for software reliability and other performance measures. IEEE Trans Reliab 28(3):206–211

Hsu CJ, Huang CY, Chang JR (2011) Enhancing software reliability modeling and prediction through the introduction of time-variable fault reduction factor. Appl Math Model 35(1):506–521

Huang CY (2005) Cost-reliability-optimal release policy for software reliability models incorporating improvements in testing efficiency. J Syst Softw 77(2):139–155

Jain M, Manjula T, Gulati TR (2014) Imperfect debugging study of SRGM with fault reduction factor and multiple change point. Int J Math Oper Res 6(2):155–175

Kapur PK, Pham H, Gupta A, Jha PC (2011) Software reliability assessment with OR applications. Springer, UK

Kim S, Kim H (2016) A new metric of absolute percentage error for intermittent demand forecasts. Int J Forecast 32:669–679

Lin CT (2011) Analyzing the effect of imperfect debugging on software fault detection and correction processes via a simulation framework. Math Comput Model 54(11–12):3046–3064

Liu Y, Li D, Wang L, Hu Q (2016) A general modeling and analysis framework for software fault detection and correction process. J Softw: Test Verif Reliab 26(5):351–365

Mudholkar GS, Srivastava DK (1993) Exponentiated weibull family for analyzing bathtub failure-rate data. IEEE Trans Reliab (42):299–302

Musa JD (2004) Software reliability engineering: more reliable software, 2nd ed. Faster and Cheaper, Authorhouse

Musa JD, Iannino A, Okumoto K (1987) Software reliability: measurement, prediction application. McGraw Hill, NY

Ohba M (1984) Inflection S-shaped software reliability growth model. In: Stochastic models in reliability theory. Springer, pp 144–162

Pachauri B, Dhar J, Kumar A (2015) Incorporating inflection S-shaped fault reduction factor to enhance software reliability growth. Appl Math Model 39(5–6):1463–1469

Peng R, Li YF, Zhang WJ, Hu QP (2014) Testing effort dependent software reliability model for imperfect debugging process considering both detection and correction. Reliab Eng Syst Saf 126:37–43

Pham H (2006) System software reliability. In: Reliability engineering series. Springer, UK

Pham H (2013) Loglog fault-detection rate and testing coverage software reliability models subject to random environments. Vietnam J Comput Sci 1:39–45

Pham H (2014) A new software reliability model with Vtub-shaped fault detection rate and the uncertainty of operating environments. Optimization 63(10):1481–1490

Pham H (2016) A generalized fault-detection software reliability model subject to random operating environments. Vietnam J Comput Sci 3:145–150

Pillai K, Nair VSS (1997) A model for software development effort and cost estimation. IEEE Trans Softw Eng 23(8):485–497

Song KY, Chang IH, Pham H (2017a) A three-parameter fault-detection software reliability model with the uncertainty of operating environments. J Syst Sci Syst Eng 26:121–132

Song KY, Chang IH, Pham H (2017b) An NHPP software reliability model with S-shaped growth curve subject to random operating environments and optimal release time. Appl Sci 7:1304

Song KY, Chang IH, Pham H (2018) Optimal release time and sensitivity analysis using a new NHPP software reliability model with probability of fault removal subject to operating environments. Appl Sci 8(5):714

Song KY, Chang IH, Pham H (2019) NHPP software reliability model with inflection factor of the fault detection rate considering the uncertainty of software operating environments and predictive analysis. Symmetry 11(521):1–21

Tjur T (2009) Coefficients of determination in logistic regression models—a new proposal: the coefficient of discrimination. Am Stat 63(4):366–372

Wu YP, Hu QP, Xie M, Ng SH (2007) Modeling and analysis of software fault detection and correction process by considering time dependency. IEEE Trans Reliab 56(4):629–642

Xie M (1991) Software reliability modelling. World Scientific Publishing

Xie M, Hu QP, Wu YP, Ng SH (2007) A study of the modeling and analysis of software fault-detection and fault-correction processes. Qual Reliab Eng Int 23:459–470

Yamada S, Osaki S (1985) Software reliability growth modeling: models and applications. IEEE Trans Softw Eng 11(12):1431–1437

Yamada S, Tokuno K, Osaki S (1992) Imperfect debugging models with fault introduction rate for software reliability assessment. Int J Syst Sci 23(12):2241–2252

Yang B, Xie M (2000) A study of operational and testing reliability in software reliability analysis. Reliab Eng Syst Saf 70(3):323–329

Zhang X, Jeske D, Pham H (2002) Calibrating software reliability models when the test environment does not match the user environment. Appl Stoch Models Bus Ind 18(1):87–99

Zhu M, Pham H (2018) A software reliability model incorporating martingale process with gamma-distributed environmental factors. Ann Oper Res. https://doi.org/10.1007/s10479-018-2951-7

Multi-objective Release Time Problem for Modular Software using Fuzzy Analytical Hierarchy Process

Neha, Anu G. Aggarwal, and Ajay Jaiswal

Abstract The revolutionary evolution of the computers and their advancements has led to an increase of human dependence on them. Their wide utilization in various aspects of everyday activities necessitates the addition of additional features and functionalities to the system. Incorporation of these features makes the software development a complicated and critical task. One approach to ease this process is to design the software in modules manner. Therefore, small modules are incorporated in the software and are evaluated separately instead of the whole software at once. This chapter proposes release time problem for modular software system considering weights of the modules and by modelling the models with two software reliability growth models (SRGMs). The effect of testing coverage function and fault reduction factor is taken into account in these SRGMs. For determine the modular weights Fuzzy Analytical Hierarchy Process (FAHP) is used. FAHP decomposes complex structure of software into different levels of hierarchy and helps in determining the relative weights of the modules. The two SRGMs are considered separately to formulate the multi-objective optimization problem by considering development cost and software reliability simultaneously. Release time problem is followed by sensitivity analysis with the purpose of understanding the importance of development cost on release time. A numerical illustration is provided to understand the proposed methodology. For the estimation of the parameters given in SRGMs; a real life fault dataset has been used.

Keywords Software reliability · Software reliability growth model · Fuzzy analytical hierarchy process · Testing coverage · Fault reduction factor · Multi-objective optimization problem · Release time problem

---

Neha · A. G. Aggarwal
Department of Operational Research, University of Delhi, New Delhi, India

A. Jaiswal (✉)
Shaheed Sukhdev College of Business Studies, University of Delhi, New Delhi, India
e-mail: ajayjaiswal@sscbsdu.ac.in

A. G. Aggarwal et al. (eds.), *Optimization Models in Software Reliability*, Springer Series in Reliability Engineering,
https://doi.org/10.1007/978-3-030-78919-0_8

1 Introduction

With rapid growth of technology, software systems have become an indispensable and efficacious part of our everyday lives. These systems demand services to its users due to reliability and stability requiremnts. Many critical operations for example, air traffic control system, national security and defence system etc. depend on reliable software. This increased belief on trustworthy software products has dragged attention of many academicians and practitioners. Hence a considerable research has been done to measure and improve the reliability during the testing or operating phase. Reliability is considered as a significant element to characterize the quality of a software system (Zhang et al. 2003). In past, numerous researches have been introduced that are focused on software reliability models. To model software faults growth process Non-homogeneous Poisson process (NHPP) is assumed as the most effective and successful mathematical tool. NHPP defines the failure phenomenon in the testing phase of software development life cycle.

A great number of NHPP based software reliability growth models (SRGM) considering different assumptions have been introduced to predict the remaining faults (Aggarwal et al. 2019; Kapur et al. 2014; Verma et al. 2019a). Goel and Okumoto (1979) reliability model was the very first attempt to incorporate NHPP as a model for failure time . Following this SRGM researchers have suggested numerous time dependent SRGMs that usually consists different distribution of fault detection rate and initial fault content up to particular time and helps in the reliability assessment process. Moreover, academicians and researchers proposed that precision of SRGMs can be improved by incorporating real life problems occurred in testing phase. For example testing coverage, fault reduction factor, fault removal efficiency, testing effort (Cai and Lyu 2007; Huang et al. 2007; Shibata et al. 2006). Effect of these factors helps in building more reliable product. Amongst these factors testing coverage and fault reduction factor are observed as the most important and effective.

From both software engineers and users perspective testing coverage is observed to be an important measure. Using this measure software engineers can assess the quality of software and can easily calculate the additional effort that is required to increase the reliability. Form users perspective, testing coverage can help them to decide when to buy a software product? (Pham and Zhang 2003). During testing phase coverage growth is characterized by coverage growth function c(t) that can be defined as "proportion of code that is covered up to time t". Hence incorporation of testing coverage to the SRGM helps developers to calculate how much extra effort is required to improve reliability.

In practice, it has been observed that the number of failure may not always be equal to the number of faults found. Such relationship between faults and failure is expressed as FRF (Musa 1975). FRFs is described as the ratio of faults and failures. FRFs are also characterised as different ratios, for example detectability, associability, fault growth rate and fault exposure ratio (Li and Malaiya 1993; Malaiya et al. 1993; Musa 1991, 2004). In proposed study we have considered SRGMs incorporating testing coverage and Fault reduction factor.

As previously mentioned, developers are bound to develop a highly stable software system. To meet these user's requirements, they put in best efforts to incorporate maximum features and functions in the software. This incorporation of features results in making of a complex and critical software structure. Therefore, in order to cope up this situation a numerous small and independent programs called components are implemented into the software system, also in module testing phase each module is tested. These are often programmed as independent software for accomplishing predefined tasks. Usually modules are developed by different programmers or sometimes at different geographical locations. Here we are using this concept of modular software system to formulate the optimization problem by assigning weights to the modules. To calculate these weights Multi Criteria Decision Making (MCDM) technique has considered.

MCDM technique is used to deal with the decision making problem when it involves multiple attributes and dimensions. One way to deal such decision making problem is Analytical Hierarchy Process (AHP). AHP was first proposed by Satty (1980), which is a top-down approach. In this technique a hierarchy is constructed where complex arrangement is decomposed into different levels. To perform AHP we need to construct pairwise comparison matrices on every level of the hierarchy for which we require linguistic scale values. Satty (1980) provided a 1 to 9 crisp scaling values for the construction of pairwise comparison matrix. Moreover, when we talk about some real life situations crisp numbers are not enough to handle the uncertainty in the decision maker's opinion. Therefore, to deal with uncertain environment fuzzy set theory was introduced (Zadeh 1965) . This study considers the FAHP technique to deal with the uncertain environment while determining module weights. The optimu release time problem is then formulated using these weights.

In the development phase of the software, the major concern of development team is to determine the optimal time to release software into the market. Such problems are commonly acknowledged as optimal release time problem. As the testing helps in improving and assessing the quality of the software but it cannot be performed indefinitely, software must be released at an optimum time to satisfy existing users and potential users as well. User's wish is to get the software at reasonable prices and also with fast delivery whereas the developers crave to minimize the development cost and maximizes the revenue and the reliability as well. If software release time is excessively delayed then the developer may undergo whipping by penalties and revenue loss. While releasing the software too early may cost heavily in terms of fault removal, which results in decrease in the reputation of developers. Hence this trade-off between these contradictory objectives is necessary. Therefore, we have considered an optimum release planning problem to minimizes the total expected cost and maximize the reliability simultaneously under the budget constraint.

Focus of the Study

Focus of the study is to provide a model for the software developers that can help to calculate release time of software. Hence, we have developed a multi-objective optimization problem that runs reliability maximization and cost minimization problem simultaneously under a budgeting constraint. We have assumed a modular software system and assigned them some weight which is calculated with the help of a MCDM technique, namely FAHP. For mean value function we have considered an SRGM including testing coverage along with FRF.

The main objectives of the proposed study can be given as follows:

(a) To analyse the significance of the software modules in the software development process by assigning them weights using FAHP method.
(b) To formulate the optimization problem that can take care of development cost and software reliability simultaneously for a modular software system under limited budget.
(c) To calculate the time to release the modular software in the market.

Organization

The chapter is organized in the following way; following the detailed introduction we have provided literature review in second section. Third section discusses the methodologically framework of the chapter. Fourth section illustrates numerical that we have done using the proposed strategy and further sensitivity analysis has been done. The last section concludes the chapter that provides the future scope and limitation of the study.

2 Related Work

In this chapter we have taken the aid of five important techniques of software reliability, namely Software Reliability Growth Model, Testing Coverage, Fault Reduction Factor, Modular Software and Release Planning. A detailed literature review of these techniques is as follows:

2.1 Software Reliability Growth Model

Reliability plays a key part in the development of software system. This vital utility of reliable software has dragged the attention of the several researchers and as a result numerous reliability models have been proposed in the last decade. Among them, time based SRGMs have pioneered the attention of researchers in assessing the reliability of software system. Some important milestone of time dependent SRGMs that have been introduced previously in literature of software reliability are as mentioned.

SRGM introduced by Goel and Okumoto (1979) bought a revolution in modelling of software growth models. The model proposed by them was a finite failure model with exponential intensity function and other parameters assumed constant. As modelling progressed, it was observed that many a time mean value function (MVF) followed S-shaped curve than the exponential functional form. Hence an improvement was made in the testing efficiency with respect to the time and a 2-stage model was given by Yamada et al. (1984). This model assumed that there exist fault detection and fault isolation phases in the testing phase software system. Authors considered delay between fault detection and eliination under the hypothesis that all faults are eliminated while conducting testing of software. Further, Kapur and Garg (1992) proposed a model in which they came up with the concept that detection or removal of faults can result in the creation of more faults in the software.

2.2 *Testing Coverage*

Based on the previous research it has been observed that SRGMs can be improved and can be more proficient if we incorporate real life issues occurred during testing phase into the SRGM (Cai and Lyu 2007; Huang et al. 2007; Shibata et al. 2006). Out of these issues testing coverage and Fault reduction factor (FRF) are considered as the most important factor. Testing coverage functions have been introduced with different distribution, for example, Logarithmic-exponential (Malaiya et al. 2002), S-shaped (Pham and Zhang 2003), Weibull-Logistic (Gokhale et al. 1996) and logistic-exponential (Chatterjee and Singh 2014).

Gokhale et al. (1996) proposed an ENHPP model by incorporating test coverage. Authors further have shown that the utility of NHPP is limited as it considers only single coverage function whereas ENHPP provides a generalized coverage function and hence ENHPP models are significant step for the unification of NHPP based models. Malaiya et al. (2002) introduced a Logarithmic-exponential model that defines the connection among testing coverage, testing time and reliability, model discusses the hypothesis that enumerable have many chances of being applied while measuring the test coverage. By proposing the model authors were able to define the relation of test coverage with the defect coverage and model considered that even at the 100% coverage, all defects might not have been found.

Further, Pham and Zhang (2003) proposed a NHPP based SRGM to address the texting coverage and compared the model with the existing SRGMs. It has been observed that model fits significantly better than the others. Authors further developed the software cost model including testing cost, fault removal cost and risk cost. Study also determined the optimal release time to find when to stop testing so that cost can be minimized and can also satisfy the reliability requirements. Further, Chatterjee and Singh (2014) proposed SRGM based on imperfect debugging by including logistic-exponential based testing coverage function and compared the results with existing SRGMs. Authors created a cost model to calculate the total expected cost. Li and Pham (2017) proposed an imperfect debugging testing coverage NHPP model along with error generation and imperfect fault removal efficiency. Authors discussed the effect of each parameter on the robustness of the model by performing sensitivity analysis. In our study testing coverage function follows Weibull distribution and Exponentiated Weibull distribution. Weibull distribution is extensively adopted distribution because of its flexibility to fit the wide variety of the datasets (Lai et al. 2006).

2.3 Fault Reduction Factor

The other important environmental factor which affects the software testing process is FRF, proposed by Musa (1975). Musa discussed FRF to explain the connection between faults and failure. A generalized definition is that "the net number of faults removed from the program is only a proportion of the failure experienced", here the corrected faults are subtracted from the introduced faults to get the net number of faults. Expression for FRF is (Musa 1975, 1980)

$$B = \frac{n}{m}, 0 < B \leq 1$$

Here, m shows the total number of failure occurred, n represents the total faults that are corrected without introducing new faults and B represents the FRF. If $B = 1$, it is assumed that failure and faults are equal in numbers. In our study we haven't assumed error generation thus the FRF is considered as one in value.

Previously, numerous researches have been done incorporating FRF to the SRGM. Later, Malaiya et al. (1993) observed the fault exposure ratio which depicts the average fault detection rate, provided FRF can be represented in terms of fault exposure ratio.

$$B = \frac{\lambda_0}{Kfm}, 0 < B \leq 1$$

where, λ_0 shows the initial failure intensity, K is the fault exposure ratio, and f is the linear execution frequency of the program. Musa, in basic execution time model defined the differential equation as,

$$\frac{dm(t)}{dt} = Bz(t) = B\varphi(a - m(t))$$

where, $z(t)$ represents hazard rate function and φ shows per fault hazard rate. Solving above equation under initial conditions $m(0) = 0$, the MVF is

$$m(t) = a(1 - e^{-B\varphi t})$$

Here, $m(t)$ shows the expected number of faults at time t.

There may be variations in FRF under different situations and environmental factor. In the proposed study, FRF is assumed to be a constant.

2.4 Modular Software

To develop highly reliable software that can meet the user's requirements many functions and features are added to them. Addition of such functions makes software system a complex structure. Hence developer integrates small and independent modules to the system and during the testing phase each module is tested independently. In past, numerous researches have been done based on modular software system. For example, Kapur et al. (2009) have proposed a study for optimal resource allocation for a modular software. Optimization problem is developed by minimizing development cost under limited resources subject to the reliability constraint. To perform the methodology, testing effort SRGM has been used which described the fault removal process for the modules and simultaneously addressed the impact of imperfect debugging and error generation process.

Further, Kapur et al. (2010) have suggested two dimensional SRGM to assess the best time and resource distribution for a modular software systems. To develop the model authors consiered Cobb Douglas function and to solve formulated problem two directional genetic algorithm have been used. Kaur et al. (2017) suggested an SRGM with change point to calculate optimal resource allocation problem for a modular software. Authors have used KKT (Karush Kuhn Tucker) conditions to work out the nonlinear optimization problem and determine the optimal resources required to build software in order to reduce evelopment costs. Here, we have used modular software system to perform release time optimization problem. Here we have considered module weights to decide the best testing time and for the weights calculation we have used MCDM technique.

2.5 Fuzzy Analytical Hierarchy Process

MCDM process is used to work with dynamic judgement senarios where the decision problem has multiple attributes and dimensions. AHP is an MCDM approach for dealing with the complicated decision-making by splitting the problem down into simpler and convenient levels Satty (1980). AHP has been applied in various problems for example, in the area of supplier selection, evaluation of experts and database system products etc. (Zahedi 1990). Due to the uncertainty and vagueness in the obtained information it was difficult to understand or express one's opinion numerically, this results in the advent of fuzzy set theory, proposed by Zadeh (1965). Numerous researches have been conducted using fuzzy set theory due to their convergence towards the genuineness. In software reliability AHP has been used as follows (Table 1).

As given in Table 1, AHP has been used several times for reliability allocation process, selection process and for comparing reliability assessment methods. To the best of our knowledge no research has been conducted to proposed a multi-objective optimization problem for release planning policies with the help of AHP. Hence this

Table 1 Software reliability literature incorporating AHP

References	Techniques used	Explanation
Zahedi and Ashrafi (1991)	Reliability allocation	The authors have discussed the reliability allocation and maximized the user's utility by calculating the reliability of module and programs
Tamura and Yamada (2005)	AHP	Authors suggested an AHP based methodology to compare the different reliability assessment methods for open source software
Li et al. (2006)	AHP, software reliability, system engineering	Authors proposed a structure for the software reliability matrices with the help of AHP and expert judgement
Roy et al. (2008)	AHP, Fault Tree Analysis (FTA)	To evaluate software allocation authors provided a combined approach using AHP-FTA
Chatterjee et al. (2015)	Reliability allocation, AHP, FAHP	Authors have proposed a software reliability allocation problem using AHP and FAHP. Further comparison has been carried out based on consistency ratio
Aggarwal et al. (2017)	Reliability allocation, FAHP, Ordered Weighted Approach	Authors discussed the reliability allocation for a software system with the help of FAHP and OWA concepts
Chatterjee et al. (2017)	Reliability allocation, Type-2 FAHP, optimization	Authors proposed the approach to determine the software components weights using type-2 FAHP and based on the efficiency of performance of each component an optimization problem has been formulated
Verma et al. (2019b)	Intuitionistic fuzzy AHP	In the proposed approach authors have discussed a reliability allocation model for a multi-software system
Neha et al. (2019)	Reliability allocation Pythagorean fuzzy AHP	Authors have discussed a reliability allocation approach by considering the Pythagorean Fuzzy AHP

(continued)

Table 1 (continued)

References	Techniques used	Explanation
Proposed approach	FAHP & Multi-objective optimization problem	Our study focused on the methodology to determine the optimal release time for a modular software system. Where FAHP is used to find the weights of the modules that are further used for multi-objective optimization

study proposed a multi-objective release time optimization problem with the help of fuzzy AHP.

2.6 *Release planning*

The major concern of the development team is when to release the software. The number of faults eliminated from the software is used to assess whether or not software should be released. Hence, a release time optimization problem considering SRGM is developed. Goel and Okumoto (1979) introduced an optimum release policies considering an exponential SRGM. Further, Yamada and Osaki (1987) proposed an optimal release time problem including cost along with reliability with the help of exponential, modified exponential and S-shaped SRGM. Huang and Lyu (2005) proposed a release time problem including testing efficiency based on reliability and cost criteria. Li et al. (2010) carried out sa release time sensitivity analysis by incorporating testing effort and multiple change point to the SRGM.

Kapur et al. (2012) have proposed a 2-dimensional multi release SRGM and developed an optimal release planning problem to calculate the optimal time to stop the testing. For modelling the SRGM Cobb Douglas function was used. Based on the proposed SRGM optimization problem has been formulated that minimizes the total development cost and optimizes the time to release the software. Kumar et al. (2018) introduced a reliable and cost-effective model comprising of patching to formulate the release policy. Kapur et al. (1994) proposed a bi-criteria optimization problem to determine software releasing time. The problem considered maximization of the reliability as well as minimization of overall expected cost. In the same way, we have formulated the multi-objective release problem by considering the development cost for each module and software reliability simultaneously to calculate the release time for a modular software system. Now considering the above discussed work we have explained our proposed methodology in the next following section.

3 Evaluation Framework

In this section we develop a mathematical model for software release planning. The methodology framework is a step by step process that consists description of the SRGM, weights calculation of the modules and formulation of the release time problem. Therefore proposed framework can be divided into four steps that can be summarised as in Fig. 1.

Notations

$m(t)$	Cumulative number of faults detected till time t.
$m_1(t), m_2(t)$	Mean value function of model 1 and model 2 respectively
$\varphi(t)$	Fault detection rate
$a(t)$	Fault content at time t
$c(t)$	Testing coverage function of time t
r	Fault reduction rate
M_i	ith module of the modular software
T	Release time of the software
$C(T)$	Total development cost
C_B	Budget for the development process
C_1	Cost of fixing faults during testing phase
C_2	Cost of fixing faults during operational phase
C_3	Fixed cost per unit time
w_i	Weight of the ith module
s, b	Shape and scale parameter of Weibull distribution
$k, v\&l$	Shape and scale paramters of Exponentiated Weibull distribution.

3.1 *Software Reliability Growth Models (SRGMs)*

The general expression for NHPP based mean value function (MVF) is as follows

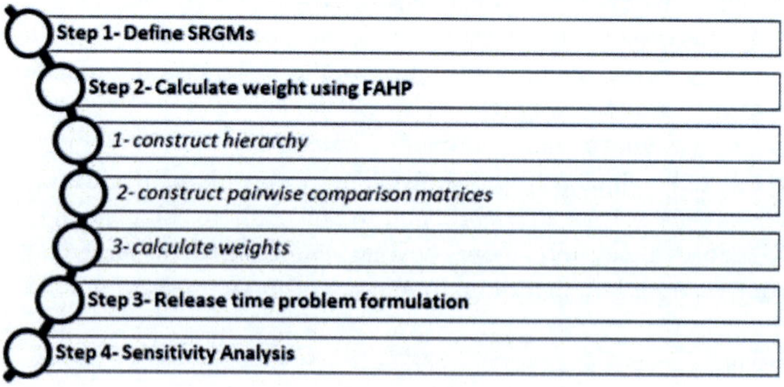

Fig. 1 Methodology framework

$$\frac{dm(t)}{dt} = \varphi(t)(a(t) - m(t)) \tag{1}$$

where, $\varphi(t)$ is the fault detection rate at time t, $a(t)$ is the faults occurred initially and $m(t)$ shows the expected number of failure by time t. Here one can obtained different MVF by substituting different values to $\varphi(t)$ & $a(t)$. We consider $\varphi(t)$ in terms of testing coverage, such as $\frac{c'(t)}{1+c(t)}$ (Pham and Zhang 2003), where $c(t)$ is the testing coverage function and initial faults content is constant as the debugging process is assumed to be perfect. Therefore MVF will be,

$$m(t) = \frac{c'(t)}{1 + c(t)}(a - m(t)) \tag{2}$$

FRF is assumed to be constant and represented by r, thus the MVF incorporating testing coverage and FRF is given as,

$$m(t) = r\frac{c'(t)}{1 + c(t)}(a - m(t)) \tag{3}$$

To evaluate the proposed methodology two NHPP based SRGMs comprising of testing coverage and FRF are considered. For 1st SRGM model testing coverage follows Weibull distribution and Exponentiated Weibull distribution function for 2nd SRGM. The mathematical expression of MVF for both the conditions is given as,

Model 1: Model 1 shows the MVF of the SRGM which incorporates testing coverage function as a Weibull distribution function.

$$m_1(t) = a\left(1 - \exp\left(-rbt^s\right)\right) \tag{4}$$

where, $c(t) = 1 - \exp(-bt^s), s > 0, t \geq 0 \& b > 0$

Model 2: Model 2 shows the SRGM which incorporates testing coverage as Exponentiated Weibull distribution and the related MVF is as follows,

$$m_2(t) = a\left[1 - \left(1 - \left(1 - \exp\left(-\left(\frac{t}{l}\right)^k\right)\right)^v\right)^{-r}\right] \tag{5}$$

where, $c(t) = \left(1 - \exp\left(-\left(\frac{t}{l}\right)^k\right)\right)^v, k > 0, l > 0, v > 0 \text{ and } t > 0.$

Further, for the validation of these models a real life dataset is used and we obtained the estimated values of the parameters. We have made a comparison of the considered SRGMs to the previously introduced SRGMs i.e, Goel-Okumoto model (Goel and Okumoto 1979) and delayed s-shaped model (Yamada et al. 1983). Their corresponding MVF are given as follows.

Goel-Okumoto Model,

$$m_3(t) = a(1 - \exp(-\varphi t)) \tag{6}$$

Here, $a(t) = a$ and $\varphi(t) = \varphi$.
And delayed s-shaped model,

$$m_4(t) = a(1 - \varphi t)\exp(-\varphi t)) \tag{7}$$

Here, $a(t) = a$ and $\varphi(t) = \frac{\varphi^2 t}{1+\varphi t}$.
Models 1 and 2 are further used to formulate the optimization problem considering different software modules and to assign weights to each the module we have used FAHP technique which is discussed in the next sub-section.

3.2 *Modular Weights calculation*

For the calculation of modular weights, we have used a fuzzy AHP approach. To understand the core idea of the fuzzy AHP we need to go through the basic theory of the fuzzy numbers that is explained in the following sub-section.

3.2.1 Fuzzy Numbers

Zadeh (1965) first proposed the fuzzy set theory to deal with the vagueness present in the datasets. According to him fuzzy set is represented as an ordered pair (D, x), here D is the set and $x : D \to [0, 1]$, i.e, membership function. Various researches have been accomplished with the help of fuzzy set theory such as in scientific environment, field of control theory and in robotics as well. There are few types of fuzzy numbers out of which triangular fuzzy number is generally used to deal the situation appeared due to the vagueness in the data. Graphical representation of a TFN is shown in Fig. 2 and membership function $\mu(x)$ for the TFN is,

$$\mu(x) = \begin{cases} \frac{x-q}{p-q} \ if \ q \leq x \leq p \\ \frac{y-x}{y-p} \ if \ p \leq x \leq y \\ 0 \quad otherwise \end{cases} \tag{8}$$

Considering the ease of expressing one's opinion, we have used Fuzzy AHP technique to get the weights of modules and to compute these weights TFN are used to express the expert's opinions mathematically.

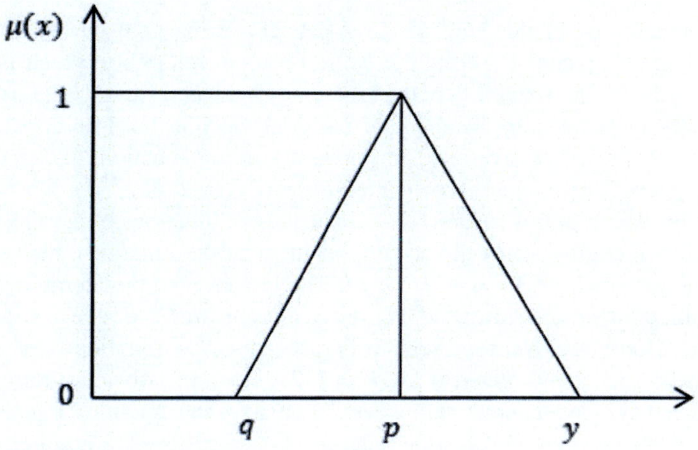

Fig. 2 Triangular fuzzy number

3.2.2 Fuzzy Analytical Hierarchy Approach (FAHP)

Previously, AHP has been used to deal with the complex or critical multi criteria based decision making problems such as component selection, supplier selection, resource allocation, etc. To perform AHP we have to develop a pairwise comparison matrix which relies on the expert's opinion using point scale (1–9). But these values used only the crisp numbers to provide the pairwise comparison matrix, whereas to incorporate real life situations we need to provide more flexibility to the expert for expressing their views. To overcome this issue fuzzy set theory was proposed that helps to define the non-clarity in the ratings provided by the experts. Incorporating fuzzy concept into AHP we construct the pairwise comparison matrices that uses TFN based linguistic scale (Table 2).

The steps involved in carrying out FAHP are as follows,

Step 1: *Construction of hierarchy*: We have developed a hierarchy based on Zahedi and Ashrafi (1991), that links the user's view to the software engineers and further

Table 2 Linguistic scale

Scale	TF scale	TF reciprocal scale
Just equal	(1,1,1)	(1,1,1)
Equally important	(1/2,1,3/2)	(2/3,1,2)
Weakly important	(1,3/2,2)	(1/2,2/3,1)
Moderately important	(3/2,2,5/2)	(2/5,1/2,2/3)
Moderately more important	(2,5/2,3)	(1/3,2/5,2)
More important	(5/2,3,7/2)	(2/7,1/3,2/5)
Strongly important	(3,7/2,4)	(1/4,2/7,1/3)
Strongly more important	(7/2,4,9/2)	(2/9,1/4,2/7)

to the programmers. In this step of the AHP, we set the goal we wish to obtain as a first level. The goal is set as the overall system reliability which is used in determining the further reliability of the functions. The second level contains multiple criteria used for the accomplishment of the goal. Second level comprises of few functions as software is developed to perform specific functions based on user's necessities, here functions are represented by $F_i(i = 1, 2, 3, \ldots f)$. Now the third level contains the programs written for the execution of functions. Programs are given by the software engineers for the accomplishment of the functions, here programs are symbolized as $P_j = (j = 1, 2, \ldots p)$. The last level of the hierarchy involves different alternatives that are used for the accomplishment of the previous level sub criteria. Here, last level represents programmer's view (modules written by the programmers) and is symbolized as $M_k(k = 1, 2, \ldots m)$. In software system modules are considered as independent unit however it may have sub-modules. Figure 3 shows the above depicted hierarchy for the software system. From the hierarchy we can state that a single program can be used for the execution of one or more functions. Similarly, a module can be used for more than one program.

Step 2: *Construction of fuzzy pairwise comparison matrices*: once we develop the hierarchy and set the target reliability of the software we than construct the fuzzy pairwise comparison matrices for each function, program and module using the linguistic scale with fuzzy numbers (Table 2).

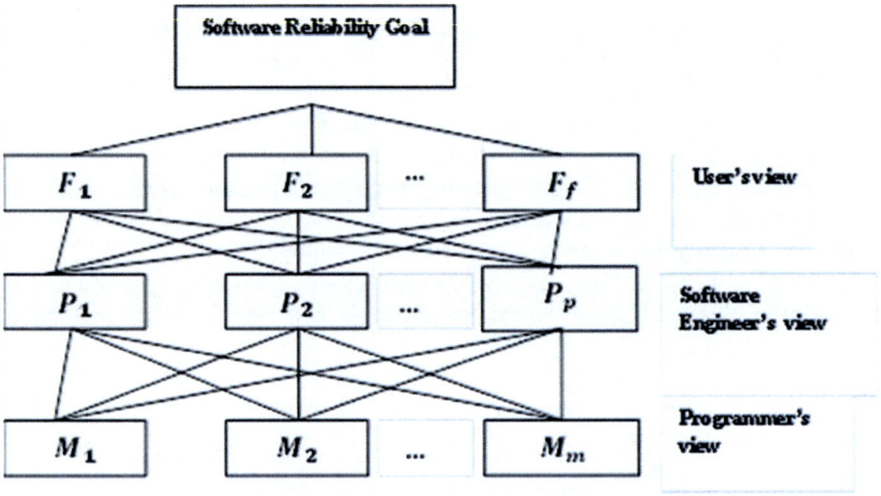

Fig. 3 Software hierarchy

$$O_e = \begin{bmatrix} (1,1,1) & (q_{12}, p_{12}, y_{12}) & \dots & (q_{1e}, p_{1e}, y_{1e}) \\ \left(\frac{1}{y_{12}}, \frac{1}{p_{12}}, \frac{1}{q_{12}}\right) & (1,1,1) & \dots & (q_{2e}, p_{2e}, y_{2e}) \\ \dots & \dots & \dots & \dots \\ \left(\frac{1}{y_{1e}}, \frac{1}{p_{e1}}, \frac{1}{q_{1e}}\right) & \left(\frac{1}{y_{2e}}, \frac{1}{p_{e2}}, \frac{1}{q_{2e}}\right) & \dots & (1,1,1) \end{bmatrix} \tag{9}$$

Let us assume that O_e is a fuzzy pairwise comparison matrix for objects $O_i (i = 1, 2, \dots e)$ of order $(e \times e)$. It is important to check the consistency of all the pairwise comparison matrix that is done using the defuzzification process. TFN are first defuzzified and then the consistency is checked as in the traditional AHP. If the consistency is less than 0.1 then we accept the matrix which can further be used for weights calculation. Consider a TFN as (q, p, y) then the defuzzified crisp number is $\frac{q+2p+y}{4}$. Now, using defuzzified number we check the consistency as $CI = (\lambda\text{max} - e)/(e - 1)$ and $CR = \frac{CI}{RI}$. Here *e* is the order of pairwise comparison matrix, *RI* is random index and *CI* is consistency index.

Step 3: *Calculation of weights*: After constructing the matrix, we calculate the relative weights using Chang's extent analysis (Chang 1996). Let us assume the relative weights for the above matrix are $(wO_1, wO_2, \dots, wO_e)$. Than the reliability assigned to them is given as (Aggarwal and Singh 1995)

$$RO_i = R^{wO_i} \tag{10}$$

Using above equation we calculate the reliability of functions, programs and modules. Therefore the relative weights for the functions, programs and modules are $wF_i = (wF_1, wF_2, \dots, wF_i)$, $wP_j = (wP_1, wP_2, \dots, wP_j)$ and $wM_k = (wM_1, wM_2, \dots, wM_k)$, respectively. The reliability assigned to the functions, programs and modules are,

$$RF_i = R^{wF_i} \tag{11}$$

$$RP_j = R^{wP_j} \tag{12}$$

$$RM_k = R^{wM_k} \tag{13}$$

respectively.

As given in the hierarchy one program (and modules) may connect to more than one function (and programs), therefore we end up getting more than one reliability to the same program (and module). In that situation it is best to select the maximum reliability.

3.3 Release Time Problem Formulation

This section determines the release time for software system using optimization technique. The problem comprises of several goals set by the development team that includes cost, reliability, intensity function etc. The optimization problem uses SRGM to find the release time and also to determine the link between the time and testing progress. In the proposed study we have used SRGM and modules weights to formulate the optimization problem. The weights obtained from the previous section are further used in the multi-objective optimization problem. We have considered that developer is interested not only in minimization of the cost but also in the maximization of reliability; the optimization is done under the budget constraint. Where, budget constraint ensures that the development cost does not go beyond the total budget assigned for the software development. The mathematical expression for the release time problem,

$$\begin{gathered}\min C(T)\\ \max R(x/T)\\ \text{Subject to}\end{gathered} \tag{14}$$

$$C(T) \le C_B$$

where, $T \ge 0$

Given problem can be rewrite as (Kapur et al. 1994)

$$\begin{gathered}\min \bar{C}(T)\\ \max R(x/T)\\ \text{Subject to}\end{gathered} \tag{15}$$

$$\bar{C}(T) \le 1$$

where, $\bar{C}(T) = C(T)/C_B$ and $T \ge 0$.

To overcome the complexity in the calculation, problem (15) is further converted into a single objective optimization problem by using priorities given by experts for the objective functions as $\theta\left(=\begin{bmatrix}\theta_1\\ \theta_2\end{bmatrix}\right) \in \mathbb{R}^2$ and $\theta_1 \ge 0, \theta_2 \ge 0 \,\&\, \sum_{i=1}^{2}\theta_i$ (Kapur et al. 1994). We have assumed reliability as $R(x/T) = w_i\frac{m_i(t)}{a_i}$. Therefore, the problem will be transformed as,

$$\min Z = \theta_1\bar{C}(T) - \theta_2 R(x/T)$$

$$\text{Subject to}$$
$$\bar{C}(T) \leq 1$$
$$T \leq 0 \quad (16)$$

By adjusting θ_1 and θ_2 according to the experts' desire we can obtain the different solutions. Using the optimization problem (Eq. 16) we calculate the release time for the software and further analyse the results by running sensitivity analysis. We have considered cost function as the composition of three cost components i.e, cost of fixing an error during testing phase (C_1), cost of fixing an error during operational phase (C_2) and testing cost per unit time (C_3).

4 Numerical Illustration

This section covers the application of proposed methodology. For the same. Parameter estimation is carried out on a real life failure dataset and the optimal release time for the modular software system is solved on an example dataset.

4.1 *Parameter Estimation*

Considering the SRGMs that are modelled using Eqs. 4 and 5, we have estimated the parameters using the Maximum Likelihood estimation. Estimation is performed on genome dataset for both the models (https://bugzilla.gnome.org). Testing process was done till 24 weeks and 85 faults were detected. Estimated values of both the models are shown in Table 3.

We have compared the models with existing SRGMs modelled by equation 6 and 7. The comparison criteria used to evaluate the performance of the proposed models with the exixting models are "Mean Square Error" (MSE), "Predictive Ratio Risk" (PRR), "Predictive Power" (PP), "Mean Absolute Percentage Error" (MAPE) and "Coefficient of Determination" (R^2). Table 4 represents the results obtained by comparing the models.

Table 3 Estimated parameters for Model 1 and Model 2

Parameters (Model 1)	Estimated value	Parameters (Model 2)	Estimated value
a	91.769	a	84.995
r	0.088	k	2.062
b	0.324	v	0.303
s	1.423	r	0.146
		l	4.881

Table 4 Performance measure

SRGMs	Performance criteria				
	MSE	PRR	PP	MAPE	R^2
G-O model	12.9104	0.2042	0.2991	7.4485	0.982
Delayed S-shaped model	8.1461	14.3861	1.0738	9.8909	0.989
Model 1	5.6438	9.4749	5.8318	7.7379	0.992
Model 2	3.8098	2.0842	2.1079	4.1972	0.991

Fig 4 shows the goodness of fit curves of both the model compared with the existing SRGMs and actual data. Here, minimum the distance between the points shows the better fit of the model.

Based on the estimated parameters form Table 3 we have determined estimated values of the parameters for each module. In the proposed study we have assumed six software modules for both the models and their estimated parameters are given in Table 5.

After calculating parameter values we proceed towards weights calculation of the modules and to get the weights we have performed FAHP technique in the next section.

4.2 AHP Weights

In order to carry release time optimization problem, apart from parametes of the models, we also need the module weights. These weights are obtained using the FAHP technique. For this, Let there are three functions in the software hierarchy say $F_1, F_2 and F_3$. Now, to accomplish these functions software engineers and programmers established four programs $(P_1, P_2, P_3 and P_4)$ and six modules $(M_1, M_2, M_3, M_4, M_5 and M_6)$. In the development process, software engineers design the programs written for the functions to fulfil the user's desire. Here, P_1, P_2 and P_4 help in accomplishing the functionF_1. For accomplishment of F_2, P_1, P_2 and P_3 are used and for F_3, P_2, P_3 and P_4. The next level of hierarchy is modules that represent the programmers view and this level comprises of six modules that helps in implementation of the programs. P_1 requires M_1, M_2 and M_5, P_2 requires M_2, M_3 and M_4, P_3 requires $M_4, M_5 and M_6$ and for P_4there are M_1, M_3 and M_6 modules.

The target reliability for the system is set as 0.90 (assumed) by considering the users requirements. Here to solve the FAHP, we have used fuzzy pairwise comparison matrices that are developed using Table 2. These matrices are constructed with the help of user's, software engineer's and programmer's opinion.

For Functions

The matrix for the system after comparing the functions is given as follows,

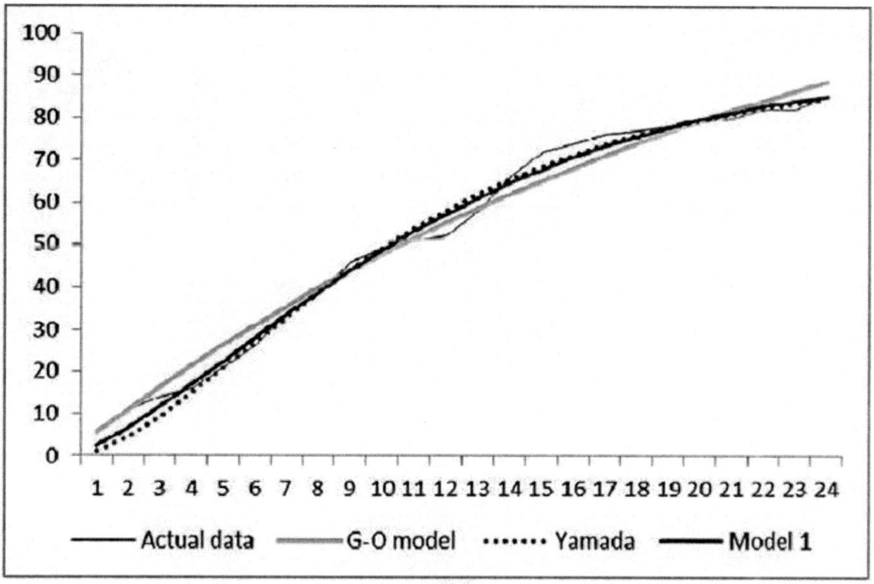

(a)

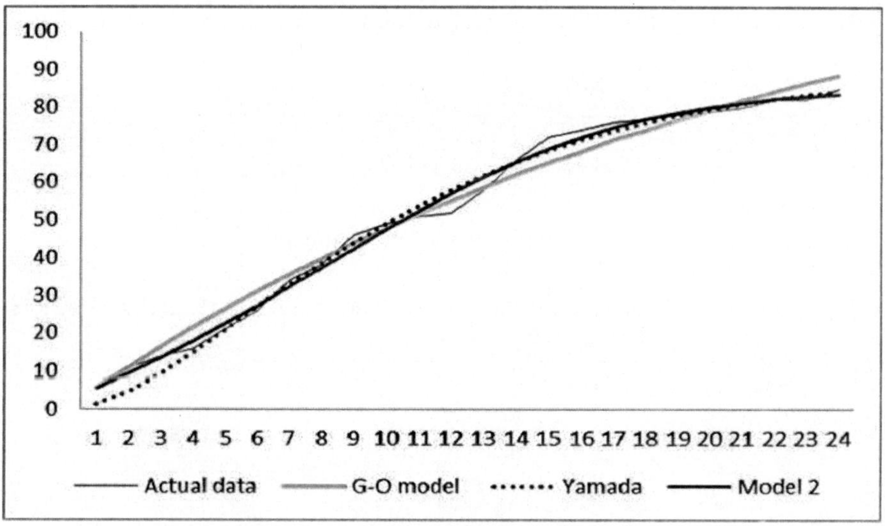

(b)

Fig. 4 Goodness of ft curve of **a** Model 1 and **b** Model 2

Table 5 Estimated parameters for modules

	Model 1				Model 2				
Modules	a	r	b	s	a	k	v	r	l
M_1	91.769	0.088	0.324	1.423	84.995	2.062	0.303	0.146	4.881
M_2	90.541	0.070	0.301	0.987	83.105	2.819	0.312	0.189	3.801
M_3	89.981	0.081	0.402	1.056	84.081	1.568	0.381	0.099	4.021
M_4	90.503	0.09	0.442	1.002	82.765	1.321	0.278	0.102	2.212
M_5	87.973	0.128	0.382	1.555	80.681	1.008	0.281	0.169	3.510
M_6	92.511	0.077	0.381	0.808	85.018	2.018	0.199	0.112	2.879

	F_1	F_2	F_3
F_1	(1, 1, 1)	(3/2,2.5/2)	(3/2,2,5/2)
F_2	(2/5,1/2,2/3)	(1, 1, 1)	(2/3,1,2)
F_3	(2/5,1/2,2/3)	(1/2,1,3/2)	(1, 1, 1)

Now we determine the relative weights for the functions and their corresponding reliabilities. The normalized weighting vector for each function is $wF = (0.43207, 0.26011, 0.155805)$ and the reliabilities for each function is as calculated,

$$RF_1 = \left(R^{wF_1}\right) = 0.95030597$$

$$RF_2 = \left(R^{wF_2}\right) = 0.97409569$$

$$RF_3 = \left(R^{wF_3}\right) = 0.97850954$$

These allocated reliability are further used in determining the reliability allocation for each programs associated to the functions.

For programs

The pairwise comparison matrix for the programs used to execute the function F_1,

	P_1	P_2	P_4
P_1	(1, 1, 1)	(3/2,2.5/2)	(1,3/2,2)
P_2	(2/5,1/2,2/3)	(1, 1, 1)	(2/5,1/2,2/3)
P_4	(1/2,2/3,1)	(3/2,2.5/2)	(1, 1, 1)

The normalized weight vector $wF_1 = (0.430988, 0.150645, 0.430988)$ and the reliabilities assigned to the programs P_1, P_2 and P_4,

$$RP_1 = \left(RF_1^{wP_1}\right) = 0.969804114$$

$$RP_2 = \left(RF_1^{wP_2}\right) = 0.988059066$$

$$RP_4 = \left(RF_1^{wP_4}\right) = 0.970645714$$

Now pairwise comparison matrix for the programs used to execute the function F_2

	P_1	P_2	P_3
P_1	(1, 1, 1)	(2/3,1,3/2)	(2/9,1/4,2/7)
P_2	(2/3,1,3/2)	(1, 1, 1)	(2/5,1/2,2/3)
P_3	(7/2,4,9/2)	(3/2,2.5/2)	(1, 1, 1)

The normalized weights are $wF_2 = (0.445249, 0.305655, 0.305655)$ and the reliabilities assigned to the programs P_1, P_2 and P_3

$$RP_1 = \left(RF_2^{wP_1}\right) = 0.98980536$$

$$RP_2 = \left(RF_2^{wP_2}\right) = 0.99563982$$

$$RP_3 = \left(RF_2^{wP_3}\right) = 0.99563982$$

The pairwise comparison matrix for the programs associated with function F_3,

	P_2	P_3	P_4
P_2	(1, 1, 1)	(2/3,1,3/2)	(2/7,1/3,2/5)
P_3	(2/3,1,3/2)	(1, 1, 1)	(2/5,1/2,2/3)
P_4	(5/2,3,7/2)	(3/2,2.5/2)	(1, 1, 1)

The normalized weights are$wF_2 = 0.508734, 0.314588, 0.314588$ and the reliabilities assigned to the P_2, P_3 and P_4,

$$RP_2 = \left(RF_3^{wP_2}\right) = 0.98945588$$

$$RP_3 = \left(RF_3^{wP_3}\right) = 0.99466523$$

$$RP_4 = \left(RF_3^{wP_4}\right) = 0.99466523$$

As we can see that each program is used for the execution of more than on function thus the final reliability assigned to the programs are as follows,

$$RP_1 = \max(0.969804114, 0.98980536) = 0.969804114$$

$$RP_2 = \max(0.988059066, 0.99563982, 0.98945588) = 0.99563982$$

$$RP_3 = \max(0.99563982, 0.99466523) = 0.99563982$$

$$RP_4 = \max(0.970645714, 0.99466523) = 0.99466523$$

Further, these reliabilities are used to determine the weights of the module attached to the software.

Weights for modules

Now, we calculate the weights for each module associated with particular program and further these weights will be used in determining the release time of software system using multi criteria optimization problem. Pairwise comparison matrix of the modules used for the accomplishment of program P_1,

	M_1	M_2	M_5
M_1	(1, 1, 1)	(2/3,1,3/2)	(2/7,1/3,2/5)
M_2	(2/3,1,3/2)	(1, 1, 1)	(2/5,1/2,2/3)
M_5	(5/2,3,7/2)	(3/2,2.5/2)	(1, 1, 1)

The normalized weights are $wP_1 = (0.355679, 0.355679, 0.223543)$.
For program P_2 the pairwise comparison matrix is,

	M_2	M_3	M_4
M_2	(1, 1, 1)	(2,5/2,3)	(5/2,3,7/2)
M_3	(1/3,2/5,1/2)	(1, 1, 1)	(1/2,2/3,1)
M_4	(2/7,1/3,2/5)	(1,3/2,2)	(1, 1, 1)

Here the normalized weights are$wP_2 = (0.2434659, 0.2434659, 0.5687433)$. Now the pairwise comparison matrix for program P_3 is,

	M_4	M_5	M_6
M_4	(1, 1, 1)	(2/7,1/3,2/5)	(1,3/2,2)
M_5	(5/2,3,7/2)	(1, 1, 1)	(7/2,4,9/2)
M_6	(1/2,2/3,1)	(2/9,1/4,2/7)	(1, 1, 1)

And the normalized weights are $wP_3 = (0.355679, 0.355679, 0.223543)$. the pairwise comparison matrix for program P_4,

Table 6 Modular weights using FAHP

Modules	P_1	P_2	P_3	P_4	Final Weights (w_i)
M_1	0.355679	…	…	0.345638	0.345638
M_2	0.355679	0.2434659	…	…	0.2434659
M_3	…	0.2434659		0.345638	0.2434659
M_4	…	0.5687433	0.355679	…	0.355679
M_5	0.223543	…	0.355679	…	0.223543
M_6	…	…	0.223543	0.308725	0.223543

	M_1	M_3	M_6
M_1	(1, 1, 1)	(2/9,1/4,2/7)	(1,3/2,2)
M_3	(7/2,4,9/2)	(1, 1, 1)	(7/2,4,9/2)
M_6	(1/2,2/3,1)	(2/9,1/4,2/7)	(1, 1, 1)

The normalized weights are $wP_4 = (0.345638, 0.345638, 0.308725)$. Weights for the modules corresponding to programs are given in Table 6. The final weights are calculated with the help of maximum reliability assigned to the modules.

These calculated weights of module will be used in formulating optimization problem to determine the release time for the modular software system.

4.3 Release Time Problem

Here, we will determined the time for releasing the software into the market. We have first assigned the estimated values of the parameters obtained by MLE (Table 4) to the MVF (Eqs. 4 and 5). Then we obtain the objective functions considering all the modules. Let's consider the cost component values as, $C_{11} = 1.5, C_{12} = 22, C_{13} = 2, C_{14} = 4, C_{15} = 3$ and $C_{16} = 6$. The cost$C_{21} = 9, C_{22} = 14, C_{23} = 10, C_{24} = 11, C_{25} = 8$ and $C_{26} = 12$. The fixed cost for all the modules is $C_{31} = 1, C_{32} = 2.5, C_{33} = 3, C_{34} = 4.5, C_{35} = 4$ and $C_{36} = 1$. (Note: cost values are taken in Dollars per unit)

Using Eq. (16), the optimization problem can be written as,

$$\min Z(T) = \theta_1 * \frac{\sum_{i=1}^{6}(C_{1i}m_i(T) + C_{2i}(m_i(\infty) - m_i(T)) + C_{3i}m_i(T))}{C_B} - \theta_2 * \prod_{i=1}^{6}\left(\frac{m_i(T)}{a_i}\right)^{w_i}$$

Subject to (17)

Table 7 Optimization results

	Model 1	Model 2
Optimal release time	36.0345 weeks	28.3087 weeks
Cost for removing faults during testing phase	872.27	967.094
Cost for removing faults during operational phase	234.27	571.08
Testing cost per unit time	277.2	452.939
Total cost	1384.63	1991.114
Reliability	0.8270	0.7163

$$\frac{\sum_{i=1}^{6}(C_{1i}m_i(T) + C_{2i}(m_i(\infty) - m_i(T)) + C_{3i}m_i(T))}{C_B} \leq 1 \tag{17}$$

where, $T \geq 0$

For the development process total budget assigned is $C_B = \$2000$. As per developers, both the objective function carries equal priorities. Thus the value assigned to θ_1 and θ_2 is 0.5. To calculate the release time, the problem (Eq. 17) is solved using MATLAB software and modules are modelled using both proposed SRGMs (Eqs. 4 and 5). The optimal time obtained to release the software and its related optimal cost is shown in Table 7.

From results we can conclude that, for Model 1, testing time has to be continued up to 36.0345 weeks or software can be released any time after 36.0345 weeks and the total cost incurred is \$1384.63 and the reliability obtained is 82.27%. Whereas on using Model 2 the optimal testing time for the software is obtained as 28.3087 weeks with total expected cost as \$1991.114 and 71.63% reliability. Results obtained from both the model support the budget assigned to them. The corresponding cost function, reliability curve and the utility curve of the models have plotted in Figs. 5, 6, 7.

4.4 Sensitivity Analysis

This section discusses the impact of budget and different values of θ_1 and θ_2 on the release time of the modualr software. To perform such study sensitivity analysis is a very helpful approach. Due to it, we are able to understand the impact of independent variables on the dependent ones in the same environmental conditions. Using this study a developer can easily identify the key component they need to focus on. We have determined the effect of modified budget on the optimal release time of the software system. We have also analyse the change in release time for the different values of θ_1 and θ_2. Analysis is done by considering two cases, in first when we increased budget by 50% and in second we decrease the budget by 50% and calculated the relative change. Relative change is defined as "the percentage change in dependent

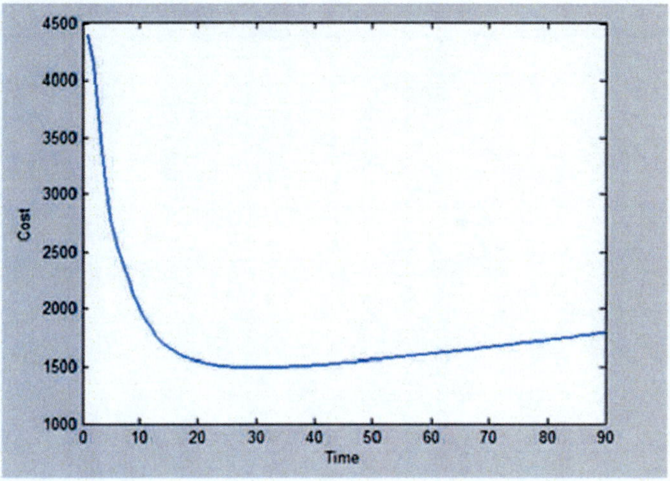

(a)

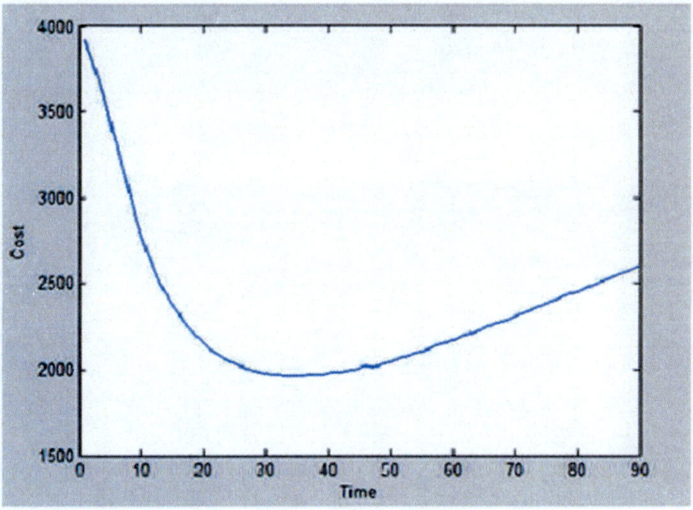

(b)

Fig. 5 Estimated cost function **a** Model 1 and **b** Model 2

quantity to the change in independent variable". The calculated value of the release time by sensitivity analysis is given in Table 8. The mathematical expression to calculate the relative change is as follows,

$$RelativeChange = \frac{newvalue - oldvalue}{oldvalue}$$

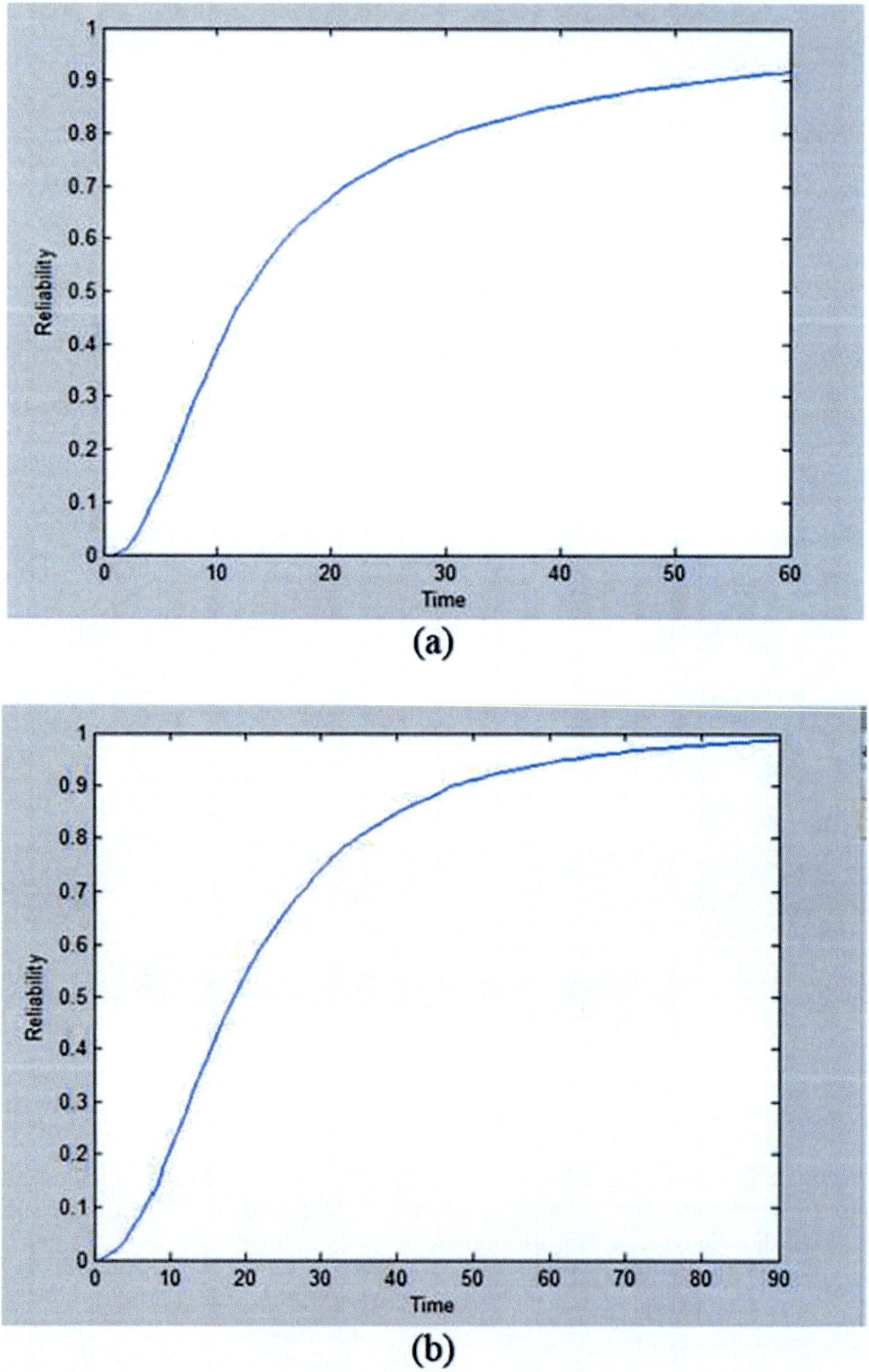

Fig. 6 Estimated reliability curve **a** Model 1 and **b** Model 2

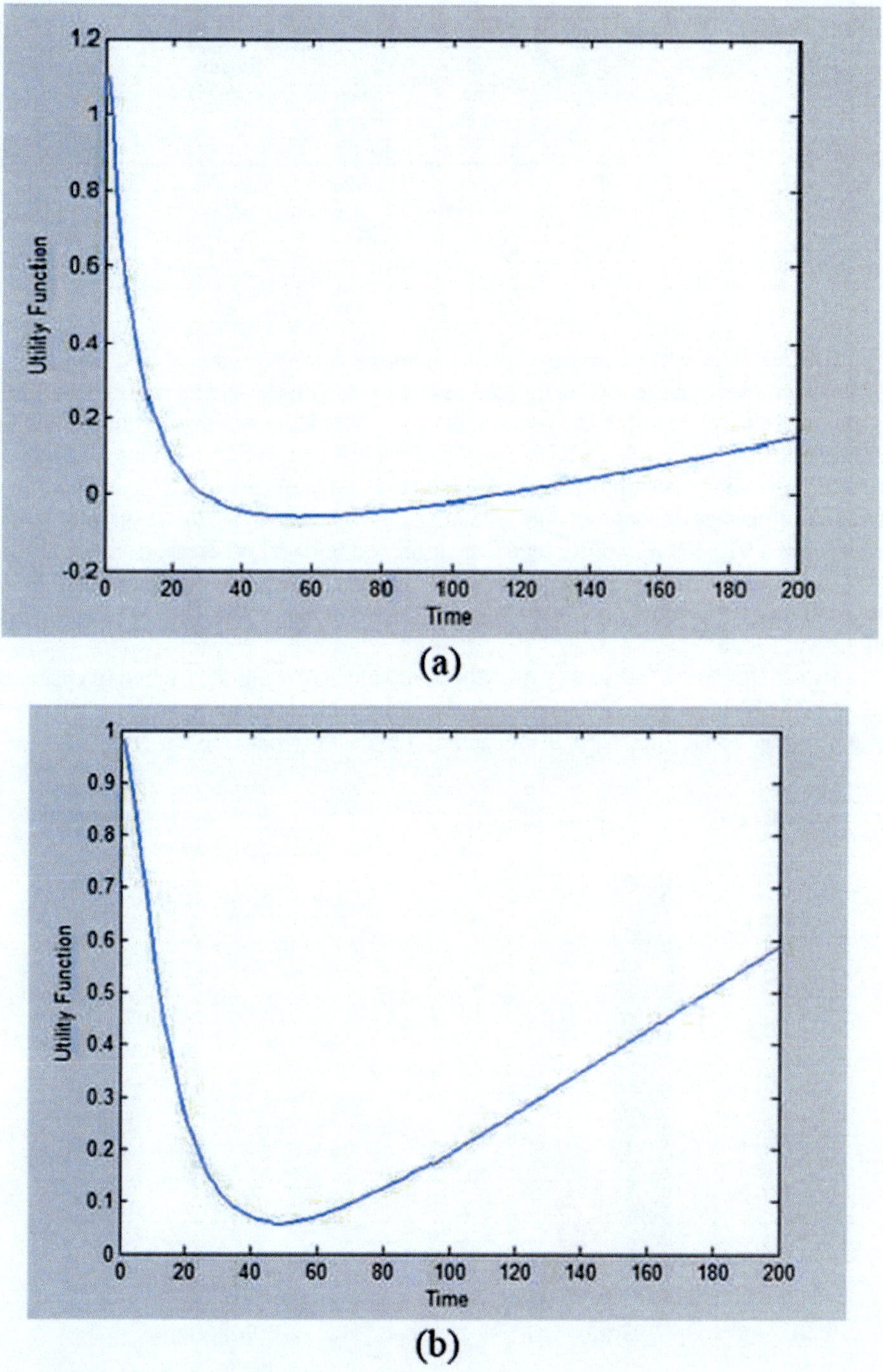

Fig. 7 Estimated utility function ($Z(T)$): **a** Model 1 and **b** Model 2

Table 8 Sensitivity analysis for modified budget

	Budget	Changed value	Testing Time (weeks)	Relative change in Testing Time	Relative change in Cost	Relative change in Reliability
Increased by 50%	C_B	3000	50.74	0.7923	0.03426	0.2774
Decreased by 50%	C_B	1000	34.59	0.2218	−0.01406	0.120

Case 1: Here we have increased the total budget by 50% and analysed the impact of this increment on the total estimated time and also on the total development cost. Same is discussed for 50% decrease in budget. The calculated testing time is given in Table 8.

From the above sensitivity analysis results we have observed that increasing the budget on perfect debugging during testing phase leads to 50.74 weeks of testing which gives 91.51% reliability under the assigned budget. Whereas, decreasing the budget implies that there is over consumption of the sources i.e., it is not possible to develop the software in such less budget. Relative change of the changed budget is plotted in Fig. 8.

Case 2: Now, case 2 discusses the different values of weight assigned to cost and reliability objective functions respectively and check the impact of these changed values on the reliability, testing time and on the development cost.

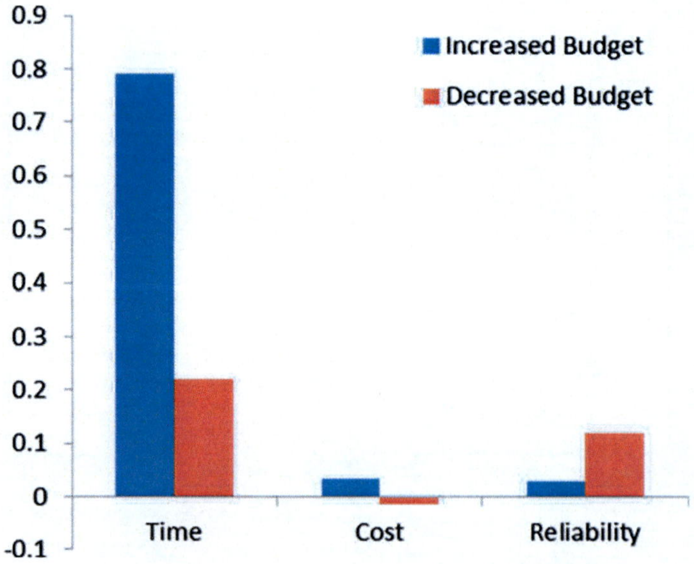

Fig. 8 Relative change in development budget

Table 9 Sensitivity analysis for different values of θ_1 and θ_2

θ_1	θ_2	Testing time (weeks)	Relative change in testing time	Relative change in cost	Relative change in reliability
0.1	0.9	28.7884	0.0169	−0.002	0.01
0.3	0.7	28.3080	−0.00002	−0.00008	−0.00001
0.5	0.5	28.3087	0	0	0
0.7	0.3	28.2175	−0.0032	0.0004	−0.0019
0.9	0.1	28.1055	−0.0071	0.00087	−0.0043

Table 9 shows that there is a slight difference in the testing time for each condition of the weights assigned and thus minimal difference in the development cost and reliability as well. Here we see that more weightage to the reliability implies higher value of the reliability. Similarly, more weight assigned to the development cost implies higher expenditure and lower reliability. Hence using θ_1 and θ_2 provides developers more freedom to set the objectives and thus he may have a trade-off between the development cost and the reliability. In the similar way the sensitivity analysis for the model 1 can be discussed.

4.5 *Implications*

The proposed study leads to major implications for software developers. The most relevant thing is that users desire software that is highly reliable and can accomplish the task it is assigned for. Moreover, users expect a software product as early as possible. On the other hand, developer is required to have reliable features in the software that are demanded by the users at the time of release of the software. At the same time the developer should make sure that the development cost must not exceed the total budget assigned to the modular testing phase of the software. Taking into consideration this trade-off, the proposed model optimizes the development cost and reliability simultaneously and it provides developer a convincing and planned process to develop software that meet the user's requirements. There must be a good communication between the users and the software developers. Hence, study provides a bridge to communicate with the users and understands their desired software. Study also provides a platform to the developers to express the views in a more flexible or wide way.

5 Conclusion

Due to the incorporation of multiple functions to meet the user's ever increasing requirements software system is getting complex and critical day by day. This creates challenging circumstances for developers to detect and correct the faults from such complex software systems. In testing phase, detection of faults from such systems is often costly and takes a lot of time. Therefore, to ease this phase, modular software systems are used, where testing is done individually for each module which is small written programs designed to perform the task, instead of testing whole software in one go. Here, we have considered a modular software system for deciding the testing time to release the software under minimization of total development cost and maximization of software reliability simultaneously.

Initially, we have estimated the parameters of the given SRGMs based on testing coverage and FRF. Where testing coverage is assumed to follow Weibull distribution for Model 1 and Exponentiated Weibull distribution for Model 2 whereas FRF is constant for both the models. The models are validated via estimation on real life dataset. To formulate the multi-objective optimization problem we have used module weights which are calculated with the help of FAHP an MCDM technique that is considered when multiple attributes are there in the decision making problem. Using FAHP we are able to get the weights by providing decision makers a wide range to express their views than expressing through crisp values. Hence obtained weights are more significant to calculate the release time.

Using these fuzzy weights of the modules and the estimated values of the parameters we have formulated the Multi-objective optimization problem to obtain the release time for the modular software system. We have solved the optimization problem by using utility approach and assigned the weights to each objective function and converted the multi-objective problem to a single objective problem under the budgetary constraint. Our study has shown significant results for optimizing the release time of a modular software that will assist the developers by providing appropriate methodology.

Limitation and Future Scope

The limitations of the proposed study that can be researched in future are:

- We have used a single version of the software for the optimization but often software comes in multiple versions. Thus to deal with this issue proposed study can be integrated by using multi-versions of the modular software system.
- The study solved the multi-objective optimization problem using a utility approach concept. This can further be solved using some meta-heuristic approaches and comparison can be made between the results of these approaches in future.
- Our modular software release planning problem involves cost components as testing cost per unit time, cost of fixing an error in testing phase and cost of fixing an error in operational phase whereas there are several costs that can influence the total development cost such as penalty cost, opportunity cost etc. Hence for further study such costs can also be included.

- The study has considered FAHP techniques to calculate the weights of the modules which deal with the uncertain information obtained by the decision makers. Whereas, many a times such obtained information can be vague or ambiguous in nature. Hence to overcome such problems extended fuzzy set theory is used. For future we can incorporate these extended from of fuzzy concept for obtaining the weights.

References

Aggarwal AG, Gandhi N, Verma V, Tandon A (2019) Multi-release software reliability growth assessment: an approach incorporating fault reduction factor and imperfect debugging. Int J Math Oper Res 15:446–463

Aggarwal AG, Verma V, Anand S (2017) Architecture-based optimal software reliability allocation under uncertain preferences. ICITKM. Delhi, India, pp 3–12

Aggarwal K, Singh Y (1995) Software reliability apportionment using the analytic hierarchy process. ACM SIGSOFT Softw Eng. Notes 20:56–61

Cai X, Lyu MR (2007) Software reliability modeling with test coverage: experimentation and measurement with a fault-tolerant software project. In: The 18th IEEE international symposium on software reliability (ISSRE'07). IEEE, pp 17–26

Chang D-Y (1996) Applications of the extent analysis method on fuzzy AHP. Eur J Oper Res 95:649–655

Chatterjee S, Chaudhuri B, Bhar C, Shukla A (2017) Estimation of software reliability and development cost using interval type-2 fuzzy AHP. In: 2017 international conference on infocom technologies and unmanned systems (trends and future directions) (ICTUS). IEEE, pp 682–688

Chatterjee S, Singh J (2014) A NHPP based software reliability model and optimal release policy with logistic–exponential test coverage under imperfect debugging. Int J Syst Assur Eng Manage 5:399–406

Chatterjee S, Singh JB, Roy A (2015) A structure-based software reliability allocation using fuzzy analytic hierarchy process. Int J Syst Sci 46:513–525

Goel AL, Okumoto K (1979) Time-dependent error-detection rate model for software reliability and other performance measures. IEEE Trans Reliab 28:206–211

Gokhale SS, Philip T, Marinos PN, Trivedi KS (1996) Unification of finite failure non-homogeneous Poisson process models through test coverage. In: Proceedings of ISSRE'96: 7th international symposium on software reliability engineering. IEEE, pp 299–307

Huang C-Y, Kuo S-Y, Lyu MR (2007) An assessment of testing-effort dependent software reliability growth models. IEEE Trans Reliab 56:198–211

Huang C-Y, Lyu MR (2005) Optimal release time for software systems considering cost, testing-effort, and test efficiency. IEEE Trans Reliab 54:583–591

Kapur P, Agarwala S, Garg R (1994) Bicriterion release policy for exponential software reliability growth model. RAIRO-Oper Res 28:165–180

Kapur P, Aggarwal AG, Kapoor K, Kaur G (2009) Optimal testing resource allocation for modular software considering cost, testing effort and reliability using genetic algorithm. Int J Reliab Qual Safety Eng 16:495–508

Kapur P, Aggarwal AG, Kaur G (2010) Simultaneous allocation of testing time and resources for a modular software. Int J Syst Assur Eng Manage 1:351–361

Kapur P, Aggarwal AG, Nijhawan N (2014) A discrete SRGM for multi release software system. Int J Indus Syst Eng 16:143–155

Kapur P, Garg R (1992) A software reliability growth model for an error-removal phenomenon. Softw Eng J 7:291–294

Kapur P, Pham H, Aggarwal AG, Kaur G (2012) Two dimensional multi-release software reliability modeling and optimal release planning. IEEE Trans Reliab 61:758–768

Kaur G, Aggarwal AG, Kedia A (2017) A study of optimal testing resource allocation problem for modular software with change point. Ann Comput Sci Inform Syst 14:77–84. https://doi.org/10.15439/2018KM11

Kumar V, Singh V, Dhamija A, Srivastav S (2018) Cost-reliability-optimal release time of software with patching considered. Int J Reliab Qual Safety Eng 25:1850018 (1850011—1850018) https://doi.org/10.1142/s0218539318500183

Lai C-D, Murthy D, Xie M (2006) Weibull distributions and their applications. https://doi.org/10.1007/978-1-84628-288-1_3

Li H, Lu M, Li Q (2006) Software reliability metrics selecting method based on analytic hierarchy process. In: 2006 sixth international conference on quality software (QSIC'06). IEEE, pp 337–346

Li N, Malaiya YK (1993) Empirical estimation of fault exposure ratio. Colorado State University, Citeseer

Li Q, Pham H (2017) A testing-coverage software reliability model considering fault removal efficiency and error generation. PLoS One 12:

Li X, Xie M, Ng SH (2010) Sensitivity analysis of release time of software reliability models incorporating testing effort with multiple change-points. Appl Math Model 34:3560–3570

Malaiya YK, Li MN, Bieman JM, Karcich R (2002) Software reliability growth with test coverage. IEEE Trans Reliab 51:420–426

Malaiya YK, Von Mayrhauser A, Srimani PK (1993) An examination of fault exposure ratio. IEEE Trans Software Eng 19:1087–1094

Musa JD (1975) A theory of software reliability and its application. IEEE Trans Software Eng 1:312–327

Musa JD (1980) The measurement and management of software reliability. Proc IEEE 68:1131–1143

Musa JD (1991) Rationale for fault exposure ratio K. ACM SIGSOFT Software Eng Notes 16:79

Musa JD (2004) Software reliability engineering: more reliable software, faster and cheaper. Tata McGraw-Hill Education

Neha, Verma V, Tandon A, Aggarwal AG (2019) Software reliability allocation incorporating pythagorean fuzzy theory and AHP. Accepted for publication in the international journal of industrial and system engineering

Pham H, Zhang X (2003) NHPP software reliability and cost models with testing coverage. Eur J Oper Res 145:443–454

Roy DS, Mohanta DK, Panda A (2008) Software reliability allocation of digital relay for transmission line protection using a combined system hierarchy and fault tree approach. IET Software 2:437–445

Satty T (1980) The analytic hierarchy process. McGraw-Hill, New York, NY

Shibata K, Rinsaka K, Dohi T (2006) Metrics-based software reliability models using non-homogeneous Poisson processes. In: 2006 17th international symposium on software reliability engineering. IEEE, pp 52–61

Tamura Y, Yamada S (2005) Comparison of software reliability assessment methods for open source software. In: 11th international conference on parallel and distributed systems (ICPADS'05). IEEE, pp 488–492

Verma V, Anand S, Aggarwal AG (2019a) Intuitionistic fuzzy AHP based reliabiltiy allocaion model for multi-software system. Accepted for publication in the international journal of services operations and informatics

Verma V, Anand S, Aggarwal AG (2019b) Software warranty cost optimization under imperfect debugging. Int J Qual Reliab Manage

Yamada S, Ohba M, Osaki S (1983) S-shaped reliability growth modeling for software error detection. IEEE Trans Reliab 32:475–484

Yamada S, Ohba M, Osaki S (1984) S-shaped software reliability growth models and their applications. IEEE Trans Reliab 33:289–292

Yamada S, Osaki S (1987) Optimal software release policies with simultaneous cost and reliability requirements. Eur J Oper Res 31:46–51
Zadeh LA (1965) Fuzzy Sets Inform Control 8:338–353
Zahedi F (1990) A method for quantitative evaluation of expert systems. Eur J Oper Res 48:136–147
Zahedi F, Ashrafi N (1991) Software reliability allocation based on structure, utility, price, and cost. IEEE Trans Software Eng:345–356
Zhang X, Teng X, Pham H (2003) Considering fault removal efficiency in software reliability assessment. IEEE Trans Syst Man Cybern Part A: Syst Humans 33:114–120

Neutrosophic AHP Approach for Budget Constrained Reliability Allocation Among Modules of Software System

Vibha Verma, Sameer Anand, and Anu G. Aggarwal

Abstract The planning and implementation of software development process should go hand in hand with the user expectations. It is crucial for firms to develop products catering user needs along with software engineers and programmers opinion. Many models have been developed in literature to allocate reliabilities to modules by setting system reliability goal based on hierarchy that inter-connects the attitudes of both developers and users. In this Chapter, we have proposed a hybrid reliability allocation model by integrating Analytical Hierarchical Process and reliability maximization problem for the software system based on the budget and reliability constraints. The comparison among functions, programs and modules by users, engineers and programmers respectively at different levels of hierarchy is performed under Neutrosophic environment to incorporate the vague and inconsistent information available. The relative weight obtained for each module is further used in optimization problem to finally allocate reliabilities to modules with the aim of maximizing the overall reliability of the software system considering the financial constraints associated with the project. The proposed methodology has been explained in detail and implemented through example problem.

Keywords Neutrosophic sets · Analytical hierarchical process · Reliability allocation · Hierarchical structure · Optimization · Budget constraint

1 Introduction

Nowadays one of the major challenges faced by Information Technology (IT) firms is to develop qualitative software systems due to its criticality in various fields like medical, defence, education, transportation, social media etc. Any kind of carelessness may result into heavy losses in monetary terms or may also result into loss of

---

V. Verma (✉) · A. G. Aggarwal
Department of Operational Research, University of Delhi, New Delhi, India

S. Anand
Shaheed Sukhdev College of Business Studies, New Delhi, India

A. G. Aggarwal et al. (eds.), *Optimization Models in Software Reliability*,
Springer Series in Reliability Engineering,
https://doi.org/10.1007/978-3-030-78919-0_9

life. Considering the importance and wide acceptability of software systems among masses has increased competition in the market which has led to remarkable improvements in methods and tools used during testing and debugging process. This is the prominent reason for the firms to continuously indulge themselves in the process of maintaining and improving the quality of the software system. The basic and important decisions regarding development of software like resource allocation, reliability allocation, code design etc. is taken by management during initial phases only. Systematically planned development process helps in successful accomplishment of the project.

It is necessary for developers to ensure that the developed software product meets the expectations of users. Developers need to achieve considerable level of reliability for their software products in order to produce a qualitative product. It is also crucial for firms to develop products that cater user needs along with taking into account the judgement of software engineers and programmers. Taking under consideration, views of all the stakeholders helps to build a better software product with higher reliability.

Defining reliability necessities for each module of a complex software system is one of the most significant tasks. Assigning goals for individual modules helps to achieve the software reliability goal. Module is considered to be smallest unit of software development process and it is defined as small piece of written code that when combined with other modules helps to execute a task. To achieve high reliability level for software system developers need to identify the amount of time and effort required on each module. Developers seek to allocate reliabilities during initial phases of Software Development Life Cycle (SDLC) in most efficient and well-organized manner.

Reliability allocation to components is a well explored area in hardware reliability (Chang et al. 2009; Elegbede et al. 2003; Mettas 2000), but not much work has been done in field of software engineering to develop methods of allocation. Few Models have been developed in literature to allocate reliabilities to modules of software system by setting software system reliability goal based on hierarchy that interconnects the opinions of developers (Software engineers and programmers) and users.

Some of the well-known software reliability allocation models discussed in the literature are Zahedi and Ashrafi (1991), Chatterjee et al. (2015), Yue et al. (2015), Aggarwal et al. (2018). Zahedi and Ashrafi (1991) proposed a model for maximization of software utility of modular software by obtaining relative weights of modules through Analytical Hierarchical Process (AHP) (Satty 1980). Chatterjee et al. (2015) used fuzzy numbers for developing preference matrix at each level of hierarchy given by Zahedi and Ashrafi (1991). Yue et al. (2015) used Dempster Shafer-theory to maximize software utility of multimedia system for customers and Aggarwal et al. (2018) used Maximum Variance Minimum Entropy-Ordered Weighted Averaging (MEMV-OWA) to assign weights to modules based on the architectural hierarchy.

Apart from the hierarchy based models, some optimization models have also been developed to allocate reliabilities among modules of the software system with objective of either maximizing the overall reliability or minimizing the software development cost. Leung (1997) developed an optimization model for minimizing cost of development and improving module reliability under uncertain operational profile. Guan et al. (2009) proposed a dynamic programming algorithm to simultaneously minimize the cost of software development and attain reliability goal. Malaiya (2014) also proposed an optimization problem to allocate reliabilities to modules while minimizing the cost and attaining the reliability objective. Mettas (2000) proposed an optimization problem to cater the reliability requirements of each component and hence attain the overall goal.

After studying the recent and the past developments in this field we wish to combine the advantages of multi criteria approach for allocation with the objective of reliability maximization considering the constraints on budget. In this Chapter, a hybrid reliability allocation model is proposed by integrating AHP and reliability maximization problem for the software system based on the budget and reliability constraints. AHP is based on the hierarchy proposed by Zahedi and Ashrafi (1991) which helps to integrate opinions of developers with the users. Comparison among functions, programs and modules by users, engineers and programmers respectively at different levels of hierarchy is performed under Neutrosophic environment to incorporate the vague, ambiguous, imprecise and inconsistent information available and hence handle indeterminacy of the available information (Ali and Smarandache 2017; Haibin et al. 2010).

Neutrosophic Sets (NS) generalize the crisp, fuzzy and intuitionistic sets and hence helps to make better decisions of real world problems. Human thinking cannot always be expressed into crisp numbers, it could be vague, incomplete, inconsistent, imprecise etc. hence transition of human preferences between truth, falsity and indeterminacy functions makes more well defined decisions. This kind of information is not embraced in the traditional reliability allocation models using Multi-Criteria Decision Making (MCDM). The relative weight obtained for each module is further used in optimization problem to finally allocate reliabilities to modules by maximizing the overall reliability of the software system. Hence, the study conducted through this chapter fulfils following objectives:

(1) To handle expert decisions through neutrosophic numbers which helps in incorporating vague, ambiguous, imprecise and inconsistent information available and hence handles indeterminacy of the available information.
(2) Obtain relative weights of modules for a software system that consists of functions, programs and modules under neutrosophic environment based on hierarchy that connects views of users, engineers and programmers.
(3) Maximization of overall reliability of software system by allocating reliabilities to modules considering the budget and reliability constraints using the relative weight obtained through Neutrosophic AHP (NAHP).
(4) Combining the benefits of MCDM technique and Non-Linear Optimization.

The rest of the chapter is organised as follows: Sect. 2 discusses the literature of all the related concepts. Section 3 presents the notations used throughout chapter followed by detailed discussion of proposed methodology in Sect. 4. The proposed methodology is implemented through an example in Sect. 5. Section 6 presents theoretical and managerial implications and then finally we conclude the study in Sect. 7.

2 Literature Review

Here, we review the literature of the related concepts and discuss the recent and past developments in the field of software reliability allocation. This helps in understanding the outlook behind the proposed methodology. Reliability Allocation during the initial design and development phase helps to effectively control the development process. Therefore, the allocation problem needs careful attention of the developers. Major work done for allocation of reliability to the modules of the software system is listed in Table 1.

It can be observed that relative weights for modules based on a structured hierarchy have been computed under fuzzy, intuitionistic and pythagorean environment but calculation of weights under neutrosophic environment helps in handling of inconsistent decisions and indeterminacy. Fuzzy concept was coined by Zadeh (1965) to incorporate imprecise decisions through a membership function. Atanassov (1999) generalised fuzzy sets through intuitionistic sets. These sets incorporate uncertain, imprecise and incomplete judgements through membership and non-membership function. Yager (2013) introduced Pythagorean fuzzy sets. It represents the general form of intuitionistic fuzzy sets and satisfies the condition where the sum of squares of the degree of belongingness and non-belongingness is less than and equal to one.

NAHP has several advantages over classical AHP, FAHP, IAHP and PFAHP and overcomes their shortcomings. This is so because to handle imprecise, vague, ambiguous, uncertain decisions NAHP considers three kinds of degree namely "membership degree", "non-membership degree" and indeterminacy degree" which helps to account for the inconsistency and falsity in decisions. NS explicitly qualify indeterminacy while the truth and false membership functions are independent. It is considered to be originated from neutrosophy and its logic was introduced by Smarandache (1995). NS came into existence to overcome the drawbacks of existing sets by considering truth, falsity and indeterminacy functions and hence are better able to represent reality. Haibin et al. (2010) described single valued NS in detail with all its properties and algebraic operations. Later Alblowi et al. (2014) discussed new concepts of these sets.

NS have been used together with AHP in various fields to record preferences of decision maker at different levels of hierarchy. Decision makers assign priorities or importance to the alternatives. Abdel-Basset et al. (2017) used neutrosophic theory to provide linguistic preferences by decision makers at various levels of AHP for selection of best candidates. Abdel-Basset et al. (2018) combined SWOT analysis with NAHP and validated their model for Starbucks company. Radwan et al. (2016) used NAHP to select the best learning management System by handling indeterminacy of information. Indeterminacy of information refers to the quality of information as uncertain or vague. Seeing the relevance of NS theory in handling the real world

Table 1 Reliability allocation models in software reliability

Researcher	Methodology
Zahedi and Ashrafi (1991)	Maximized customer utility using weights of modules obtained from hierarchy of functions, programs and modules that connects user and developers. AHP approach was used for weight calculation
Aggarwal and Singh (1995)	Reliability allocation through AHP
Leung (1997)	Developed a optimization model for minimizing cost of development and improving module reliability under uncertain operational profile
Lyu et al. (1997)	Applied reliability growth models to guide modules with multiple applications and proposed reliability allocation procedure for the single application component
Mettas (2000)	Proposed an optimization problem to cater the reliability requirements of each component and hence attain the overall goal
Kapur et al. (2003)	Reliability maximization problem incorporating redundancy at module level with budgetary constraints
Misra (2005)	Optimization problem to obtain failure intensities of modules by minimizing the development cost subject to reliability constraints. The weight of importance is given by usage time of module
Guan et al. (2009)	Proposed a dynamic programming algorithm to simultaneously minimize the cost of software development and attain reliability goal
Tian et al. (2009)	Software utility maximized by developing nonlinear optimization within the framework of genetic algorithm for multi-user software system considering development cost using software fault tree analysis to analyse relationship between modules and their respective reliability requirements
Pietrantuono et al. (2010)	Developed tool for identification of most important modules and then to assign resources and allocate reliability efficiently
Li et al. (2012)	Petri network approach is used to develop a model that can account the changes in software system with time. Also reliability is assessed using the dependency graph of modules. At last Petri network and Analytical Network Process (ANP) are combined for allocation purpose
Malaiya (2014)	Proposed an optimization problem to allocate reliabilities to modules while minimizing the cost and attaining the reliability objective
Chatterjee et al. (2015)	Used fuzzy numbers for developing preference matrix at each level of hierarchy given by Zahedi and Ashrafi (1991). This approach is known as Fuzzy AHP (FAHP)
Yue et al. (2015)	Used DS-theory to maximize software utility for customers of multimedia system. The weights for modules was obtained through AHP
Aggarwal et al. (2018)	Integrated FAHP and MEMV-OWA approach to assign weights to modules based on the architectural hierarchy

(continued)

Table 1 (continued)

Researcher	Methodology
Verma et al. (2019)	Intuitionistic AHP (IAHP) to allocate reliability to modules based on structured hierarchy
Neha et al. (2019)	Integrated Pythagorean Fuzzy Theory and AHP (PFAHP) to allocate reliability to modules based on hierarchy given by Zahedi and Ashrafi (1991)
Proposed approach	Relative weight calculation of modules using NAHP and allocate reliability to modules using obtained weights by reliability maximization subject to budget constraints

problems through linguistic preferences it has been extended to be applied in the field of software reliability for reliability allocation among modules of software system.

3 Notations

The Notations used throughout this chapter are listed as follows:

$\tilde{a_{ij}}$	Single valued triangular neutrosophic number
F_j	jth function of software system
P_k	kth program
M_i	ith independent module
I	Number of modules in the software system
J	Number of functions in the software system
K	Number of programs in the software system
w_i	Final relative weight of ith module
w_j	Final weight of jth function
w_k	Final weight of kth program
w_{ik}	Weight of ith module connected to kth program
w_{kj}	Weight of kth program connected to jth Function
Z	Objective function
R_i	Reliability of ith module obtained by NAHP
a_i	Fixed cost incurred during development of ith module
b_i	Variable cost required to achieve the considerable reliability for ith module
c_i	Penalty cost for ith module
$u_i(l_i)$	Upper (lower) limit for reliability of ith module
R^*	Reliability target for software system
B	Budget available for software development

4 Reliability Allocation Methodology

The objective of this chapter is to allocate reliability among modules of software system with the aim of maximizing the overall reliability of the software system based on budgetary and reliability constraints. The objective is attained in two stages:

1. In first stage relative weight of each module is obtained using NAHP based on a structural hierarchy of software system consisting of functions, programs and modules that relates users, software engineers and programmers opinion for software development. The hierarchy was initially developed by Zahedi and Ashrafi (1991). The elements at each level of hierarchy are compared through neutrosophic numbers which helps to incorporate inconsistent decisions and incomplete information.
2. Then the relative weights of modules obtained in previous stage are used into the optimization problem to allocate reliabilities to modules. The proposed problem maximizes software system reliability while allocating reliabilities to the modules of software system subject to budget available and target reliability of the system. The problem consists of non-linear objective function with non-linear constraints where the parameters can be easily set based on past literature and industry practices (Elegbede et al. 2003; Kuo and Wan 2007; Leung 1997).

This section describes the proposed model that integrates NAHP with reliability maximization problem for allocation of reliability among modules. The proposed methodology is discussed in following 6 steps:

Step 1: Hierarchy Development

The proposed allocation problem is based on the structured hierarchy proposed by Zahedi and Ashrafi (1991) for reliability apportionment among the modules of a modular software during initial phases of software development. This hierarchy has four levels that connect views and opinions of users with developers. The first level represents the goal of the problem defined by the firm. The second level consists of functions that are defined based on requirement and expectation of users from the software. Next level has the programs designed by software engineers for accomplishment of desired functions. Last level comprises of independent modules i.e. the piece of written code which combines to form programs. These codes are written by programmers.

Here, it is possible that a particular module is connected to more than one program and a program is connected to more than one function. This hierarchy has been used in many research models for the reliability allocation purpose (Chatterjee et al. 2015; Aggarwal and Singh 1995; Neha et al. 2019; Verma et al. 2019). The hierarchy has been also used for allocation purpose in multi software systems considering reliability goal of software at the first level. The general hierarchy for a single software system is presented in Fig. 1.

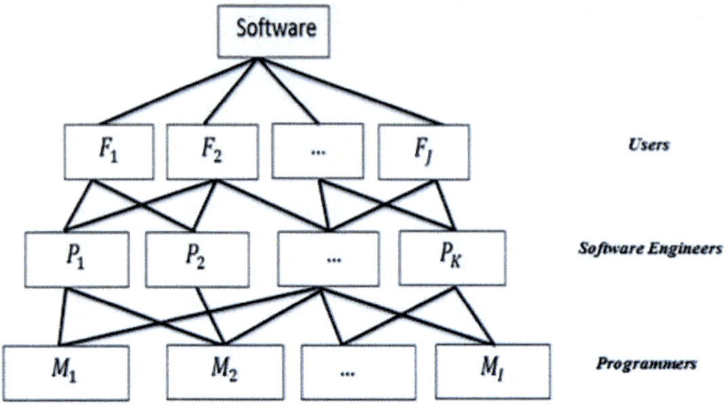

Fig. 1 Hierarchy for reliability allocation

Step 2: Neutrosophic Fuzzy Comparative Judgements

After decomposing the problem into different hierarchical levels, the neutrosophic pair-wise comparison matrices are developed for evaluating functions, programs and modules for their importance in the software system by the respective decision-makers. These comparison matrices are developed under neutrosophic environment using neutrosophic set (N) which is given by truth ($T_N(x)$), falsity ($F_N(x)$), and indeterminacy ($I_N(x)$) membership functions for any point x in the space of points. The function values are subset of $[0^-, 1^+]$ but there is no restriction on the sum of these functions. Mathematically, $0^- \leq \sup\ T_N(x) + \sup\ F_N(x) + \sup\ I_N(x) \leq 3^+$.

The basic arithmetic operations like addition, subtraction, division, multiplication, complimentary etc. performed on NS has been explained by (Alblowi et al. 2014; Ali and Smarandache 2017; Haibin et al. 2010). Neutrosophic sets generalise classical sets, fuzzy sets, intuitionistic sets because it not only considers truth and falsity membership functions but also considered indeterminacy function which makes is more viable for real world problems. The scale used for comparison using neutrosophic fuzzy numbers has been given by Abdel-Basset et al. (2018) is presented in Table 2. Neutrosophic numbers has been combined with AHP for assigning priorities or weights to alternatives and selection purposes (Abdel-Basset et al. 2018, 2017; Radwan et al. 2016). This helps in strategic judgements for development of products.

Step 3: Consistency Check of judgement Matrix

Before calculating relative weight of each module, it is must to ensure the consistency of judgement matrices at each level of hierarchy given by the respective decision-makers. The consistency for neutrosophic judgement matrix is checked through evaluation of Consistency Index (CI) and Consistency Ratio (CR) given by AHP methodology (Satty 1980; Franek and Kresta 2014). If CR is less than 0.1 then the matrix is said to be consistent. Consistency of the comparative judgements matrix is calculated after converting the neutrosophic numbers to crisp values. This is done using the score values given in Eq. 2. After obtaining comparison matrix with crisp

Table 2 NAHP triangular linguistic scale (Abdel-Basset et al. 2018)

Scale	Explanation
$\tilde{1} = \langle(1, 1, 1); 0.50, 0.50, 0.50\rangle$	Equally preferable
$\tilde{2} = \langle(1, 2, 3); 0.40, 0.65, 0.60\rangle$	Moderately lowly preferable
$\tilde{3} = \langle(2, 3, 4); 0.30, 0.75, 0.70\rangle$	Slightly preferable
$\tilde{4} = \langle(3, 4, 5); 0.60, 0.35, 0.40\rangle$	Moderately highly preferable
$\tilde{5} = \langle(4, 5, 6); 0.80, 0.15, 0.20\rangle$	Strongly preferable
$\tilde{6} = \langle(5, 6, 7); 0.70, 0.25, 0.30\rangle$	Very strongly preferable
$\tilde{7} = \langle(6, 7, 8); 0.90, 0.10, 0.10\rangle$	Extremely preferable
$\tilde{8} = \langle(7, 8, 9); 0.85, 0.10, 0.15\rangle$	Extremely highly preferable
$\tilde{9} = \langle(9, 9, 9); 1.00, 0.00, 0.00\rangle$	Absolutely preferable

values, consistency is checked by following the steps followed in AHP for checking consistency of judgement matrix. Franek and Kresta (2014) gave detailed view of judgement scale and consistency check in AHP.

Step 4: Calculation of Module's relative weights using NAHP Method

The weights for functions, programs and modules can be calculated from the consistent neutrosophic pair-wise comparison matrix. First the matrix is converted to the deterministic form using score and accuracy degree values. The score and accuracy degree is obtained from Eqs. 1, 2 respectively. For a neutrosophic number $\widetilde{a_{ij}}=\langle(a_1, b_1, c_1), \alpha_{\tilde{a}}, \theta_{\tilde{a}}, \beta_{\tilde{a}}\rangle$ Score and accuracy degree are given as (Abdel-Basset et al. 2018):

$$\text{Score}(\widetilde{a_{ij}}) = \frac{1}{8}[a_1 + b_1 + c_1] \times (2 + \alpha_{\tilde{a}} - \theta_{\tilde{a}} - \beta_{\tilde{a}}),\ \text{Score}(\widetilde{a_{ij}}) = 1/\text{Score}(\widetilde{a_{ij}}) \tag{1}$$

$$\text{Accu}(\widetilde{a_{ij}}) = \frac{1}{8}[a_1 + b_1 + c_1] \times (2 + \alpha_{\tilde{a}} - \theta_{\tilde{a}} + \beta_{\tilde{a}}),\ \text{Accu}(\widetilde{a_{ij}}) = 1/\text{Accu}(\widetilde{a_{ij}}). \tag{2}$$

Using the score values the deterministic matrix is obtained corresponding to each neutrosophic value in the pairwise comparison matrices. Next the column entries of the deterministic matrix are normalised by dividing each and every entry in the column with the corresponding column sum. At last the total of row averages is obtained. In this way, relative weights for alternatives at each level are obtained.

At first functions of the software system are compared to each other by users based on their knowledge and requirement. The remainder of this step is relative weight of functions. Then corresponding to each function the programs attached are compared by software engineers. Here weight for each program is obtained for the functions it is connected with. As a program may be attached to one or more functions. Then at the last level of hierarchy modules are compared by programmers. A module may be connected to more than one program and hence weight for each module corresponding to every program is obtained. Then at the end the final relative weights of modules is calculated using the weights obtained for functions and programs. The final relative weights are given by Eq. 3.

$$w_i = \sum_{i=1}^{I} w_k * w_{ik} \quad \text{for } k = 1, 2, \ldots K \tag{3}$$

where

$$w_k = \sum_{k=1}^{K} w_j * P_{kj} \quad \text{for } j = 1, 2 \ldots J.$$

Step 5: Formulation of Optimization (Reliability Maximization) Problem

After obtaining final relative weights for the modules of the software system, we move towards our aim of allocating reliability to each of the module such that the overall reliability of the software system is maximized. Hence, we propose an optimization problem (Eqs. 4–7) to maximize reliability subject to budget and reliability constraints. This optimization problem uses the final weights of modules to quantify reliability of the corresponding module.

$$\text{Max } Z = \prod_{i=1}^{I} R_i^{w_i} \tag{4}$$

s.t.

$$\sum_{i=1}^{I} a_i + b_i R_i + c_i(1 - R_i) \leq B \tag{5}$$

$$\prod_{i=1}^{I} R_i \geq R^* \tag{6}$$

$$u_i \geq R_i \geq l_i \quad \text{for } i = 1, 2, \ldots I \tag{7}$$

Equation 4 represents the objective of maximizing the overall reliability; the Eq. 5 denotes the budget constraint that considers fixed cost of testing and debugging, variable cost involved for improving the reliability and unreliability (penalty) cost due to failures that may occur in operational phase. The combined cost of developing all the modules for particular software system should not exceed the budget. The Eq. 6 presents the reliability constraint for the software system. The target is set by developers based on requirement analysis. At last the bound constraints are given through Eq. 7. This ensures that reliability of each module is within the acceptable limit.

Step 6: Solution of Optimization Problem

The proposed optimization problem is solved using the module weights obtained from NAHP, setting development budget and reliability goal and assuming the cost parameter values of the budget constraints based on the literature and industry practices. The formulated non-linear constrained optimization problem to maximize reliability of modular software is solved in MATLAB.

5 Numerical Illustration

The proposed methodology for allocating reliability to the modules of software system based on budget and reliability constraints has been implemented using an example case. As already discussed in previous section the methodology is divided into two major steps. First, the relative weights for modules is calculated using NAHP methodology then these relative weights of importance are used in optimization problem for maximizing software system reliability and allocating reliability to modules.

5.1 Relative Weights of Modules

The module weights are obtained by following the step by step process discussed in Sect. 4. The structural hierarchy of the software system considered in this study consists of 4 functions (F_1, F_2, F_3, F_4), 5 programs (P_1, P_2, P_3, P_4, P_5) and 7 modules ($M_1, M_2, M_3, M_4, M_5, M_6, M_7$) at each subsequent level respectively. The decision maker comparing the corresponding functions, programs and modules in their ability to influence the software system reliability is different for each level.

Users are responsible for assigning importance to functions, while software engineers provide importance for programs and finally at the last level of hierarchy programmers compare the modules based on their importance in the software system. The linguistic preferences of decision makers are expressed in terms of neutrosophic numbers. The scale used for comparison is due to Abdel-Basset et al. (2018). The scale has been defined in Table 2. The authors explained neutrosophic triangular scale corresponding to AHP scale given by Satty (1980). The authors also discussed the basic algebraic operations for neutrosophic sets. Using the neutrosophic triangular scale (Table 2) the pairwise comparison matrix for functions given by user is as follows:

Functions	F_1	F_2	F_3	F_4
F_1	$\tilde{1}$	$\tilde{4}$	$\tilde{3}$	$\tilde{5}$
F_2	$\tilde{4}^{-1}$	$\tilde{1}$	$\tilde{2}$	$\tilde{4}$
F_3	$\tilde{3}^{-1}$	$\tilde{2}^{-1}$	$\tilde{1}$	$\tilde{5}$
F_4	$\tilde{5}^{-1}$	$\tilde{4}^{-1}$	$\tilde{5}^{-1}$	$\tilde{1}$

Users compared the four functions of the software system for its importance based on their demand and expectations. The linguistic decisions are expressed in the form of neutrosophic numbers. To obtain the weights of functions the NAHP methodology discussed in Sect. 4 is implemented. At first consistency of the matrix is checked and then the weights are computed for the functions. The consistency results are presented in Table 4. The weights obtained for functions based on pairwise comparison matrix are $F_1 = 0.4454$, $F_2 = 0.1916$, $F_3 = 0.1715$ and $F_4 = 0.0585$. This represents the weight matrix of functions in the software system. It is observed that first function has the highest weight of importance while the fourth function has least weight of importance among the four functions of the example case.

After obtaining weights for each function there is need to calculate weights of each program connected to functions at the upper level. First function comprises of three programs, second one consists of two programs while third function works on execution of three programs and fourth one requires 4 programs for working. The comparison matrix given by software engineer corresponding to each function as follows:

F_1	P_1	P_2	P_3
P_1	$\tilde{1}$	$\tilde{5}^{-1}$	$\tilde{3}^{-1}$
P_2	$\tilde{5}$	$\tilde{1}$	$\tilde{5}$
P_3	$\tilde{3}$	$\tilde{5}^{-1}$	$\tilde{1}$

F_2	P_2	P_4
P_1	$\tilde{1}$	$\tilde{3}^{-1}$
P_2	$\tilde{3}$	$\tilde{1}$

F_3	P_3	P_4	P_5
P_3	$\tilde{1}$	$\tilde{8}$	$\tilde{4}$
P_4	$\tilde{8}^{-1}$	$\tilde{1}$	$\tilde{2}$
P_5	$\tilde{4}^{-1}$	$\tilde{2}^{-1}$	$\tilde{1}$

F_4	P_1	P_3	P_4	P_5
P_1	$\tilde{1}$	$\tilde{1}$	$\tilde{5}$	$\tilde{7}$
P_3	$\tilde{1}$	$\tilde{1}$	$\tilde{5}$	$\tilde{7}$
P_4	$\tilde{5}^{-1}$	$\tilde{5}^{-1}$	$\tilde{1}$	$\tilde{3}^{-1}$
P_5	$\tilde{7}^{-1}$	$\tilde{7}^{-1}$	$\tilde{3}$	$\tilde{1}$

Some programs are connected to more than one function for example we can observe that program P_1 is connected to first (F_1) and fourth (F_4) functions. These matrices represent importance of programs for the execution of particular function. After checking the consistency of these matrices relative weights are calculated. The result of consistency for each function is given in Table 4. The weights obtained for each program connected to one or more than one function is given in Table 3. The Table 3 represents the program weight matrix for each function. For further calculations it is assumed that if a program is not connected to that function than its weight is zero. This implies that the program is not important for executing that

Table 3 Weights allocated to programs

Program	F_1	F_2	F_3	F_4
P_1	0.06754	–	–	0.3734
P_2	0.45712	0.30304	–	–
P_3	0.13553	–	0.4699	0.3734
P_4	–	0.10711	0.10187	0.0593
P_5	–	–	0.08461	0.0896

Table 4 Consistency result for pairwise matrix

Matrix	λ_{max}	n	CI	(Random index) RI	CR*
S	4.0768	4	0.0256	0.89	0.02
F_1	3.0632	3	0.0316	0.58	0.05448
F_2	2 × 2 Matrix is always consistent				
F_3	3.0843	3	0.0421	0.58	0.072
F_4	4.0498	4	0.0166	0.89	0.0187
P_1	3.0205	3	0.01027	0.58	0.0177
P_2	2 × 2 Matrix is always consistent				
P_3	3.06041	3	0.0302	0.58	0.052
P_4	4.1908	4	0.06362	0.89	0.07149
P_5	3.0443	3	0.02216	0.58	0.0382

*CR < 0.10 = > Matrix consistent

function. From Table 3 it can be observed that P_2 has highest weight corresponding to first (F_1) and second (F_2) function while program P_3 has highest weight for function F_3. But in case of function F_4 program P_1 and P_3 have equal weight of importance.

Now, there is need to determine the overall relative weight of the program since some programs are connected to more than one function. The overall relative weight for program is obtained by multiplying the weight matrix of programs (Table 3) with the weight matrix of functions as shown in Q_1. The final weight of programs are $P_1 = 0.051926$, $P_2 = 0.261664$, $P_3 = 0.162797$, $P_4 = 0.041462$ and $P_5 = 0.019752$. The final weights of programs show that program P_2 has highest weight.

$$\underset{\text{Program Weight Matrix}}{\begin{bmatrix} 0.06754 & 0 & 0 & 0.3734 \\ 0.45712 & 0.30304 & 0 & 0 \\ 0.13553 & 0 & 0.4699 & 0.3734 \\ 0 & 0.10711 & 0.10187 & 0.0593 \\ 0 & 0 & 0.08461 & 0.0896 \end{bmatrix}} \times \underset{\text{Function Weight Matrix}}{\begin{bmatrix} 0.4454 \\ 0.1916 \\ 0.1715 \\ 0.0595 \end{bmatrix}} = \underset{\text{Final Program Weights}}{\begin{bmatrix} 0.051926 \\ 0.261664 \\ 0.162797 \\ 0.041462 \\ 0.019752 \end{bmatrix}} Q_1$$

Software development hierarchy considered for the study consists of modules at the last level. These modules are independent in nature i.e. they are not further connected. Module can be described as a written piece of code that helps in execution of the program. Here, we assume that first program consists of 3 modules, second program is formed from 2 modules, third program requires 3 modules while fourth program needs 4 modules and fifth program can be accomplished through 3 modules. The neutrosophic pairwise comparison matrix given by the programmers for importance of module in influencing the reliability of the system are as follows:

P_1	M_1	M_3	M_5
M_1	$\tilde{1}$	$\tilde{7}$	$\tilde{5}$
M_3	$\tilde{7}^{-1}$	$\tilde{1}$	$\tilde{3}$
M_5	$\tilde{5}^{-1}$	$\tilde{3}^{-1}$	$\tilde{1}$

P_2	M_2	M_4
M_2	$\tilde{1}$	$\tilde{5}$
M_4	$\tilde{5}^{-1}$	$\tilde{1}$

P_3	M_4	M_6	M_7
M_4	$\tilde{1}$	$\tilde{8}$	$\tilde{5}$
M_6	$\tilde{8}^{-1}$	$\tilde{1}$	$\tilde{3}$
M_7	$\tilde{5}^{-1}$	$\tilde{3}^{-1}$	$\tilde{1}$

P_4	M_2	M_4	M_6	M_7
M_2	$\tilde{1}$	$\tilde{2}$	$\tilde{3}$	$\tilde{4}$
M_4	$\tilde{2}^{-1}$	$\tilde{1}$	$\tilde{2}$	$\tilde{3}$
M_6	$\tilde{3}^{-1}$	$\tilde{2}^{-1}$	$\tilde{1}$	$\tilde{2}$
M_7	$\tilde{4}^{-1}$	$\tilde{3}^{-1}$	$\tilde{2}^{-1}$	$\tilde{1}$

P_5	M_1	M_4	M_7
M_1	$\tilde{1}$	$\tilde{3}$	$\tilde{5}$
M_4	$\tilde{3}^{-1}$	$\tilde{1}$	$\tilde{4}$
M_7	$\tilde{5}^{-1}$	$\tilde{4}^{-1}$	$\tilde{1}$

One module may be connected to one or more program for example M_2 is connected to second (P_2) and fourth (P_4) programs. Before computing the relative weights, it is important to ensure that the pairwise comparison matrixes are consistent. The consistency values for all the matrices are listed in Table 4.

Table 5 Relative weights obtained for modules

Modules connected to programs	P_1	P_2	P_3	P_4	P_5
M_1	0.47986	–	–	–	0.39508
M_2	–	0.34721	–	0.39027	–
M_3	0.1202	–	–	–	–
M_4	–	0.0710	0.48391	0.23895	0.18357
M_5	0.065	–	–	–	–
M_6	–	–	0.12347	0.1407	–
M_7	–	–	0.0759	0.08693	0.06456

The relative weights for modules connected to programs at upper level are determined by methodology discussed in Sect. 4 and are given in Table 5. Table 5 represents the module weight matrix corresponding to each program. For final weight calculation, it is considered that the modules that are not connected to the program will be assigned weight as zero. This signifies that the module is not part of that program.

Now, the final weights of modules are computed by multiplying the weight matrix modules (Table 5) and final weights obtained for programs in Q_1 is shown in Q_2. The final weights of modules are $M_1 = 0.032721$, $M_2 = 0.107034$, $M_3 = 0.0066242$, $M_4 = 0.11089$, $M_5 = 0.003375$, $M_6 = 0.024934$ and $M_7 = 0.017236$. We observe that M_4 has maximum weight among all the seven modules of the software system. Mathematically, it is given as:

$$\underbrace{\begin{bmatrix} 0.47986 & 0 & 0 & 0 & 0.39508 \\ 0 & 0.34721 & 0 & 0.39027 & 0 \\ 0.1202 & 0 & 0 & 0 & 0 \\ 0 & 0.0710 & 0.48391 & 0.23895 & 0.18357 \\ 0.065 & 0 & 0 & 0 & 0 \\ 0 & 0 & 0.12347 & 0.1407 & 0 \\ 0 & 0 & 0.0759 & 0.08693 & 0.06456 \end{bmatrix}}_{\text{Module Weight Matrix}} \times \underbrace{\begin{bmatrix} 0.051926 \\ 0.261664 \\ 0.162797 \\ 0.041462 \\ 0.019752 \end{bmatrix}}_{\text{Final Program Weights}} = \underbrace{\begin{bmatrix} 0.032721 \\ 0.107034 \\ 0.0066242 \\ 0.11089 \\ 0.003375 \\ 0.024934 \\ 0.017236 \end{bmatrix}}_{\text{Final Module Weights}} \quad Q_2$$

5.2 Optimization Problem Solution

The final relative weights of modules obtained from Q_2 are used in the optimization problem to allocate reliabilities to each module. These weights have been obtained after integrating the users, software engineers and programmer's expectations and anticipations for software product. Going through the discussed procedure helps in reliability allocation incorporating views of all the stakeholders. For our example problem the optimization problem to maximize software system reliability by allocating reliability to modules is as given in Eqs. (4)–(7). The optimization problem for the example problem is given by Q_3.

$$\left.\begin{array}{c} \text{Max } Z = R_1{}^{w_1} * R_2{}^{w_2} * R_3{}^{w_3} * R_4{}^{w_4} * R_5{}^{w_5} * R_6{}^{w_6} * R_7{}^{w_7} \\ \text{s.t} \\ \sum_{i=1}^{7} a_i + b_i R_i + c_i(1 - R_i) \le 1000 \\ \prod_{i=1}^{7} R_i \ge 0.90 \\ 0 \le R_i \le 1 \quad i = 1, 2 \ldots 7 \end{array}\right\} Q_3$$

where w_i's are relative weights of modules obtained through NAHP and all the other parameters have been assumed. The budget of the software development has been assumed to be 1000 units and reliability goal has been set to be 0.90. The assumed values for budget constraint are as follows:

$$a_i = 80 \text{ for } i = 1, 2 \ldots 7$$

$$b_1 = 40, b_2 = 50, b_3 = 60, b_4 = 20, b_5 = 70, b_6 = 50, b_7 = 35$$

$$c_1 = 20, c_2 = 30, c_3 = 35, c_4 = 10, c_5 = 40, c_6 = 15, c_7 = 20$$

a_i is fixed cost of testing and debugging. So, it is assumed to be same for every module. R_i's are the decision variable representing reliability allocated to ith Module.

Solving the non-linear optimization problem Q_3 in MATLAB, we obtain the reliability allocated to each module as $R_1 = 0.9822$, $R_2 = 0.9976$, $R_3 = 0.9899$, $R_4 = 0.985$, $R_5 = 0.9833$,$R_6 = 0.9924$ and $R_7 = 0.9746$. Maximum reliability is allocated to the second (M_2) module while minimum reliability is allocated to seventh (M_7) module. These reliability values not only depend on final relative weights of modules but also depend on the assumed cost parameters. So, by changing the parameters values various scenarios can be experimented. The overall reliability of the software system is computed by multiplying reliability of all the seven modules. The software system reliability is computed to be 0.9086. The budget utilized for development of software is 883.0455 units.

6 Implications

6.1 Theoretical Implication

The study states several theoretical and managerial implications. The proposed methodology for reliability allocation among modules of the software system is a straightforward and effective approach that uses neutrosophic numbers to compare functions, programs and modules based on their importance to users, software engineers and programmers respectively. Decision making under neutrosophic environment provides more flexibility to the ones involved in the process. A neutrosophic set includes in itself all other existing sets like classical, fuzzy and intuitionistic sets because it not only consist of truth and false membership functions but also considers indeterminacy function. These three membership functions help numbers to represent real life situations by expressing support and objection evidences together. In this way neutrosophic sets provide a broader aspect for making judgements. Due to the importance of these sets it has been widely used in various fields (Abdel-Basset et al. 2017; Radwan et al. 2016). Here, in this chapter the weights are obtained for modules based on a structured hierarchy for reliability allocation helps to develop a product that can cater the demands of user along with meeting the expectations of developers.

Hence, the study highlights the importance of structured hierarchy that connects opinions and thoughts of all the stakeholders of SDP. This hierarchy has been discussed and illustrated in literature by many researchers (Aggarwal et al. 2018; Chatterjee et al. 2015; Yue et al. 2015; Zahedi and Ashrafi 1991). The optimization problem considers the cost of achieving the considerable reliability for each module and also the penalty cost for any failure due to unreliability of the modules. This makes optimization problem more realistic and appropriate for SDP. Based on the total cost incurred management can take important decisions regarding testing of the modules. Several other researches has considered the effect of reliability on the overall cost of development (Chatterjee and Shukla 2017; Huang and Lyu 2005; Kapur and Garg 1990).

Functions have been assigned importance by users based on their requirements, programs are given importance by software engineers based on its relevance for function and modules are compared by programmers created on the need of module for execution of a program. The final weights of modules are obtained by following top-down hierarchy of software development given by experts for reliability apportionment (Zahedi and Ashrafi 1991).

6.2 Managerial Implications

The final reliability allocated to the modules based on the weights obtained using NAHP are $R_1 = 0.9822$, $R_2 = 0.9976$, $R_3 = 0.9899$, $R_4 = 0.985$, $R_5 = 0.9833$, $R_6 =$

0.9924 and $R_7 = 0.9746$. This result goes well with the literature of reliability allocation among modules of software system (Chatterjee et al. 2015; Yue et al. 2015; Zahedi and Ashrafi 1991). The methodology combines the useful effect of judgements using neutrosophic numbers and AHP with the optimization problem to maximize reliability of the software system subject to budget and reliability constraint. The optimization problem consisted of non-linear objective and constraints with simple parameters that can be easily quantified. The developer sets the reliability goal that can be achieved by allocating reliabilities to the modules. The parameter of the budget constraint can be altered, allowing the developer to investigate various possible scenarios. Therefore, the IT firms can decide how to achieve the reliability goals for software system. The proposed optimization problem uses the relative weights of modules computed from NAHP and hence is able to capture the views of both experts and users for software development.

Successful implementation of the proposed methodology has been shown through an example problem. The example exhibits reliability allocation to modules by going through all the intermediary levels showing its ability to capture the software systems complexity. In this study it was found that relative weight of modules, functions and programs, cost parameters of optimization problem affect the reliability of the software system. The lower values of reliability signify that those modules need more attention to improve the overall reliability. This attention can be in terms of time and resource employed to develop that particular module.

Another attention-grabbing observation is that the calculation of weights using NAHP based on the hierarchy considered for the study builds a linkage between expectations, views and thoughts of users, programmers and software engineers. The decisions using NS helps to incorporate vague, ambiguous, uncertain, biased judgements. This also helps to tackle the decisions taken under scarcity of information or incomplete information available to decisions makers. Including customers/users in the process help the management to increase their customer base in the market.

7 Conclusions, Limitations and Future Scope

Allocating reliabilities to modules during initial design and development phase is very critical task of Software Development Process. This is so because the reliability of individual modules determines the reliability of the software system and helps in reliability goal attainment. This helps developer to learn about the module requirements and their importance in the software system in maximizing overall reliability. After knowing about the reliability necessity for each module the developer plans about resources and time needed for developing the module and hence the software system. In this chapter, we proposed a hybrid approach by combining NAHP methodology and proposed optimization problem for maximizing overall software reliability to allocate reliabilities among the modules of the software system. Hence, this chapter amalgamates the subjective and objective views about the software from different stakeholders of the Software development project i.e. users, engineers and programmers present at different levels of hierarchy. The hierarchy evaluates the importance of functions to users, importance of programs to software engineers and importance of modules to programmers and finally the judgements at all the levels are combined to obtain relative weights of modules. Hence, the final weights are resulting from combination of decisions at all the levels.

Comparison of various alternatives using NAHP and incorporating relative weights of module into the optimization problem seems to be very promising in the determining the reliabilities of individual modules. This is so because the past work in this field either uses AHP to prioritise modules and allocate reliability or researchers have proposed optimization problems with the objective to either maximize the overall reliability of software system or minimize the software development cost. These were the two prominent ways of reliability goal attainment for software system. The simplicity and effectiveness of the methodology to comprehend the relationship between modules, programs and functions makes it more efficient to be widely accepted by the IT community. The methodology is of great help for software developers during decision making process.

There are some limitations to the study; which opens path for future research. The present study is based on preferences of single decision maker at each level. In future the proposed model can be extended for more than one decision maker at a particular level. The demographic diversities of the group members can positively affect the decision making. Properly managed groups tend to produce quality decisions. The weights can be calculated using various methods available in literature like Ordered Weighted Average, Principal Component Analysis, distance methods and MCDM techniques etc. NS can be used with different MCDM techniques. In particular it can be employed with ANP to handle interdependencies of functions, programs and modules. Also interval valued neutrosophic numbers can be used for comparisons at different levels of hierarchy. Interval neutrosophic sets generalize all the other sets such as fuzzy, interval valued fuzzy, interval valued intuitionistic and hence will be able to create better decisions. The proposed study can also be conducted for multi-software or multimedia software systems.

References

Abdel-Basset M, Mohamed M, Zhou Y, Hezam I (2017) Multi-criteria group decision making based on neutrosophic analytic hierarchy process. J Intell Fuzzy Syst 33(6):4055–4066

Abdel-Basset M, Mohamed M, Smarandache F (2018) An extension of neutrosophic AHP–SWOT analysis for strategic planning and decision-making. Symmetry 10(4):116

Aggarwal K, Singh Y (1995) Software reliability apportionment using the analytic hierarchy process. ACM SIGSOFT Softw Eng Notes 20(5):56–61

Aggarwal AG, Verma V, Anand S (2018) Architecture-based optimal software reliability allocation under uncertain preferences. In: Proceedings of the first international conference on information technology and knowledge management annals of computer science and information systems, vol 14, pp 3–14

Alblowi S, Salama A, Eisa M (2014) New concepts of neutrosophic sets. Infinite Study,

Ali M, Smarandache F (2017) Complex neutrosophic set. Neural Comput Appl 28(7):1817–1834

Atanassov KT (1999) Intuitionistic fuzzy sets. In: Intuitionistic fuzzy sets. Springer, pp 1–137

Chang Y-C, Chang K-H, Liaw C-S (2009) Innovative reliability allocation using the maximal entropy ordered weighted averaging method. Comput Ind Eng 57(4):1274–1281

Chatterjee S, Shukla A (2017) An ideal software release policy for an improved software reliability growth model incorporating imperfect debugging with fault removal efficiency and change point. Asia-Pacific J Oper Res 34(03):1740017

Chatterjee S, Singh JB, Roy A (2015) A structure-based software reliability allocation using fuzzy analytic hierarchy process. Int J Syst Sci 46(3):513–525

Elegbede AC, Chu C, Adjallah KH, Yalaoui F (2003) Reliability allocation through cost minimization. IEEE Trans Reliab 52(1):106–111

Franek J, Kresta A (2014) Judgment scales and consistency measure in AHP. Procedia Econ Financ 12:164–173

Guan H, Wang T, Chen W (2009) Exploring architecture-based software reliability allocation using a dynamic programming algorithm. In: Proceedings of the 2009 international symposium on computer science and computational technology (ISCSCI 2009), 2009. Citeseer, p 106

Haibin W, Smarandache F, Zhang Y, Sunderraman R (2010) Single valued neutrosophic sets. Infinite Study

Huang C-Y, Lyu MR (2005) Optimal release time for software systems considering cost, testing-effort, and test efficiency. IEEE Trans Reliab 54(4):583–591

Kapur P, Garg R (1990) Optimal sofware release policies for software reliability growth models under imperfect debugging. RAIRO—Oper Res 24(3):295–305

Kapur P, Bardhan A, Jha P (2003) Optimal reliability allocation problem for a modular software system. Opsearch 40(2):138–148

Kuo W, Wan R (2007) Recent advances in optimal reliability allocation. In: Computational intelligence in reliability engineering. Springer, pp 1–36

Leung Y (1997) Software reliability allocation under an uncertain operational profile. J Oper Res Soc 48(4):401–411

Li Q, Yang H, Wang H (2012) Component-based software system reliability allocation and assessment based on ANP and petri. In: 2012 International conference on quality, reliability, risk, maintenance, and safety engineering, 2012. IEEE, pp 227–231

Lyu MR, Rangarajan S, van Moorsel AP (1997) Optimization of reliability allocation and testing schedule for software systems. In: Proceedings of the eighth international symposium on software reliability engineering, 1997. IEEE, pp 336–347

Malaiya YK (2014) Reliability allocation. Wiley StatsRef: Statistics Reference Online

Mettas A (2000) Reliability allocation and optimization for complex systems. In: Annual reliability and maintainability symposium. 2000 Proceedings. International symposium on product quality and integrity (Cat. No. 00CH37055), 2000. IEEE, pp 216–221

Misra R (2005) Economic allocation of target reliability in modular software systems. In: Annual reliability and maintainability symposium, 2005. Proceedings, 2005. IEEE, pp 428–432

Neha, Verma V, Tandon A, Aggarwal AG (2019) Software reliability allocation incorporating pythogorean fuzzy theory and AHP. Int J Ind Syst Eng. Accepted for Publication

Pietrantuono R, Russo S, Trivedi KS (2010) Software reliability and testing time allocation: an architecture-based approach. IEEE Trans Softw Eng 36(3):323–337

Radwan NM, Senousy MB, Alaa El Din MR (2016) Neutrosophic AHP multi criteria decision making method applied on the selection of learning management system. Infinite Study

Satty TL (1980) The analytical hierarchy process: planning, priority setting, resource allocation. RWS Publication, Pittsburg

Smarandache F (1995) Neutrosophic logic and set, mss

Tian P, Wang J, Zhang W, Liu J A fault tree analysis based software system reliability allocation using genetic algorithm optimization. In: 2009 WRI world congress on software engineering, 2009. IEEE, pp 194–198

Verma V, Anand S, Aggarwal AG (2019) Intuitionistic fuzzy AHP based reliability allocation model for multi-software system. Int J Serv Oper Inform. Accpeted for Publication

Yager RR (2013) Pythagorean membership grades in multicriteria decision making. IEEE Trans Fuzzy Syst 22(4):958–965

Yue F, Zhang G, Su Z, Lu Y, Zhang T (2015) Multi-software reliability allocation in multimedia systems with budget constraints using Dempster-Shafer theory and improved differential evolution. Neurocomputing 169:13–22

Zadeh LA (1965) Fuzzy sets. Inf Control 8(3):338–353

Zahedi F, Ashrafi N (1991) Software reliability allocation based on structure, utility, price, and cost. IEEE Trans Softw Eng 17(4):345–356

Testing Resource Allocation for Software System: An Approach Integrating MEMV-OWA and DEMATEL

Rubina Mittal and Rajat Arora

Abstract In the process of resource allocation for software development, specific amount of resources need to be assigned to different modules of a software application from the available resources. To achieve this, it is essential to consider the various quality characteristics of a software. The key criteria for gauging the quality attributes of a software has been taken from the previous studies, practice and theory. The present study, one of the multi-criteria decision making technique, Decision Making Trial and Evaluation Laboratory (DEMATEL) is applied to determine the cause and effect relationship between quality characteristics. This helps in selecting the important characteristics and discarding the ones that are not so significant. The outcomes of this study can be used to divide criteria into two groups namely cause and effect groups and hence construct an Impact Relationship Map. Later, the maximal entropy minimum variance ordered weighted averaging (MEMV-OWA) method has been used to determine the module weights for resource allocation and ranking of various modules based on conflicting nature of characteristics in order to allocate the resources in a competent way. The detailed methodology has been illustrated through a numerical example.

Keywords Resource allocation · SRGM · MCDM · DEMATEL · MEMV-OWA

1 Introduction

In today's competent world, it is a challenge for the developers to use the resources in an efficient manner so that the quality of the product can be improved. Hence, resource allocation is a very crucial step of software development. In Resource allocation, the given fixed amount of resources are assigned to different modules of

R. Mittal (✉)
Keshav Mahavidyalaya, University of Delhi, New Delhi, Delhi, India
e-mail: drmittal@keshav.du.ac.in

R. Arora
Department of Operational Research, University of Delhi, New Delhi, Delhi, India

A. G. Aggarwal et al. (eds.), *Optimization Models in Software Reliability*, Springer Series in Reliability Engineering,
https://doi.org/10.1007/978-3-030-78919-0_10

a software application according to its need and importance. The process involves apportioning the available amount of resources among different modules by taking into consideration the quality characteristics of software system. It is an essential topic of concern for both the developer and the ultimate user. It is one of the crucial aspect to be considered during the development of a software application.

The major challenge faced while apportioning resources is that in what manner the uncertain preferences of a decision maker should be handled for several attributes of interest. Various development resources are expended during testing phase of a software. The testing phase primarily consists of module testing, integration testing and system testing. Therefore, it is extremely important for the development team to decide how to effectively employ the available resources during software testing for developing a reliable and quality software.

So far, resource allocation process has been done by either maximizing the number of faults removed or minimizing the development cost subject to reliability and resource constraints (Huang and Lo 2006; Khan et al. 2008; Kapur et al. 2009). But, none of these studies focused on the importance of quality characteristics while apportioning resources. They did not take into consideration the biased or subjective behavior of decision maker and the amount of information available. To bridge this gap, the present study fulfills the following objectives (1) to shortlist the most dominant/critical criteria influencing software quality (2) to exploit cause and effect relationship between different quality characteristics using DEMATEL technique (3) to obtain the relative weights of modules using MEMV-OWA approach and the results obtained from DEMATEL.

The major challenge for developers is to identify the relative importance of modules while considering various characteristics of interest. First, it is crucial to understand the relationship between various characteristics such as functionality, reliability, usability, efficiency, maintainability and portability and then based on that relationship, the relative importance of modules is obtained. It is necessary for developers to incorporate the biased and uncertain preferences of the experts while making decision. To identify the relationship, a well-known multi-criteria decision making technique called DEMATEL is used. To handle the biased nature of experts, MEMV-OWA operator is used. This helps in maximizing the use of available information and minimizing the variance.

In this way, maximizing the entropy can effectively utilize the uncertain information of decision maker's experience and minimizing the variance of the weight vector is a potential way to avoid overestimation of the decision maker's preferences. Various optimization problems have been studied in context of allocation viz. traditional reliability redundancy allocation, resource allocation, life percentile optimization using different techniques. Later, using expert's opinion, the importance of each of the shortlisted criteria in each of the four modules of software application is obtained and a numerical is illustrated to demonstrate how a pre-specified amount of resources are allocated to different modules.

Rest of the paper is organized as follows. Section 2 presents the review of literature of DEMATEL, MEMV-OWA and resource allocation. Section 3 discusses the detailed resource allocation methodology, a numerical example is illustrated in Sect. 4 followed by conclusions and future scope in Sect. 5.

2 Literature Review

This section discusses the rationale behind the proposed study by discussing the current and past studies in the area. Vast research has been done in context of resource allocation problems using different techniques. Ohtera and Yamada (1990) introduced software reliability growth model (SRGM) based on Non-Homogeneous Poisson Process (NHPP) and discussed the time-dependent behavior of resource expenditure spent during testing. They discussed two types of managerial problems in software testing viz. testing resource allocation that deals with the best use of available testing resources during module testing and testing resource-control problem. Yamada et al. (1995) considered two kinds of software testing-resource allocation problems to make the best use of the specified testing-resources during module testing and also introduced a NHPP based SRGM for describing the time-dependent behavior of detected software faults and testing-resource expenditures spent during the testing.

Hou et al. (1996) investigated the efficient allocation of resources for modular software based on Hyper-geometric distribution based SRGM. Lo et al. (2002) studied optimal testing resource allocation and reliability analysis of component based software application. He proposed an analytical approach for estimating the reliability of a component-based software under the assumption that the components of a software are heterogeneous and the transfers of control between components follow a discrete time Markov process. Apart from this, they also formulated and solved two resource allocation problems. Lyu et al. (2002) studied how to allocate the test resources in an optimal manner for software development. Different types of reliability growth curves were used to model the relation between the component's failure rates and the cost required to reduce this rate.

Dai et al. (2003) presented a genetic algorithm for testing-resource allocation that can be used in case of complex systems and also when there are multiple objectives viz. minimizing cost and maximizing reliability while allocating finite resources. Huang and Lo (2006) studied optimal resource allocation problem in modular software system under reliability constraint. Further, an optimization problem is discussed which minimizes the cost of software development when a fixed amount of testing effort and a desired reliability objective are given using Lagrange multiplier method. Khan et al. (2008) used a Dynamic Programming approach for optimal allocation of testing resources for modular software based on inflection S-shaped SRGM with Exponentiated Weibull testing effort function. Kapur et al. (2009) considered cost, testing effort and reliability for optimal resource allocation during module testing using Genetic algorithm. Pietrantuono et al. (2010) used architecture-based approach for testing time allocation.

Kapur et al. (2010) proposed Two-dimensional allocation problem for modular software. In order to take into consideration simultaneous effects, Cobb Douglas production function is used. Akhtar et al. (2011) proposed an imperfect debugging SRGM during testing and allocation of resources is done based on optimizing the effort and reliability. Fiondella and Gokhale (2012) used architecture-based optimization in resource allocation process. Nasar et al. (2014) investigated various software release policies and resource allocation based on dual constraints of cost and reliability. Kumar and Yadav (2017) developed a model for optimally allocating effort between detection and correction process during testing phase of software development life cycle (SDLC). For this, flexible SRGM with testing effort is considered under dynamic environment. Okamura and Dohi (2018) allocated the testing resources optimally using architecture-based SRGM. In this chapter, we considered the relative modular weights obtained using integration of DEMATEL and MEMV-OWA and software development cost in the testing-resource allocation problems while maximizing number of faults removed.

The DEMATEL method was proposed by the Battelle Memorial Institute Fontela and Gabus (1976). This multi-criteria decision making (MCDM) technique is applied to analyze the relationship, impact of criteria on each other and to select the critical criteria among the set of criteria. The assessment of criteria is based on expert decision. This methodology has been applied in multiple disciplines. Tzeng et al. (2007) proposed hybrid MCDM model combining factor analysis and DEMATEL to evaluate intertwined effects in e-learning programs by addressing the independent relations among criteria with the help of factor analysis. Tsai and Chou (2009) used DEMATEL in combination with ANP in selecting management systems for sustainable development in SME's. They used this approach to construct inter-relations among diverse criteria and obtain their weights through ANP. Further ANP is integrated with Zero–one goal programming to achieve the organization goals while simultaneously considering goals on resources. Chang et al. (2009) integrated OWGA operator and DEMATEL for prioritization of failures in a product Failure modes and effects analysis to rank risk for failure problems.

Chang et al. (2011) used Fuzzy DEMATEL method for developing supplier selection criteria. Zhou et al. (2011) used Fuzzy DEMATEL method for identifying critical success factors in emergency management. Büyüközkan and Çifçi (2012) integrated Fuzzy DEMATEL, fuzzy ANP and fuzzy TOPSIS to evaluate green suppliers. BaykasoğLu et al. (2013) integrated fuzzy DEMATEL and TOPSIS for truck selection problem of a land transportation company. Rochimah (2013) integrated DEMATEL and ANP methods to calculate the weights of software quality characteristics. Sadehnezhad et al. (2014) used combination of DEMATEL and ANP with fuzzy approach to evaluate business intelligence performance. Govindan et al. (2015) proposed Intuitionistic fuzzy based DEMATEL method in the domain of supply chain. Su et al. (2016) used a hierarchical grey DEMATEL approach to identify and analyze criteria and alternatives in incomplete information in the domain of supply chain. Si et al. (2018) briefed about the various methodologies and applications of DEMATEL technique But it has not been applied for allocating testing resources during software development process. Integrating it with OWA operator helps to incorporate subjective behavior of decision maker. It also helps in maximizing the available information and minimizing the variance.

OWA technique has been widely applied for determining weights. O'Hagan (1990) used the concept of maximum entropy in fuzzy neuron. Yager (1993) first introduced the term OWA for finding the weights. He later combined it with the concept of maximum entropy or minimization of dispersion. Cheng et al. (2009) applied OWA operator to combine multiple criteria into aggregated value of a single criteria and further used these values for classification tasks. Later, Ahn (2012) proposed a new weighing method known as minimizing distances from the extreme points (MDP) wherein the OWA operator weights are so chosen that minimize the expected quadratic distance with respect to the set of extreme points. Aggarwal et al. (2017) proposed architectural based software reliability allocation using OWA.

All the research work on resource allocation in literature has developed an optimization problem with the objective of minimizing cost or delivery time based on a single criterion and maximizing number of faults removed or reliability (Lo et al. 2002; Kapur et al. 2010). They didn't consider the multiple characteristics in a single problem. There are different aspects of software quality. The six important quality attributes of the software considered in literature are Functionality, Reliability, Usability, Efficiency, Maintainability and Portability (Rochimah 2013). Each of these attributes can influence each other (Table 1).

In this paper, we allocate the available resources to different modules in a software application based on multiple attributes of software quality. Here, we will propose a novel hybrid model integrating MEMV-OWA and DEMATEL. This model addresses the dependent and independent relationships among criteria through DEMATEL and thus select the most dominant ones. This is graphically represented through Impact Relationship map or digraph. In order to address the decision maker's uncertain preferences in resource allocation, the proposed OWA operator takes into consideration a bi-objective mathematical programming model incorporating both maximal entropy and minimal variance. These are further used for computing the orness weights of evaluation criteria (quality characteristics). The advantage of maximal entropy minimal variance OWA operator is that it incorporates all the available information and avoids over-estimation of preferences of a decision maker. The resulting scores obtained for the characteristics represent their relative importance while evaluating

Table 1 Brief description of quality attributes

Code no.	Quality attributes	Definition
V1	Functionality	Essential purpose of any product or service
V3	Usability	Ability to use the program
V4	Efficiency	Concerned with system resources used to satisfy desired functionality
V5	Maintainability	Effort needed to locate and fix a bug in an operational software application
V6	Portability	Capacity of a system to be used in diverse platforms

different modules in software application. The proposed model is able to produce effective evaluations for models particularly when there are intertwined and diverse evaluation criteria.

3 Resource Allocation Methodology

Resource allocation among modules of the software system has been accomplished integrating two techniques namely DEMATEL and MEMV-OWA. In this section, we discuss the techniques and step-wise methodology.

3.1 DEMATEL Method

The DEMATEL method, originally proposed at The Battelle Memorial Institute through its Geneva Research Centre (Fontela and Gabus 1976) is a logical way for capturing cause and effect relationships between criteria and visualizing them through matrices, impact relation map or digraphs. The numerical value in the digraph is indicative of the strength of the influence. This method helps in constructing a comprehensible prototypical model which not only captures the direct influences but also the indirect influences between multiple attributes (Wu and Lee 2007). This method splits the relevant criteria into cause group and effect group to enable efficient and accurate decision making. The procedure of DEMATEL can be summarized into steps as follows (Seyed-Hosseini et al. 2006):

Step 1. First we need to build the hierarchy for resource allocation among modules of the software based on quality characteristics.

Step 2. A pairwise-comparison matrix is constructed for different quality characteristics of software based on expert judgement as follows:

- The pairwise comparison measure may bifurcates into four echelons where the score of 0 represent "no influence", 1 indicates "low influence", 2 indicates "high influence" and 4 is indicating "very high influence".
- The initial direct relation matrix M is a matrix of order n obtained by pairwise comparison of different criteria with respect to their influence by the decision maker and the elements of matrix m_{ij} represent the degree to which criterion V_i affects criterion V_j. Here, all entries in principal diagonal are zero.

Step 3. Let g denote the greatest element among row sums. Mathematically, it can be written as:

$$g = \max\left(\sum_{j=1}^{n} m_{ij}\right) \tag{1}$$

Step 4. The normalized direct relation matrix D can be obtained as

$$D = M/g \tag{2}$$

This is also called by the name Direct and Indirect Relative Severity Matrix (DIRSM).

1. The Total relation matrix TR can be obtained as

$$TR = D(I - D)^{-1} \tag{3}$$

   where, I denotes the identity matrix of order n.
   Let t_{ij} represents the elements of Total relation matrix TR and let R_i and C_j represent row sum and column sum of this matrix for all $i, j = 1, 2, \ldots, n$.
2. A causal diagram can be constructed by plotting the ordered pairs (R + C, R − C) where the horizontal axis (R + C) is called "prominence" and the vertical axis (R – C) is termed as "relation". Using the values of R + C and R – C, where C is the sum of the columns and R is the sum of the rows of the DIRSM, a level of influence and a level of relationship are defined.

The prominence is indicative of the importance of each criteria. On the other hand, relation separates the criteria into cause and effect groups. The value of (R − C) represents the severity of influence of each alternative criteria whereas the measure (R + C) indicates the degree of relation between each alternative with one another. The positive value of (R − C) represents that the criteria falls in the cause group and its negative value indicates the belongingness to the effect group. The criteria which is having greater effect on another criteria are under cause group and those which are more influenced from the other criteria are listed under effect group.

The accurate decisions can be made with the help of causal diagram and by identifying the difference between cause and effect criteria.

3.2 *Ordered Weighted Averaging (OWA) Operator*

The Ordered weighted averaging operator is of great use in solving Multi-criteria decision making problems. Recently, it has been widely used in studying software quality. Researchers have worked on various OWA operators for finding the associated weights. Yager (1993) was the first to introduce the concept of OWA operator in 1988. Later, O'Hagan proposed the ME-OWA operator to determine the weighing vector under the non -linear constraint of maximum entropy with the pre-specified orness level. The similar problem with minimum variance constraint was proposed by Fullér and Majlender (2003).

The objective of using this operator is to take complete benefits of the existing information and to remove the biasness of decision maker's priorities.

Definition An OWA operator of dimension n is a mapping $F:R^n \rightarrow R$ with associated weighting vector $W = (w_1, w_2, w_{3,\ldots,} w_n)^T$ $W = (w_1, w_2, w_{3,\ldots,} w_n)^T$ with the properties.

$$\sum_{i=1}^{n} w_i = 1; \quad 0 \leq w_i \leq 1, \quad i = 1, 2, \ldots, n \tag{4}$$

and

$$F(a_1, a_2, a_3, \ldots a_n) = \sum_{i=1}^{n} w_i a_i \tag{5}$$

where a_1 is the largest element in the collection (Han and Deng 2018).

- **Orness associated with OWA Operator**: Let F be an OWA operator and W be the weight vector given by $W = (w_1, w_2, w_{3,\ldots,} w_n)^T$ then, the degree of "Orness" associated with F is defined as

$$\text{Orness}(W) = \frac{1}{n-1}\sum_{i=1}^{n}(n-i)w_i = \mu \quad \text{where,} \quad \mu \in [0, 1]. \tag{6}$$

  The measure of Orness indicates whether the relationship between multiple attributes shows an and-like or an or-like level. An Orness value μ close to zero, then the multiple attributes are related by a higher and-like value indicating that the decision maker is maximally non-committal. On the other hand, if orness is closer to one, then the multiple attributes are related by a higher or-like value/behavior indicating that the decision maker is maximally optimistic. If all the weights are equal, in that case orness value is 0.5 indicating that the decision maker faces a restrained assessment.
- **Entropy (or dispersion) measure**: This measure accounts for, to what extent the information in an uncertain environment is utilized. For a given weight vector

$$W = (w_1, w_2, w_3, \ldots, w_n)^T, \quad \text{dispersion}(W) = -\sum_{i=1}^{n} w_i \ln w_i \tag{7}$$

If only one component of weight vector is 1 and rest all are 0. It shows minimum dispersion and disp(W) is zero. It represents that only one attribute is accountable in the aggregation process.

When all the weights are equally likely, then disp(W) is maximum and equals ln(n). It represents that all the attributes are accountable in the aggregation process.

- **Measure of Variance**: This measure determines the variation in components of weight vector for a given orness level and is defined as $V^2(W) = E\big(W^2\big) - (E(W))^2$. The variance of weighting vector should be controlled in the process of decision-making in order to avoid the overestimation of a single attribute.
- **Maximum entropy-ordered weighted averaging (ME-OWA) Operator**: ME-OWA Operator was introduced by O'Hagan (1990) that maximizes the entropy under pre-defined orness level. This operator has also been used by other researchers for various purposes. Cheng et al. (2008), Fullér and Majlender (2001), Chang et al. (2009) used this operator for reliability allocation. Mathematically, it can be formulated as a mathematical programming problem as follows:

$$\text{Maximize } -\sum_{i=1}^{n} w_i \ln w_i \tag{8}$$

$$\text{subject to } \frac{1}{n-1}\sum_{i=1}^{n}(n-i)w_i = \mu;\quad \mu \in [0, 1] \tag{9}$$

$$\sum_{i=1}^{n} w_i = 1;\quad 0 \le w_i \le 1, i = 1, 2, \ldots, n \tag{10}$$

This problem was solved by Fuller and Majlender using Lagrange multipliers to find optimal weighting vector under maximum entropy. After solving, the obtained weight vectors corresponding to the multiple attributes are given below:

$$\ln w_j = \frac{j-1}{n-1}\ln w_n + \frac{n-j}{n-1}\ln w_1 \tag{11}$$

$$\text{or } w_j = \sqrt[n-1]{w_1^{n-j} w_n^{j-1}},\quad \text{for } 1 \le j \le n \tag{12}$$

where n is the total no. of attributes, and

$$w_n = \frac{((n-1)\alpha - n)w_1 + 1}{(n-1)\alpha + 1 - nw_1} \tag{13}$$

Minimum variance-ordered weighted averaging (MV-OWA) Operator.

This variant of OWA was proposed by Fuller and Majlender that minimizes the dispersion of weighting vector under the constraint of orness level. Mathematically, the MV-OWA problem can be formulated as:

$$\text{Minimize } V^2(W) = \frac{1}{n}\sum_{i=1}^{n} w_i^2 - \frac{1}{n^2} \tag{14}$$

$$\text{subject to} \sum_{i=1}^{n} \frac{n-i}{n-1} w_i = \mu, \quad 0 \le \mu \le 1 \tag{15}$$

$$\sum_{i=1}^{n} w_i = 1; \ 0 \le w_i \le 1, \quad i = 1, 2, \ldots, n \tag{16}$$

which can be solved analytically.

3.3 Maximum Entropy Minimum Variance-Ordered Weighted Averaging (MEMV-OWA) Operator

Based on the above two concepts, a bi-objective programming problem is formulated in order to find the weighting vector of MEMV-OWA Operator. It allows for getting the advantages of both ME-OWA and MV-OWA Operators i.e. uncertain information can be utilized to its maximum and at the same time over-estimation can be avoided. The corresponding bi-objective programming model is given as follows:

$$\text{Maximize} \ -\sum_{i=1}^{n} w_i \ln w_i \tag{17}$$

$$\text{Minimize} \ V^2(W) = \frac{1}{n} \sum_{i=1}^{n} w_i^2 - \frac{1}{n^2} \tag{18}$$

$$\text{subject to} \ \frac{1}{n-1} \sum_{i=1}^{n} (n-i) w_i = \mu; \quad \mu \in [0, 1] \tag{19}$$

$$\sum_{i=1}^{n} w_i = 1; \quad 0 \le w_i \le 1, \quad i = 1, 2, \ldots, n. \tag{20}$$

The solution of the above programming model can be obtained by the Ideal-Point Method (IPM) by converting the bi-objective model into single objective one. This method works by evaluating the non-inferior alternatives based on their absolute distances from the ideal point that is the hypothetical alternative closest to the optimal solution. Under IPM method there's no need to treat dependence among multiple objectives as separable and hence this method is helpful in overcoming some of the difficulties associated with interdependence between multiple objectives. In order to arrive at solution, Lagrange multiplier method is used. The working of the method is illustrated as follows:

Consider the following two objectives

$$Z_1 = \text{Maximize } -\sum_{i=1}^{n} w_i \ln w_i \tag{21}$$

and

$$Z_2 = \text{Minimize } V^2(W) = \frac{1}{n}\sum_{i=1}^{n} w_i^2 - \frac{1}{n^2} \tag{22}$$

and their optimal values are z_1^0 *and* z_2^0.
The ideal point method is defined as

$$\text{Minimize } \|Z - Z^0\| = |Z_1^0 - Z_1| + |Z_2^0 - Z_2| = \left(Z_1^0 - \left(-\sum_{i=1}^{n} w_i \ln w_i\right)\right) + \left(Z_2^0 - \frac{1}{n}\sum_{i=1}^{n} w_i^2 - \frac{1}{n^2}\right) \tag{23}$$

The above constrained optimization problem will be solved by Lagrange multiplier method and the corresponding Lagrange function is:

$$L(w,\ \lambda_{1,},\lambda_2 = \left(Z_1^0 - \left(-\sum_{i=1}^{n} w_i \ln w_i\right)\right) + \left(Z_2^0 - \frac{1}{n}\sum_{i=1}^{n} w_i^2 - \frac{1}{n^2}\right) + \lambda_1\left(\frac{1}{n-1}\sum_{i=1}^{n}(n-i)w_i - \mu\right) + \left(\sum_{i=1}^{n} w_i - 1\right) \tag{24}$$

where, λ_1 and λ_2 are real numbers.

Taking the partial derivatives of the above Eq. (21) with respect to w_i, λ_1 and λ_2 and equating them to zero gives:

$$\frac{\partial L}{\partial w_i} = \frac{-2w_i}{n} - (1 + \ln\ w_i) + \frac{n-i}{n-1}\lambda_1 + \lambda_2 = 0 \tag{25}$$

$$\frac{\partial L}{\partial \lambda_1} = \frac{1}{n-1}\sum_{i=1}^{n}(n-i)w_i - \mu = 0 \tag{26}$$

$$\frac{\partial L}{\partial \lambda_2} = \sum_{i=1}^{n} w_i - 1 = 0 \tag{27}$$

For $i = 1$ and $i = n$, Eq. (22) becomes

$$\frac{-2w_1}{n} - (1 + \ln\ \ w_1)\lambda_1 + \lambda_2 = 0 \tag{28}$$

$$\frac{-2w_n}{n} - (1 + \ln \quad w_n) + \lambda_2 = 0 \tag{29}$$

This yields

$$\lambda_1 = \frac{2w_1}{n} + \ln w_1 - \frac{2w_n}{n} - \ln w_n \tag{30}$$

And

$$\lambda_2 = \frac{2w_n}{n} + 1 + \ln w_n \tag{31}$$

Incorporating the above values of λ in Lagrange function yield a single objective model with the objective function as

$$\frac{2}{n}w_i + \ln w_i = \frac{n-i}{n-1}\left(\frac{2}{n}w_1 + \ln w_1\right) + \frac{i-1}{n-1}\left(\frac{2}{n}w_n + \ln w_n\right) \tag{32}$$

The weighting vector of MEMV-OWA operator can be obtained by solving the above equations.

The step-wise methodology for testing resource allocation among modules of the software systems has been demonstrated through flow chart (Fig. 1). The resources are allocated in proportion to the weights obtained for the modules.

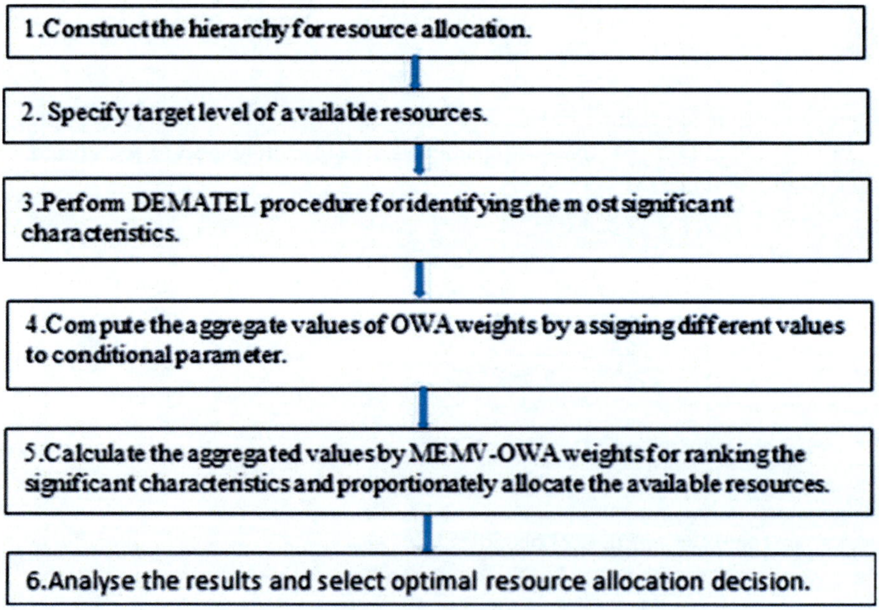

Fig. 1 Procedure of integrated MEMV-OWA and DEMATEL

4 Numerical Example

In this section, we will demonstrate the implementation of the proposed methodology with the help of an example.

At first, DEMATEL technique is used for determining cause and effect relationship between different software characteristics. The hierarchy used for applying DEMATEL has been demonstrated through Fig. 2. In the initial stage, we need to obtain pairwise comparison matrices for quality characteristics from experts (Table 2). The six criteria viz. Functionality, Reliability, Usability, Efficiency Maintainability and Portability are denoted by V1,V2,V3,V4,V5 and V6 respectively.

The pairwise comparison measure may bifurcates into four echelons where the score of 0 represent "no influence", 1 indicates "low influence", 2 indicates "moderate influence", 3 indicates "high influence" and 4 is indicating "very high influence". The scores in the matrix indicate the level of direct influence that one characteristic exert on another. The higher score indicates greater influence.

Initial direct influence matrix D is obtained by normalizing the pair-wise comparison matrix or average matrix by dividing the pair-wise comparison matrix by the greatest row sum. Here all the principle diagonal elements are zero. The elements

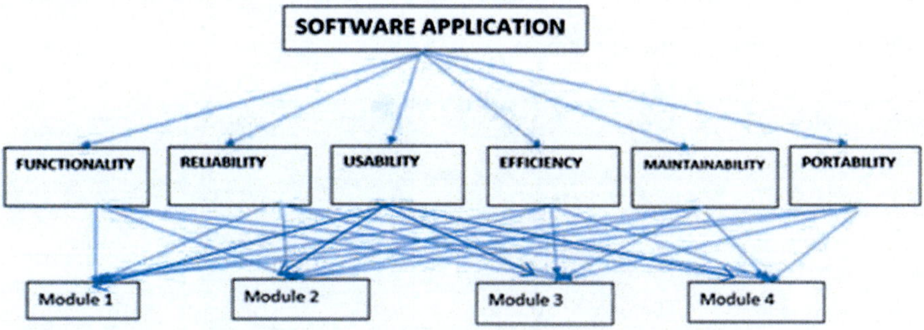

Fig. 2 Hierarchy of a software application before applying DEMATEL

Table 2 Pairwise comparison of quality characteristics (relationship matrix)

	V1	V2	V3	V4	V5	V6	Row sum
V1	0	1	3	4	4	3	15
V2	4	0	1	1	4	1	11
V3	1	2	0	3	2	4	12
V4	3	3	2	0	2	3	13
V5	1	1	3	2	0	4	11
V6	2	1	2	1	3	0	9
col_sum	11	8	11	11	15	15	

of matrix D represent the initial influence that an element exert and receive from another.

The greatest row sum is denoted by g. In this example, the value of g is computed as g = max{15, 7, 12, 9, 15, 11} = 15. The normalized direct relation matrix D can be obtained using Eq. 2. The resulting matrix is given through Table 3.

The Total relation matrix is obtained by infinite series of direct and indirect effects of each element. It is computed by the formula $TR = D(I - D)^{(-1)}$ where D is the normalized direct relation matrix obtained in Table 3. The entries in total relation matrix represents how element *i* influence element *j*. Corresponding to our numerical, the total relationship matrix is given in Table 4.

Table 5 presents the criticality of software quality characteristics and classifies them into cause and effect group. The row sum R indicates the sum of direct and indirect influences dispatching from element *i* to other elements and the column sum C represents the sum of influence that the element receives from other elements. Here, $R + C$ and $R - C$ values are computed for each characteristic. Higher (positive) the $R - C$ value of the characteristic, higher will be its influence on other characteristics and will be considered in cause group and hence classified as critical. According to this, the priority order obtained for our examples is $V1 > V2 > V3 > V4 > V5 > V6$.

Table 3 Normalized relationship matrix

	V1	V2	V3	V4	V5	V6
V1	0	0.0666	0.2	0.2667	0.2667	0.2
V2	0.2667	0	0.0667	0.0667	0.2667	0.0667
V3	0.0667	0.1333	0	0.2	0.1333	0.2667
V4	0.2	0.2	0.1333	0	0.1333	0.2
V5	0.0667	0.0667	0.2	0.1333	0	0.2667
V6	0.1333	0.0667	0.1333	0.0667	0.2	0

Table 4 Total relationship matrix

	V1	V2	V3	V4	V5	V6	R_Row sum	C_Col sum
V1	0.518007	0.478976	0.74749	0.771511	0.922073	0.935052	4.373109	3.164662
V2	0.628258	0.311245	0.527651	0.510063	0.788789	0.666131	3.432137	2.449921
V3	0.490367	0.446401	0.453994	0.597472	0.686985	0.818255	3.493474	3.431876
V4	0.642174	0.529945	0.625126	0.488407	0.760562	0.830963	3.877178	3.301288
V5	0.444481	0.364354	0.58936	0.51506	0.51861	0.778324	3.210189	4.298572
V6	0.441375	0.319	0.488255	0.418774	0.621553	0.489245	2.778203	4.517971

Table 5 Cause-effect group and criticality of quality attributes

Characteristics	R	C	R + C	R–C	CAUSE/EFFECT GROUP	CRITICALITY
V1	4.373109	3.164662	7.537772	1.208447	CAUSE GROUP	CRITICAL
V2	3.432137	2.449921	5.882058	0.982215	CAUSE GROUP	CRITICAL
V3	3.493474	3.431876	6.92535	0.061598	CAUSE GROUP	CRITICAL
V4	3.877178	3.301288	7.178466	0.575891	CAUSE GROUP	CRITICAL
V5	3.210189	4.298572	7.508761	−1.08838	EFFECT GROUP	NOT-CRITICAL
V6	2.778203	4.517971	7.296174	−1.73977	EFFECT GROUP	NOT-CRITICAL

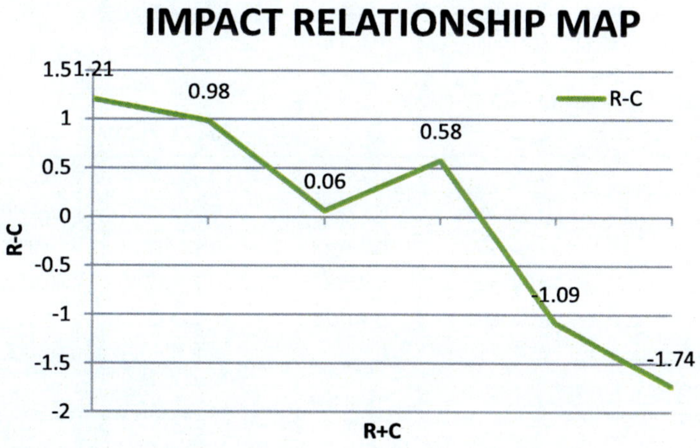

Fig. 3 Cause-Effect diagram for each criteria

From the Impact Relationship map obtained in Fig. 3, those criteria whose R − C value is negative are falling under effect group and are discarded for further analysis since they are not critical (Han and Deng 2018). The above graph illustrates that the first characteristic, functionality has the greatest R − C value viz. 1.21 and hence has maximum influence on other factors. Therefore it is given the highest priority during the resource allocation among modules. On the other hand, portability has the least R − C value viz. -1.74 and given least preference during module allocation.

The hierarchy of software application after the critical criteria are shortlisted through the application of DEMATEL technique has been shown in Fig. 4. The criteria which are part of cause group are classified as critical and the others in effect group are non-critical.

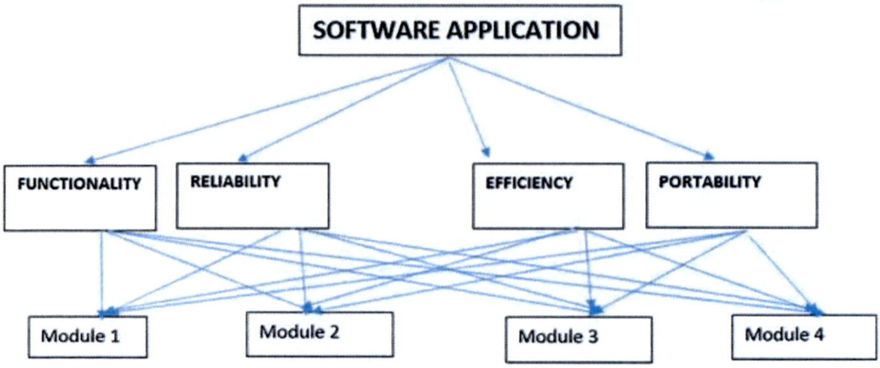

Fig. 4 Hierarchy considering shortlisted criteria after application of DEMATEL

Table 6 Weighing vector of MEMV-OWA

Orness level (μ)	w_1	w_2	w_3	w_4
0.5	0.25	0.25	0.25	0.25
0.6	0.34621	0.27196	0.21556	0.16703
0.7	0.44156	0.29748	0.17763	0.08379
0.8	0.59358	0.25727	0.10467	0.04509
0.9	0.76089	0.18528	0.04432	0.01040
1	1	0	0	0

Next, we use MEMV-OWA for determining optimal weights of modules. The MEMV-OWA technique has been used to find the weights of the shortlisted criteria (through DEMATEL) at different level of Orness viz. 0.5, 0.6, 0.7, 0.8, 0.9 and 1. The results of weights at different orness level are shown in Table 6 that have been obtained using Genetic Algorithm in MATLAB by solving Eqs. 17–20.

In our study, a software application is considered with four independent modules viz. M1, M2, M3 and M4 respectively (Fig. 4). The relative importance of each characteristic is different for different modules. In order to cater this, the ratings are allocated corresponding to each characteristic for given modules using an Expert's opinion based on a Likert scale of 1 to 5 where 1 indicates least significant and 5 is indicative of the most significant attribute with respect to a module. Consequently, the weights of the modules are computed by taking the weighted average of characteristic rating and characteristic weights obtained corresponding to orness level of 0.8.

Let us consider an XYZ software company which has fixed amount of resources worth Rs. 10 lakhs to be spent on the software application consisting of four modules. The amount to be allocated to each module can be computed in proportion to the normalized weights of modules as shown in Table 8 using the normalized weights calculated in Table 7.

Table 7 Normalised weights of modules

	M1	M2	M3	M4	Criteria weights ($\mu = 0.8$)
V1	5	4	5	5	0.594
V2	5	4	3	3	0.257
V3	4	5	3	2	0.105
V4	3	5	4	2	0.045
Normalised weights of modules	4.81	4.15	4.23	4.04	

Table 8 Resource allocation to modules

Modules	Normalized weights	Amount allocated from 10 lakhs
M1	4.805263	4,805,263
M2	4.149671	4,149,671
M3	4.231506	4,231,506
M4	4.036768	4,036,768

The above Table 8 tells us that out of total resources worth Rs. 10 lakhs, the maximum amount was allocated to module M1 viz. 4,805,263 and the minimum was allocated to module M4.

5 Conclusion and Future Scope

The optimal apportionment of the resources is very essential due to their limited availability. In this paper, a hybrid method integrating MEMV-OWA and DEMATEL has been proposed to first identify critical quality attributes for the software application and then using multi-objective ordered weighted-averaging technique to rank the critical attributes for optimal allocation of resources. DEMATEL is also implemented to make out the cause and effect categories and the quality criteria in the cause group are identified as critical success factors. In this paper, four quality attributes are identified as critical ones for the software application viz. Functionality, Reliability, Usability and Efficiency. Through application of MEMV-OWA to rank the critical attributes we obtained weighing vector that minimized the dispersion from equal waits along with maximizing the entropy. In our results, maximum weightage is given for functionality and the least weightage is given to usability for resource allocation. This method provides a well- organized approach for resource allocation. This study can be further extended to deal with uncertainty in expert's judgement by application of DEMATEL in fuzzy environs. Also for ranking it may be integrated with other MCDM techniques viz. TOPSIS, VIKOR and PROMETHEE.

References

Aggarwal AG, Verma V, Anand S (2017) Architecture-based optimal software reliability allocation under uncertain preferences. Ann Comput Sci Inf Syst 14:3–12

Ahn BS (2012) Programming-based OWA operator weights with quadratic objective function. IEEE Trans Fuzzy Syst 20(5):986–992

Akhtar MS, Rafi U, Usmani MK, Dey D (2011) A review of aphid parasitoids (Hymenoptera: Braconidae) of Uttar Pradesh and Uttarakhand, India. Biol Med 3(2):320–323

Baykasoğlu A, Kaplanoğlu V, Durmuşoğlu ZD, Şahin C (2013) Integrating fuzzy DEMATEL and fuzzy hierarchical TOPSIS methods for truck selection. Expert Syst Appl 40(3):899–907

Büyüközkan G, Çifçi G (2012) A novel hybrid MCDM approach based on fuzzy DEMATEL, fuzzy ANP and fuzzy TOPSIS to evaluate green suppliers. Expert Syst Appl 39(3):3000–3011

Chang Y-C, Chang K-H, Liaw C-S (2009) Innovative reliability allocation using the maximal entropy ordered weighted averaging method. Comput Ind Eng 57(4):1274–1281

Chang B, Chang C-W, Wu C-H (2011) Fuzzy DEMATEL method for developing supplier selection criteria. Expert Syst Appl 38(3):1850–1858

Cheng C-H, Chang K-H, Chang Y-C (2008) Using ME-OWA approach to modify prioritization of failures in the IPM system. 國防管理學報 29(2):97–110

Cheng C-H, Wang J-W, Wu M-C (2009) OWA-weighted based clustering method for classification problem. Expert Syst Appl 36(3):4988–4995

Dai Y-S, Xie M, Poh K-L, Yang B (2003) Optimal testing-resource allocation with genetic algorithm for modular software systems. J Syst Softw 66(1):47–55

Fiondella L, Gokhale SS (2012) Optimal allocation of testing effort considering software architecture. IEEE Trans Reliab 61(2):580–589

Fontela E, Gabus A (1976) The DEMATEL observer, DEMATEL 1976 report. Battelle Geneva Research Center, Switzerland, Geneva

Fullér R, Majlender P (2001) An analytic approach for obtaining maximal entropy OWA operator weights. Fuzzy Sets Syst 124(1):53–57

Fullér R, Majlender P (2003) On obtaining minimal variability OWA operator weights. Fuzzy Sets Syst 136(2):203–215

Govindan K, Khodaverdi R, Vafadarnikjoo A (2015) Intuitionistic fuzzy based DEMATEL method for developing green practices and performances in a green supply chain. Expert Syst Appl 42(20):7207–7220

Han Y, Deng Y (2018) An enhanced fuzzy evidential DEMATEL method with its application to identify critical success factors. Soft Comput 22(15):5073–5090

Hou R-H, Kuo S-Y, Chang Y-P (1996) Needed resources for software module test, using the hypergeometric software reliability growth model. IEEE Trans Reliab 45(4):541–549

Huang C-Y, Lo J-H (2006) Optimal resource allocation for cost and reliability of modular software systems in the testing phase. J Syst Softw 79(5):653–664

Kapur P, Aggarwal AG, Kapoor K, Kaur G (2009) Optimal testing resource allocation for modular software considering cost, testing effort and reliability using genetic algorithm. Int J Reliab Qual Saf Eng 16(06):495–508

Kapur P, Aggarwal AG, Kaur G (2010) Simultaneous allocation of testing time and resources for a modular software. Int J Syst Assur Eng Manag 1(4):351–361

Khan MG, Ahmad N, Rafi L (2008) Optimal testing resource allocation for modular software based on a software reliability growth model: a dynamic programming approach. In: 2008 International conference on computer science and software engineering. IEEE, pp 759–762

Kumar C, Yadav DK (2017) Software defects estimation using metrics of early phases of software development life cycle. Int J Syst Assur Eng Manag 8(4):2109–2117

Lo J-H, Kuo S-Y, Lyu MR, Huang C-Y (2002) Optimal resource allocation and reliability analysis for component-based software applications. In: Proceedings 26th annual international computer software and applications, 2002. IEEE, pp 7–12

Lyu MR, Rangarajan S, Van Moorsel AP (2002) Optimal allocation of test resources for software reliability growth modeling in software development. IEEE Trans Reliab 51(2):183–192
Nasar M, Johri P, Chanda U (2014) Software testing resourcc allocation and release time problem: a review. Int J Mod Educ Comput Sci 6(2):48
O'Hagan M (1990) A fuzzy neuron based upon maximum entropy ordered weighted averaging. In: International conference on information processing and management of uncertainty in knowledge-based systems, 1990. Springer, pp 598–609
Ohtera H, Yamada S (1990) Optimal allocation and control problems for software-testing resources. IEEE Trans Reliab 39(2):171–176
Okamura H, Dohi T (2018) Optimizing testing-resource allocation using architecture-based software reliability model. J Optim
Pietrantuono R, Russo S, Trivedi KS (2010) Software reliability and testing time allocation: an architecture-based approach. IEEE Trans Softw Eng 36(3):323–337
Rochimah S (2013) Integration of DEMATEL and ANP methods for calculate the weight of characteristics software quality based model ISO 9126. In: 2013 International conference on information technology and electrical engineering (ICITEE). IEEE, pp 143–148
Sadehnezhad F, Zaranejad M, Gheitani A (2014) Using combinational method DEMATEL and ANP with fuzzy approach to evaluate business intelligence performance. Eur Online J Nat Soc Sci Proc 2(3):1374–1386
Seyed-Hosseini SM, Safaei N, Asgharpour M (2006) Reprioritization of failures in a system failure mode and effects analysis by decision making trial and evaluation laboratory technique. Reliab Eng Syst Saf 91(8):872–881
Si S-L, You X-Y, Liu H-C, Zhang P (2018) DEMATEL technique: a systematic review of the state-of-the-art literature on methodologies and applications. Math Probl Eng
Su C-M, Horng D-J, Tseng M-L, Chiu AS, Wu K-J, Chen H-P (2016) Improving sustainable supply chain management using a novel hierarchical grey-DEMATEL approach. J Clean Prod 134:469–481
Tsai W-H, Chou W-C (2009) Selecting management systems for sustainable development in SMEs: a novel hybrid model based on DEMATEL, ANP, and ZOGP. Expert Syst Appl 36(2):1444–1458
Tzeng G-H, Chiang C-H, Li C-W (2007) Evaluating intertwined effects in e-learning programs: a novel hybrid MCDM model based on factor analysis and DEMATEL. Expert Syst Appl 32(4):1028–1044
Wu W-W, Lee Y-T (2007) Developing global managers' competencies using the fuzzy DEMATEL method. Expert Syst Appl 32(2):499–507
Yager RR (1993) Families of OWA operators. Fuzzy Sets Syst 59(2):125–148
Yamada S, Ichimori T, Nishiwaki M (1995) Optimal allocation policies for testing-resource based on a software reliability growth model. Math Comput Model 22(10–12):295–301
Zhou Q, Huang W, Zhang Y (2011) Identifying critical success factors in emergency management using a fuzzy DEMATEL method. Saf Sci 49(2):243–252

Modeling Allocation Problem for Software with Varied Levels of Fault Severity

Gurjeet Kaur

Abstract The resource allocation problem considered in this chapter focuses on the objective of maximization of removal of fault content from a modular software. The allocation problem accounts the view wherein there is an ordering of the severity of the faults on continuity of levels varying from the lowest level to the highest level; lowest representing the easiest traceable and detectable fault in terms of time and resources; while the highest severity level indicating a long time gap and more resources for getting traced and rectified. For modeling the allocation problem, aid of Software Reliability Growth Models is taken. The proposed Software Reliability Growth model takes into consideration two important factors. First, the effect of time distribution of testing-resource function is considered as reliability growth curve depends strongly on this factor. Second, the growth model incorporates a novel idea of faults being categorized under varied levels (1-Level being simplest severity level and k-Level being hardest severity level) of fault severity. After the parameter estimation and analyses of the goodness of fit criterions; the proposed testing resource, varied severity fault modeling framework is used in resource allocation problem. The allocation problem is subjected to availability of resources and budget with an aspired level of reliability for each module. Also, for devising the optimal allocation problem, it is taken that the cost of removing the fault from each module is dependent on its severity. The formulated problem is a complex nonlinear programming problem and is solved by a meta-heuristic technique of genetic algorithm. Numerical illustrations are also taken in the chapter. Managerial implications, conclusion and limitations of the proposed allocation modeling are highlighted at the end of the chapter.

Keywords Resource allocation · Fault severity · Modular software · Software reliability growth model

G. Kaur (✉)
Shaheed Sukhdev College of Business Studies, University of Delhi, Delhi, India
e-mail: gurjeetkaur@sscbsdu.ac.in

A. G. Aggarwal et al. (eds.), *Optimization Models in Software Reliability*,
Springer Series in Reliability Engineering,
https://doi.org/10.1007/978-3-030-78919-0_11

1 Introduction

Software Reliability rests on the concept of error, fault, and failure. Error is the cause of the fault. Fault implies that error is lying in the code, but unless and until it is executed, it doesn't turn into failure. And, failure refers to the case when the expected outcome doesn't come as per the input test case (Pham 2006). During the execution of testing of software, the emphasis is on digging out maximum faults and rectifying them so that the commitment of reliable software is well accomplished by the software developer. So considering this objective of software developers, the chapter focuses on the modeling of the optimization problem which maximizes the total faults removed from a modular software system.

The modeling of optimization problem grounds on the fact that each fault lying in the code of the software is not equally severe; that is to say that there may be faults that are easily traceable to time or resources, while others may not be that simple to get detected and resolved. This concept of severity of faults in software reliability growth models (SRGMs) is modeled in literature by putting faults under discrete groups of simple, hard and complex faults (Aggarwal et al. 2011; Kapur et al. 2010a, b). However one can view the severity based on continuity of levels from simplest to hardest. This chapter takes into account this view wherein there is an ordering of the severity of the faults varying from the lowest level to the highest level; lowest representing the easiest traceable and detectable fault in terms of time and fault of highest severity level indicates a long time gap of getting traced and rectified. This varied level of severity of faults is modeled for exponential SRGM.

With the modeling of the Non-Homogenous Poisson Process (NHPP) based Software Reliability Growth Model by Goel and Okumoto (1979), several SRGMs are modeled till date. One line of modeling SRGMs relates fault detection concerning the testing time consumed (Musa 1975; Yamada and Osaki 1987; Kapur and Garg 1990; Huang 2005; Aggarwal et al. 2019). But the finding of faults may be narrowly related to the amount of effort or resources spent on testing and not just the amount of testing time elapsed. Kapur et al. (2007) cited that the resources consumed can be personnel based like manpower or it can be process based like CPU time. And because of the complex dependence of testing-effort, SRGMs which implicitly try to relate fault detection to testing-effort through the elapsed testing time may not adequately characterize the fault detection process. Musa and Okumoto (1984), Trachtenberg (1990), Musa et al. (1987) and Obha (1984a) showed that the effort or testing resource index is a better exposure indicator for software reliability modeling than calendar time because of the shape of the observed reliability growth curve depends strongly on the time distribution of the testing-resource too. As a result, another line of modeling SRGMs relates to testing resource-dependent SRGMs (Huang 2005; Kapur et al. 2011). This chapter too takes into consideration testing resource-dependent SRGM with faults at varying severity levels.

This testing resource SRGM is then employed for modular software systems wherein the huge complex software consisting of thousands of lines of code is broken into small packets of programs called modules. When such modular software is tested then they are tested in three stages. The first stage is unit testing or module testing, the second is integration testing and the third stage is system testing. In the first stage of this testing process, one of the primary concerns is to allocate resources to modules in an effective manner so that the concerned aim is met. These resource allocation problems are extensively modeled in software reliability. A detailed literature work on the resource allocation problem is presented in the next section of the chapter. We are modeling one such resource allocation problem wherein the aim is to maximize the total faults removed in a modular software system. This optimal resource allocation problem is subject to the availability of total testing resource expenditure and budget constraints. Also, it is expected that a certain reliability level is achieved for each module while determining the effective allocation of resources to the modules. The formulated problem is a complex nonlinear programming problem that is therefore solved by a meta-heuristic technique of genetic algorithm (Sastry 2007).

So, with the motive of modeling an optimal testing resource allocation problem considering the varied severity level of faults in the software, the chapter workings are highlighted as follows-

a. Modeling Time-Dependent SRGM with faults of varied severity level (based on Model framework given by Shatnawi and Kapur (2008).
b. Modeling Resource-Dependent SRGM with faults of varied severity level.
c. Modeling Maximization of total fault removal from modular software constrained to budget, total resource expenditure available, and aspired reliability level.

The chapter is organized as follows-Sect. 2 highlights the related research literature work. Section 3 considers the modeling framework of SRGM and the resource allocation optimization problem. Section 4 briefs on the procedure adopted for model validation and solving optimal allocation problem formulated. In Sect. 5 numerical illustration of validation of model on real data example and numerical illustration on framed data for the optimization problem is taken. The chapter brings to a close with Theoretical and Managerial Implications of the study, conclusion, and, limitations and future scope of the study with Sects. 6, 7, and 8 respectively.

2 Related Research Work

2.1 NHPP Based SRGMs

Software reliability growth models are used to evaluate modular software quantitatively and also to forecast the reliability of each of the modules during modular testing. Various SRGMs, which considers the number of failures (faults recognized) and the execution time (CPU time/Calendar time) have been talked about in the literature (Musa and Okumoto 1984; Kapur et al. 1999, 2011; Pham 2006). At the outset, in 1979, an SRGM was proposed by Goel and Okumoto where reliability growth was observed under a perfect debugging environment. With this advancement, plethora of SRGMs now exists with modifications in the assumption of the Goel and Okumoto model (Obha and Yamada 1984; Kapur and Garg 1990; Kapur et al. 2012).

SRGMs with delay in the fault removal process

The above stated SRGMs speak about the immediate removal of a fault when it is found, but in fact, circumstances are very far from this. The removal of any fault is entirely dependent on several circumstances such as fault sophistication, methodologies for testing, efforts for testing, environment, etc. So, because of these variables, there could be a delay in removing the fault. Yamada et al. (1985) proposed a modified exponential model assuming that the software contains two types of faults namely, simple and hard. An S-shaped Reliability Growth Model with Two Types of Errors was proposed by Kareer et al. (1990). A three-stage Erlang model proposed by Kapur and Garg (1990) was used to model the third type of fault. Shatnawi and Kapur (2008) later proposed a generalized model based on the classification of the faults in the software system according to their removal complexity. Aggarwal et al. (2012) proposed a generalized framework for software reliability growth modeling concerning testing effort expenditure considering the faults types as simple, hard, and complex. Zhu and Pham (2018) recently conducted experiments on two forms of faults (dependent and independent) and showed that for both styles, certain faults are not removable in the removal process. For the fault removal process, Aggarwal et al. (2018) proposed a discrete software reliability growth model (SRGM) where faults present in the software are not of the same form and can be categorized as easy and hard faults depending on the effort and time taken to remove them. This chapter proposes testing resource-dependent SRGM with faults at varying severity levels (1-Level to k-Level).

2.2 *Resource Allocation Problems*

Many authors have investigated the problem of resource allocation in software reliability (Ohetera and Yamda 1990, Kapur et al. 2011; Huang et al. 2004; Aggarwal et al. 2012; Kaur et al. 2017). The main purpose of these resource allocation problems varied from the minimization of software development cost when the number of remaining faults and the desired reliability objective is given, minimizing the mean number of remaining faults in the software modules when the amount of available testing resources is specified and vice versa, to maximizing the remaining number of fault in modular software under budget and availability of resources constraint. Xie and Yang (2001) studied the problem of optimal testing-time allocation for modular software systems intending to maximize the operational reliability of a simple software system. Optimal Testing Resource Allocation for Modular Software Considering Cost, Testing Effort and Reliability using Genetic Algorithm was proposed by Kapur et al. (2009), Aggarwal et al. (2010) proposed optimization problem of simultaneously allocating of time and resources Further, Aggarwal et al. (2011)

developed the Testing Time and Resource Dependent Two Dimensional Software Reliability Model for Faults of Simple, Hard, and Complex Severity and gave the Related Optimal Allocation Problem. This chapter proposes a resource allocation problem where faults are modeled with SRGMs taking into consideration varied levels of severity of faults.

The modeling framework of the SRGM and optimal resource allocation problem is presented in the next section.

3 Modeling Framework

This section takes into consideration the modeling framework of resource allocation problems considering faults of varying severity levels. For this, first, time-dependent SRGM with faults of varied severity level is provided. This modeling is based on testing time SRGM with faults of varying severity levels by Shatnawi and Kapur (2008). Then on the basis of this model, testing resource dependent SRGM with faults of varied severity level is proposed.

3.1 Time-Dependent SRGM with Faults of Varied Severity Level

The notations used in the modeling framework for time dependent SRGM with faults of varied severity level are:

k	Total Number of Levels of Fault Severity
j	Index counter for levels of severity of faults (j = 1, 2, ... k)
aj	Constant, representing the number of faults lying dormant at the beginning of testing at jth level of fault severity
a	Constant, representing the total number of faults lying dormant at the beginning of testing
i	Step counter at jth level of fault severity
$m_{ji}(t)$	Mean number of fault removed for level j at step i
$m_j(t)$	The cumulative number of faults removed by time t at jth severity level
bj	The cumulative number of faults removed by time t at jth severity level
j	Constant, fault detection rate per remaining fault at the *j*th level of fault severity
pj	Proportion of total number of faults lying dormant at the beginning of testing at *j*th level of fault severity
$m(t)$	Cumulative number of faults removed by time t

Apart from general assumptions of Non-Homogeneous Poisson Process SRGM models (Pham 2006), the proposed model is based upon the following basic assumptions:

1. A software system is subject to failure during execution caused by faults remaining in the system.
2. The number of faults detected at any time instant is proportional to the remaining number of faults in the software.
3. The faults existing in the software are of varying severity levels. The level varies from 1 to k; 1 being simplest to k being hardest.
4. The fault removal process is perfect and failure observation/fault isolation/ fault removal rate is constant.
5. Each time a failure occurs, an immediate effort takes place to decide the cause of the failure to remove it. The time delay between the failure observation and its subsequent fault removal is assumed to represent the severity of the faults. The more severe the fault more is the time delay.

Taking into consideration these assumptions, in this SRGM, the testing phase consists of k different level of processing the faults–

1-Level: In this level, the fault is simplest to debug and there is a negligible time gap in detection and correction of the fault. Therefore, one step differential equation resulting from assumption 2 at 1-Level is–

$$\frac{d}{dt}m_{11}(t) = b_1[a_1 - m_{11}(t)] \tag{1}$$

Solving this linear differential equation with the initial condition $m_{11}(0) = 0$ we obtain

$$m_1(t) = m_{11}(t) = a_1(1 - e^{-b_1 t}) \tag{2}$$

(Fault severity 1-Level)

This model is known as Goel-Okumoto Model (1979).

2-Level: In this, the level of fault severity raises high from 1-level in debug and removal. So there is a gap time gap between detection and execution. Considering this, the two step processed differential equation for 2-level will be

$$\frac{d}{dt}m_{21}(t) = b_2[a_2 - m_{21}(t)] \tag{3}$$

and

$$\frac{d}{dt}m_{22}(t) = b_2[m_{21}(t) - m_{22}(t)] \tag{4}$$

Solving this linear differential equation with the initial condition $m_{21}(0) = 0, m_{22}(0) = 0$ we obtain

$$m_2(t) = m_{22}(t) = a_2(1 - (1 + b_2 t)e^{-b_2 t}) \tag{5}$$

(Fault severity 2-Level)

This model form is due to Yamada et al. (1983).

3-Level: The level of fault severity further raises high from the second level to the third level. So there is more time gap between detection and execution. Considering this, the three step processed differential equation for 3-level 3 will be

$$\frac{d}{dt}m_{31}(t) = b_3[a_3 - m_{31}(t)], \tag{6}$$

$$\frac{d}{dt}m_{32}(t) = b_3[m_{31}(t) - m_{32}(t)] \tag{7}$$

and

$$\frac{d}{dt}m_{33}(t) = b_3[m_{32}(t) - m_{33}(t)] \tag{8}$$

Solving this linear differential equation with the initial condition $m_{31}(0) = 0, m_{32}(0) = 0, m_{33}(0) = 0$ we obtain

$$m_3(t) = m_{33}(t) = a_3(1 - (1 + b_3t + \frac{{b_3}^2t^2}{2!})e^{-b_3t}) \tag{9}$$

(Fault severity 3-Level)

The existing literature of SRGM term these three phases together as SRGM with faults classification as simple, hard, and complex faults (Kapur et al. 1995).

4-Level: Further extending the level of severity from 3-level to 4-level, there is much gap being observed between detection and removal of fault; thereby the four step processed differential equation for 4-level will be-

$$\frac{d}{dt}m_{41}(t) = b_4[a_4 - m_{41}(t)], \tag{10}$$

$$\frac{d}{dt}m_{42}(t) = b_4[m_{41}(t) - m_{42}(t)], \tag{11}$$

$$\frac{d}{dt}m_{43}(t) = b_4[m_{42}(t) - m_{43}(t)] \tag{12}$$

and

$$\frac{d}{dt}m_{44}(t) = b_4[m_{43}(t) - m_{44}(t)] \tag{13}$$

Solving this linear differential equation with the initial condition

$m_{41}(0) = 0, m_{42}(0) = 0, m_{43}(0) = 0, m_{44}(0) = 0$ we obtain

$$m_4(t) = m_{44}(t) = a_4(1 - (1 + b_4t + \frac{{b_4}^2t^2}{2!} + \frac{{b_4}^3t^3}{3!})e^{-b_4t}) \tag{14}$$

(Fault severity 4-Level)

Going with a similar concept we can have k level of severity being modeled as-

k-Level:

$$m_k(t) = m_{kk}(t) = a_k(1 - (1 + b_kt + \frac{{b_k}^2t^2}{2!} + \frac{{b_k}^3t^3}{3!} + \frac{{b_k}^4t^4}{4!} + \cdots + \frac{{b_k}^{k-1}t^{k-1}}{(k-1)!})e^{-b_kt}) \tag{15}$$

(Fault severity k-Level)

Incorporating the k-level of severity, the time dependent exponential SRGM with varied severity levels of fault is given by-

$$\begin{aligned} m(t) =& m_1(t) + m_2(t) + m_3(t) + m_4(t) + \ldots m_k(t) = ap_1\left(1 - e^{-b_1t}\right) \\ &+ ap_2\left(1 - (1 + b_2t)e^{-b_2t}\right) + ap_3\left(1 - \left(1 + b_3t + \frac{{b_3}^2t^2}{2!}\right)e^{-b_3t}\right) \\ &+ ap_4(1 - (1 + b_4t + \frac{{b_4}^2t^2}{2!} + \frac{{b_4}^3t^3}{3!})e^{-b_4t}) + \cdots + ap_k(1 - (1 + b_kt + \frac{{b_k}^2t^2}{2!} \\ &+ \frac{{b_k}^3t^3}{3!} + \frac{{b_k}^4t^4}{4!} + \cdots + \frac{{b_k}^{k-1}t^{k-1}}{(k-1)!})e^{-b_kt}) \end{aligned} \tag{16}$$

(Time-Dependent SRGM with faults of varying severity)

Where a is the total fault content in the software and$a_1 = ap_1; a_2 = ap_2; a_3 = ap_3; a_4 = ap_4; \ldots a_k = ap_k; \sum_{j=1}^{k} p_j = 1$.

3.2 *Testing Resource Dependent SRGM with Faults of Varied Severity Level*

The notations used in the modeling framework for testing resource dependent SRGM with faults of varied severity level are:

k	Total Number of Levels of Fault Severity
j	Index counter for levels of severity of faults (j = 1, 2, … k)
aj	Constant, representing the number of faults lying dormant at the beginning of testing at jth level of fault severity
a	Constant, representing the total number of faults lying dormant at the beginning of testing
i	Step counter at jth level of fault severity
W (t)	Cumulative testing effort in the time interval (0, t]

w (t)	Testing resource intensity; $w(t) = \frac{d}{dt} W(t)$
$m_j(W(t))$	The cumulative number of faults removed with consumption of time–dependent testing resource by time t at jth severity level
bj	Constant, fault detection rate per remaining fault at the *j*th level of fault severity
pj	Proportion of total number of faults lying dormant at the beginning of testing at *j*th level of fault severity
$m(W(t))$	Cumulative number of faults removed with consumption of time–dependent testing resource by time t

Apart from assumptions stated in Sect. 3.1, testing resource-dependent SRGM with varied faults of varied severity level rests on:

1. The fault isolation/removal rate with respect to testing effort intensity is proportional to the number of observed failures.
2. To describe the behavior of testing resource Exponential, Rayleigh, and Weibull function has been used given by the following equations respectively-

$$\boldsymbol{W(t) = \alpha \cdot \left[1 - \exp(-\beta \cdot t)\right]} \tag{17}$$

(Exponential function)

$$\boldsymbol{W(t) = \alpha \cdot \left(1 - \exp\left[-\frac{\beta}{2} \cdot t^2\right]\right)} \tag{18}$$

(Rayleigh function)

$$\boldsymbol{W(t) = \alpha \cdot \left(1 - \exp\left[-\beta \cdot t^{\gamma}\right]\right)} \tag{19}$$

(Weibull function)

Yamada et al. (1986) presented a testing resource-dependent SRGM based on the GO model assuming that the number of faults detected in $(t, t + \Delta t)$ per unit testing effort expenditure is proportional to the remaining faults given as:

$$\frac{\frac{dm(t)}{dt}}{w(t)} = b\,[a - m(t)] \tag{20}$$

Solving the above equation with the initial condition $w(0) = 0$ and $m(0) = 0$ we get

$$m\,(W(t)) = a\,(1 - e^{-bW(t)}) \tag{21}$$

Based on this concept and considering the time dependent SRGM with faults of varied severity level given in Sect. 3.1, we have testing resource dependent SRGM with faults of varied (i.e. k) severity level given is given by-

$$
\begin{aligned}
m(W(t)) &= m_1(W(t)) + m_2(W(t)) + m_3(W(t)) + m_4(W(t)) + \ldots m_k(W(t)) \\
&= ap_1\left(1 - e^{-b_1 W(t)}\right) + ap_2\left(1 - (1 + b_2 W(t))e^{-b_2 W(t)}\right) \\
&+ ap_3\left(1 - \left(1 + b_3 W(t) + \frac{b_3{}^2 W(t)^2}{2!}\right)e^{-b_3 W(t)}\right) \\
&+ ap_4(1 - (1 + b_4 W(t) + \frac{b_4{}^2 W(t)^2}{2!} + \frac{b_4{}^3 W(t)^3}{3!})e^{-b_4 W(t)}) + \ldots \\
&+ ap_k(1 - (1 + b_k W(t) + \frac{b_k{}^2 W(t)^2}{2!} + \frac{b_k{}^3 W(t)^3}{3!} + \frac{b_k{}^4 W(t)^4}{4!} + \cdots + \frac{b_k{}^{k-1} W(t)^{k-1}}{(k-1)!})e^{-b_k W(t)})
\end{aligned} \tag{22}
$$

(Testing Resource dependent SRGM with faults of varied severity level)

Where a is the total fault content in the software and$a_1 = ap_1$; $a_2 = ap_2$; $a_3 = ap_3$; $a_4 = ap_4$; $\ldots a_k = ap_k$; $\sum_{j=1}^{k} p_j = 1$.

3.3 *Modeling Resource Allocation Problem for Maximizing the Total Fault Removal from Software*

The notations that are used for the resource allocation problem are:

N	Number of modules
K	Level of severity
J	1, 2, 3 … k; Simplest faults i.e. 1-Level to Hardest fault i.e. k–Level
I	Module, 1, 2, … N
$m_i(W_t)$	Mean value function for *i*th module
b_{ji}	Constant fault detection rate for *j*th fault type in the *i*th module
a_{ji}	Constant, representing the number of *j*th fault type lying dormant in the *i*th module at the beginning of testing,
c_{ji}	Cost of removing *j*th fault from *i*th module
W_i	Testing effort for *i*th module
R_i	Reliability of *i*th module
B	Total cost of removing different types of faults
W	Total testing resource expenditure

The resource allocation problem considered in this chapter focuses on the objective of maximization of removal of fault content from a modular software. For devising the optimal allocation problem it is taken that the cost of removing the fault from each module is dependent on its severity. That is to say, if the fault is of 1-Level then the cost of removing it will be less than the cost of removing the fault at 2-Level and so on; therefore the cost will be highest for the *k*th level of fault severity.

Therefore, the problem of maximizing the faults of each of the N independent modules whose failure process is modeled by the SRGM proposed in the Sect. 3.2 such that reliability of each module is at least R_0 is formulated as:

Maximize

$$
\begin{aligned}
m(W) = \sum_{i=1}^{N} m_i(W_i) = \sum_{1=1}^{N} a_{1i}\left(1 - e^{-b_{1i}W_i}\right) + \sum_{i=1}^{N} a_{2i}\left(1 - (1 + b_{2i}W_i)e^{-b_{2i}W_i}\right) \\
+ \sum_{i=1}^{N} a_{3i}\left(1 - \left(1 + b_{3i}W_i + \frac{{b_{3i}}^2{W_i}^2}{2!}\right)e^{-b_{3i}W_i}\right) \\
+ \sum_{i=1}^{N} a_{4i}\left(1 - \left(1 + b_{4i}W_i + \frac{{b_{4i}}^2{W_i}^2}{2!} + \frac{{b_{4i}}^3{W_i}^3}{3!}\right)e^{-b_{4i}W_i}\right) + \ldots \\
+ \sum_{i=1}^{N} a_{ki}(1 - (1 + b_{ki}W_i + \frac{{b_{ki}}^2{W_i}^2}{2!} + \frac{{b_{ki}}^3{W_i}^3}{3!} + \frac{{b_{ki}}^4{W_i}^4}{4!} + \cdots + \frac{{b_{ki}}^{k-1}{W_i}^{k-1}}{(k-1)!})e^{-b_{ki}W_i})
\end{aligned} \tag{23}
$$

Subject to:

$$
\begin{aligned}
\sum_{i=1}^{N} (C_{1i}m_{1i}(W_i) + C_{2i}m_{2i}(W_i) + C_{3i}m_{3i}(W_i) + C_{4i}m_{4i}(W_i) + \ldots \\
+ C_{ki}m_{ki}(W_i)) \le B\mathrm{i} = 1, 2\ldots \mathrm{N}
\end{aligned} \tag{24}
$$

$$\sum_{i=1}^{N} W_i \le W \quad i = 1, 2, \ldots N \tag{25}$$

$$R_i \ge R_0 \quad i = 1, 2, \ldots N \tag{26}$$

$$W_i \ge 0 \quad i = 1, 2, \ldots N \tag{27}$$

(-Resource Allocation Problem with Faults of Varied Severity Levels)

For evaluating the reliability, we are using the measure of defining software reliability at time t as given by Huang et al. (2004); that states "the ratio of the cumulative number of detected faults at time t to the expected number of initial fault content of the software.", i,e. the reliability expression for the solving Optimal Allocation problem is taken by the expression of the form

$$\text{Reliability} = R(W(t)) = \frac{m(W(t))}{a} \tag{28}$$

The above-formulated problem is a complex nonlinear problem otherwise not solvable by analytical mathematical tools and is solved by the Genetic Algorithm described in the next section. Considering the case of five modules we have solved the above problem by taking 4-levels and 5-levels of fault severity via numerical illustration shown in Sect. 5.2.

4 Solution Methodologies

4.1 Parameter Estimation

It should be noted that software reliability growth models become useful only when if it is possible to estimate their parameters. For validating the model described in the above section, parameter estimation is done on actual software failure data. The proposed models presented in the chapter are non-linear and presents extra problems in estimating the parameters. SPSS Regression Models enables the user to apply more sophisticated models to the data using its wide range of nonlinear regression models. For the estimation of the parameters of the proposed models, the Method of Least Square (Non-Linear Regression method) has been used. To determine how good the model is, the goodness of fit measures of Mean Square Error (MSE) and Coefficient of Multiple Determination (R^2) is used. R^2 measures the percentage of the total variation about the mean accounted for by the fitted curve. It ranges in value from 0 to 1. Small values indicate that the model does not fit the data well. The larger, the better the model explains the variation in the data (Kapur et al. 1999). The lower MSE indicates less fitting error, thus better goodness of fit.

4.2 Genetic Algorithm

We are using Genetic Algorithm (GA) to solve the resource allocation problem presented in this research paper. GA stands up as a powerful tool for solving search & optimization problems (Goldberg 1989). The reason behind choosing the genetic algorithm as the solving tool is that the optimal effort allocation problem is of complex nonlinear nature.

GA always takes into consideration a population of solutions. Since GA can be applied to solve any kind of problem, therefore, before using GA's, there is no particular requirement on the problem. Genetic Algorithm begins with the initial population of solutions, which are represented as chromosomes. For the total testing time given as W, GA generates the initial population randomly. It initializes to random values within the limits of each variable. When we talk about the Fitness of a Chromosome, then, it implies measuring the quality of the solution, which it represents in terms of several optimization parameters of the solution. A fit chromosome means a better solution. The objective of allocation problem along with the penalties of the constraints, which are not met, forms the fitness function of GA methodology used in this chapter. Selection refers to the process of choosing two parents out of the population for crossover. A higher fitness function suggests that the chance an individual has to be selected is more. Crossover refers to the process of considering two-parent solutions and producing two similar chromosomes. This is done by swapping sets of genes, with the hope that at least one child will have genes that improve its fitness. Mutation helps in preventing the algorithm from getting trapped in a local minimum.

Further, mutation serves the role of recovering the lost genetic materials, along with randomly disturbing genetic information.

GA is a meta-heuristic iterative algorithm. The procedure taken for solving the optimal resource allocation problem is as follows (Sastry 2007; Aggarwal et al. 2012).

a. Start the genetic algorithm procedure by coding the objective function along with the constraints stated in Sect. 3.3 (Eqs. (23) to (27)) as a Matlab file.
(Step No. 1)
b. Generate a random population of chromosomes. This is obtained by the heuristic method of GA by assigning random values within the limits specified in the constraints of allocation problem.
(Step No. 2)
c. For each chromosome in the population, evaluate the fitness function coded in Step No.1
(Step No. 3)
d. By repeating the following steps, create a new population until the new population is complete:
   (i) From a population, select two parent chromosomes randomly as per their fitness function. This is called as selection procedure in GA.
   (ii) With a crossover probability, cross-over the parents, to form new offspring (children). In case, no crossover is performed, then, offspring is the exact copy of parents. This crossover process is vital for producing the next iteration of GA.
   (iii) With a mutation probability, at each locus (position in chromosome) mutate the offspring. Mutation probability is kept low in GA.
   (iv) Place new offspring in the new population. A new iteration is on its way by the process of selection, crossover and mutation.
   (v) For the next iterations, use this newly generated population.
   (vi) Test the case that if the termination condition of the iterative process is reached or not. In case the end condition is satisfied (that can be the number of generations satisfied or negligible improvement in fitness function of the subsequent generated populations etc.), stop, and return the best solution in the current population.

(Step No. 4)

e. Continue performing step 3 and Step 4 till an optimal solution is obtained.

(Step No. 5)

With this description, the numerical illustration of the proposed model of SRGM (as given by Eq. (22)) and optimal allocation (given by Eqs. (23) to (28)) is put forward in the next section.

5 Numerical Illustration

In this section, the mathematically derived SRGM with varied levels of fault severity is validated on real data example. Further, the resource allocation optimization problem is solved on a numerically framed case data example.

5.1 *Model Validation*

Software reliability growth models become useful only when if it is possible to estimate their parameters. For validating the model described in Sect. 3, parameter estimation is done on actual software failure data.

The parameter estimation is carried out as on a data set considering three cases which are as follows-

Case 1: It takes into consideration estimation of time dependent SRGM as given by Eq. (16). We have carried out estimation of Eq. (16) by putting k = 1, 2, 3 and 4. In this we have presented a comparative analysis on the basis of level of severity of faults. That is to say we have compared four time dependent SRGMs obtained by Eq. (16) at 1-Level, 2-Level, 3-Level and 4-Level respectively on the Data Set.

Case 2: It takes into consideration estimation of time and resource dependent SRGM with faults of varied severity as given by Eqs. (16) and (22). A comparative analysis on the basis of time and resource dependent SRGM with faults of varied level severity is obtained in this case by putting k = 3 in Eqs. (16) and (22) the data set.

Case 3: It takes into consideration estimation of resource dependent SRGM with faults of varied level severity as given by Eq. (22). We have carried out estimation of Eq. (22) by putting k = 1, 2, 3, 4 and 5 respectively. In this we have presented a comparative analysis on the basis of level of severity of faults. That is to say we have compared five resource dependent SRGMs obtained by Eq. (22) at 1-Level, 2-Level, 3-Level, 4-Level and 5-level respectively on a data set.

We have carried out the parameter estimation on the data set cited in Ohba (1984b) (DS-1). The software was tested for 19 weeks during which 47.65 computer hours were used and 328 faults were removed.

Case 1: Estimation of Time-Dependent SRGM with faults of varied level of severity

We have estimated the dataset by considering time-dependent SRGM, as stated in Sect. 3.1 with levels of fault severity as four levels (i.e. taking k = 4). That is, we are considering that faults are classified from severity 1-level to 4-level; 1-Level being the simplest level of fault to be identified and 4-Level indicating the hardest level of fault detection and removal. The comparative analysis is done concerning levels of severity of faults; i.e. the four software reliability growth models are compared considering the estimation of the dataset as SRGM with only 1 level of fault severity,

SRGM with 2 levels of fault severity, SRGM with 3 levels of fault severity, and SRGM with 4 levels of fault severity.

The parameters are estimated by the non-linear Least Square technique with the aid of the software package of SPSS. The results are shown in Table 1.

The goodness of fit measures used is Mean Square Error (MSE) and Coefficient of multiple determination (R^2). A high value of R^2 and a low value of MSE is an indication of a good fit. The comparison results of the time-dependent SRGMs considering varied levels of fault severity are shown in Table 2 and Figs. 1, 2, 3, 4.

Based on the value of R^2 and MSE it is evident that modeling SRGM on this dataset with 4 levels of fault severity gives better estimation results as compared with 1, 2, or 3 levels of fault severity.

Table 1 Parameter Estimates for DS-1; Time dependent SRGM with varied levels of fault severity ($k = 1$ to 4)

Time dependent SRGM	A	b_1	b_2	b_3	b_4
1-Level of fault severity ($k = 1$)	760.5342	0.0310			
2-Level of fault severity ($k = 2$)	431.2673	0.1375	0.1386		
3-Level of fault severity ($k = 3$)	378.6730	0.5090	0.0630	0.2580	
4-Level of fault severity ($k = 4$)	380.4156	0.3278	0.1882	0.0010	0.3450
Time dependent SRGM		p_1	p_2	p_3	p_4
1-Level of fault severity ($k = 1$)					
2-Level of fault severity ($k = 2$)		0.2709	0.7291		
3-Level of fault severity ($k = 3$)		0.1670	0.0010	0.8320	
4-Level of fault severity ($k = 4$)		0.2480	0.0010	0.0410	0.7100

Table 2 Goodness of fit measures for DS-1; Time dependent SRGM with varied levels of fault severity ($k = 1$ to 4)

SRGM with varied levels of fault severity	R^2	MSE
1 Level of fault severity ($k = 1$)	0.986	158.264
2 Level of fault severity ($k = 2$)	0.991	123.562
3 Level of fault severity ($k = 3$)	0.993	114.702
4 Level of fault severity ($k = 4$)	0.994	109.287

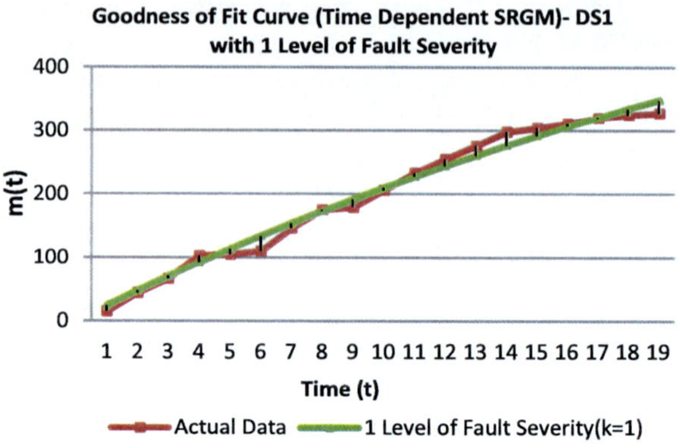

Fig. 1 Goodness of fit curve (Time dependent SRGM)-DS1 with 1 level of fault severity

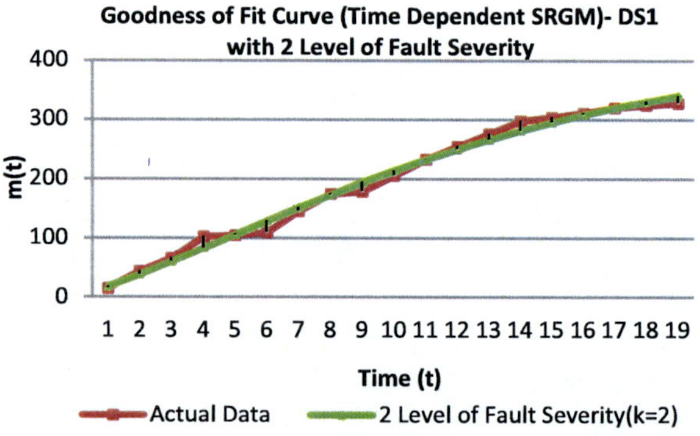

Fig. 2 Goodness of fit curve (Time dependent SRGM)-DS1 with 2 level of fault severity

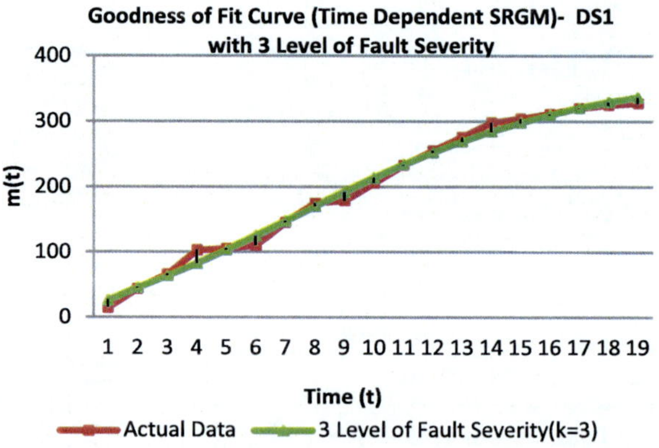

Fig. 3 Goodness of fit curve (Time dependent SRGM)-DS1 with 3 level of fault severity

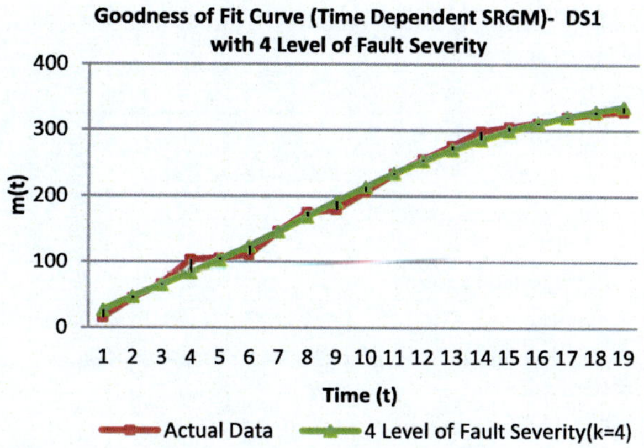

Fig. 4 Goodness of fit curve (Time dependent SRGM)-DS1 with 4 level of fault severity

Case 2: Estimation of Time-Dependent SRGM and Testing Effort Dependent SRGM with faults of varied level of severity

We have estimated the dataset by considering levels of fault severity as three levels (i.e. taking k = 3). That is we are considering that faults are classified from severity 1-level to 3-level; 1 Level being the simplest level of fault to be identified and 4 Level indicating the hardest level of fault detection and removal. The comparative analysis is done based on testing time SRGM and testing resource SRGM, i.e. the models are compared considering the estimation of the dataset as SRGM with 1, 2, and 3 levels of fault severity related to time and testing resource respectively. We have used the exponential distribution function for parameter estimation of the testing resource. The values of the parameters obtained by the non-linear Least Square technique are $\alpha = 503.993$; $\beta = 0.005$. Using these estimates, the parameters of effort based SRGM are obtained. The parameters estimated results are shown in Table 3. The comparison results of the time and resource-dependent SRGM for data set DS-1 considering varied levels of fault are shown in Table 4 and Figs. 5 and 6.

Based on the results of the goodness of fit test for DS-1, it is obtained that the accuracy of estimation of time-dependent SRGM improves with the level of fault severity. This goodness of fit gives better results when testing resource-based SRGM is considered.

Case 3: Estimation of Testing Resource Dependent SRGM with faults of varied level of severity

In this case, to measure the performance of the testing resource model with faults of varying levels of severity given in Sect. 3.2, Eq. (22).

We have estimated the dataset by considering resource-dependent SRGM with levels of fault severity as five levels (i.e. taking k = 5). That is we are considering that faults are classified from severity level 1 to 5; 1-Level being the simplest level of fault to be identified and 5-Level indicating the hardest level of fault detection and removal. The comparative analysis is done with respect to levels of severity of faults;

Table 3 Parameter dstimates for DS-1; testing time SRGM and testing resource SRGM with faults of varied severity level

	a	b_1	b_2	b_3	p_1	p_2	p_3
1-Level of fault severity							
Time-dependent SRGM	760.5342	0.0310					
testing resource dependent SRGM	871.9494	0.0110					
2-Level of fault severity							
Time-dependent SRGM	431.2673	0.1375	0.1386		0.2709	0.7291	
testing resource dependent SRGM	423.2818	0.3380	0.0629		0.0747	0.9253	
3-Level of fault severity							
Time-dependent SRGM	378.6730	0.5090	0.0630	0.2580	0.1670	0.0010	0.8320
testing resource dependent SRGM	384.8269	0.0165	0.4671	0.1031	0.0010	0.1520	0.8470

Table 4 Goodness of fit measures for DS-1; Time versus resource dependent SRGM with faults of 1, 2, and 3 levels of severity

Time-dependent SRGM versus resource dependent SRGM with varied levels of fault severity		MSE	R^2
1 Level of fault severity	Time-dependent SRGM	0.986	158.264
	testing resource dependent SRGM	0.986	157.631
2 Level of fault severity	Time-dependent SRGM	0.991	123.562
	testing resource dependent SRGM	0.991	113.517
3 Level of fault severity	Time-dependent SRGM	0.993	114.702
	testing resource dependent SRGM	0.994	98.335

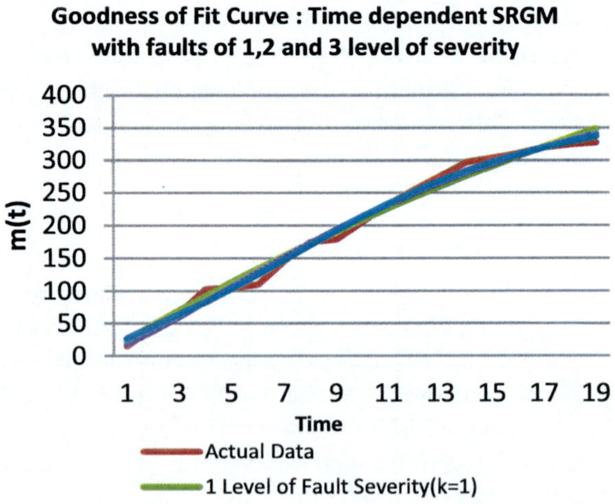

Fig. 5 Goodness of fit curve for DS-1; Time dependent SRGM with faults of 1, 2, and 3 levels of severity

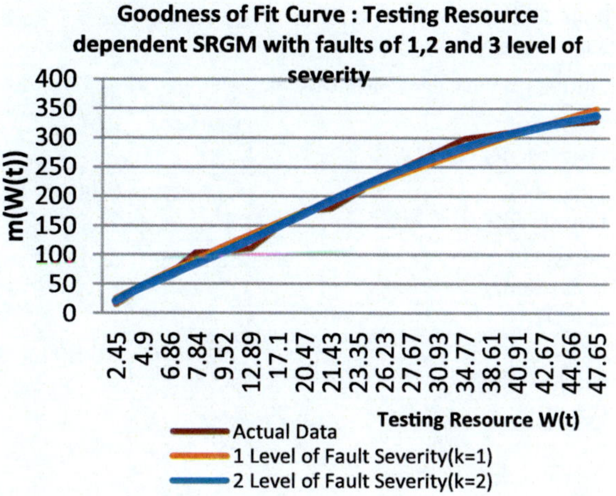

Fig. 6 Goodness of fit curve for DS-1; Testing resource dependent SRGM with faults of 1, 2, and 3 levels of severity

i.e. the models are compared considering the estimation of the dataset as resource-dependent SRGM with only 1 level of fault severity, resource-dependent SRGM with 2 levels of fault severity, resource-dependent SRGM with 3 levels of fault severity, resource-dependent SRGM with 4 levels of fault severity and resource-dependent SRGM with 5 levels of fault severity. The parameters estimation results are shown in Table 5. We have used the Exponential distribution function for parameter estimation of the testing resource. The values of the parameters obtained by the non-linear Least Square technique are $\alpha = 503.993$; $\beta = 0.005$.

The comparison results of the testing resource-dependent SRGMs considering varied levels of fault severity are shown in Table 6 and Figs. 7, 8, 9, 10, 11.

Based on the value of R^2 and MSE with graphical view support, it is evident that testing resource-based SRGMs on this dataset with 5 levels of fault severity gives better estimation results as compared with 1,2 3, or 4 levels of fault severity.

5.2 *Resource Allocation Numerical Illustration*

Case 1: Resource Allocation Problem with 4 levels of Fault Severity

We are taking a framed data case wherein we are taking into consideration that the software consists of five modules. These assumed parameter estimates for each module is shown in Table 7. The total testing resources available is assumed to be 5000 units. The total cost of removing the different types of faults is 40000 units. Also, it is desired that the reliability of each module is at least 0.90.

Using the parameters given in the Table 7, the problem stated in Sect. 3.3, i.e. Resource Allocation problem for Maximizing the Total Fault Removal from Software is considered. The modular software optimization problem is for five modules with 4 levels of Fault Severity in each module was coded in Matlab and solved using

Table 5 Parameter estimates for DS-1; Testing resource dependent SRGM with varied levels of fault severity (k = 1 to 5)

Testing resource dependent SRGM	a	b_1	b_2	b_3	b_4	b_5
1-Level of fault severity (k = 1)	871.9494	0.0110				
2-Level of fault severity (k = 2)	423.2818	0.3380	0.0629			
3-Level of fault severity (k = 3)	384.8269	0.0165	0.4671	0.1031		
4-Level of fault severity (k = 4)	392.9696	0.0010	0.3932	0.0649	0.1421	
5-Level of fault severity (k = 5)	354.5444	0.6071	0.4375	0.5586	0.1774	0.1818
Testing resource dependent SRGM		p_1	p_2	p_3	p_4	p_5
1-Level of fault severity (k = 1)						
2-Level of fault severity (k = 2)		0.0747	0.9253			
3-Level of fault severity (k = 3)		0.0010	0.1520	0.8470		
4-Level of fault severity (k = 4)		0.0676	0.1911	0.0010	0.7403	
5-Level of fault severity (k = 5)		0.0010	0.0010	0.2346	0.0020	0.7614

Table 6 Goodness of fit measures for DS-1; Testing resource dependent SRGM with varied levels of fault severity (k = 1 to 5)

Testing resource dependent SRGM with varied levels of fault severity	R^2	MSE
1 Level of fault severity (k = 1)	0.986	157.631
2 Level of fault severity (k = 2)	0.991	113.517
3 Level of fault severity (k = 3)	0.994	98.335
4 Level of fault severity (k = 4)	0.995	96.099
5 Level of fault severity (k = 4)	0.996	91.671

GA. Several runs of GA were made with Population Size 50, several generations 50, Selection Method as Tournament without Replacement, Crossover Probability as 0.8, and mutation probability as 0.1; and the solution was stabilized with these GA parameters. It is taken that cost of removing Level 1 severe fault is 5 units, Level 2 severe fault is 10 units, Level 3 severe fault is 15 units and Level 4 severe fault is 20 units. The optimal testing time allocation to each type of fault in the module and hence

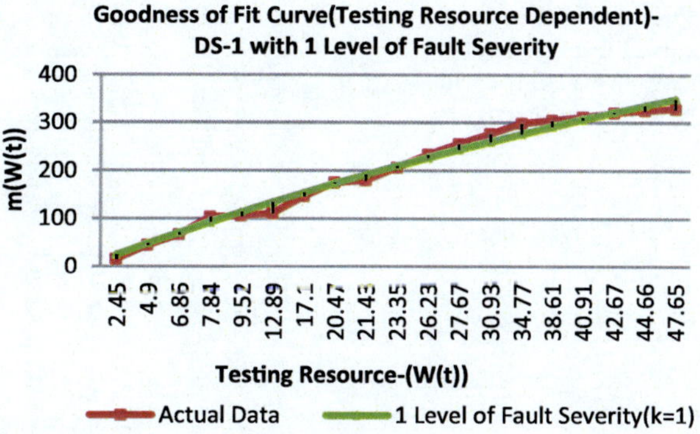

Fig. 7 Goodness of fit curve (Testing resource dependent)—DS-1 with 1 level of fault severity

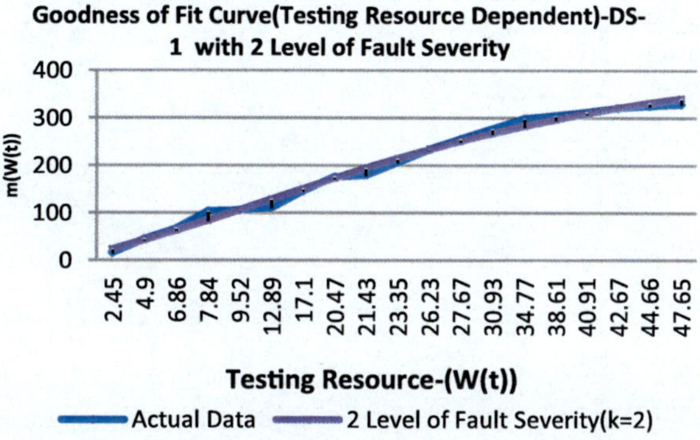

Fig. 8 Goodness of fit curve (Testing resource dependent)—DS-1 with 2 level of fault severity

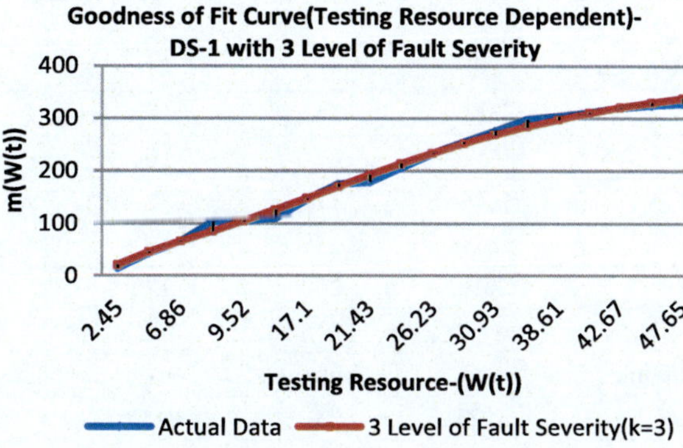

Fig. 9 Goodness of fit curve (Testing resource dependent)—DS-1 with 3 level of fault severity

total fault removed (in integral values) from each module and their corresponding cost of removing is shown in Table 8.

From Table 8, we have that, from the total cost available for removing faults of 40,000 units 32,110 is used in attaining this optimal allocation of resources among modules.

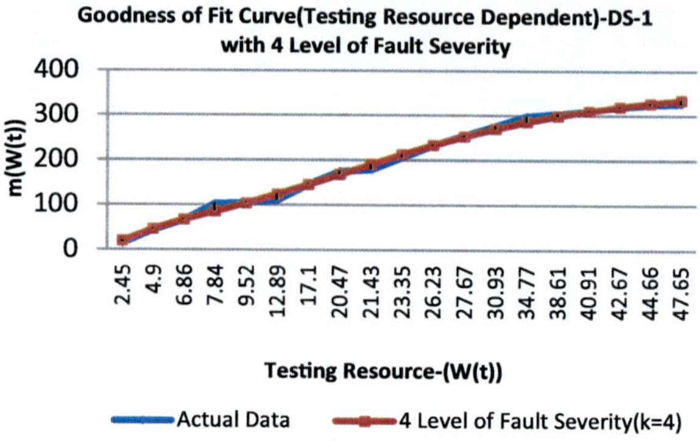

Fig. 10 Goodness of fit curve (Testing resource dependent)—DS-1 with 4 level of fault severity

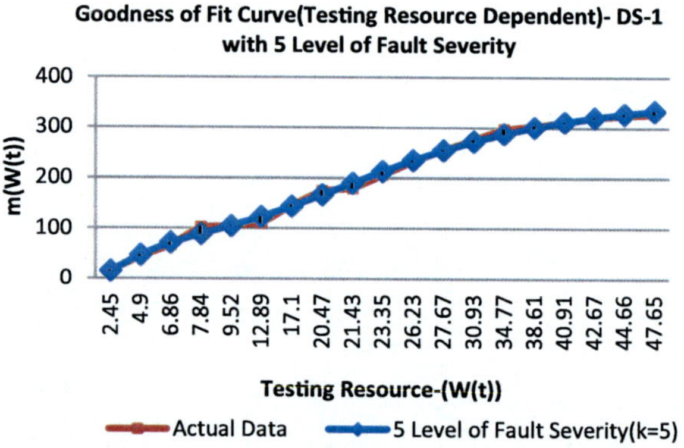

Fig. 11 Goodness of fit curve (Testing resource dependent)—DS-1 with 5 level of fault severity

Table 7 Parameter estimates for resource allocation problem with 4 levels of fault severity in each module

Module	a_1	a_2	a_3	a_4	Total faults
1	313	243	112	58	726
2	332	87	100	210	729
3	135	170	82	178	565
4	145	215	85	62	507
5	301	200	172	34	707
Total	1226	915	551	542	3234
Module	b_1	b_2	b_3	b_4	
1	0.23	0.019	0.11	0.012	
2	0.25	0.072	0.07	0.003	
3	0.39	0.043	0.005	0.012	
4	0.34	0.034	0.17	0.017	
5	0.27	0.018	0.019	0.024	

Table 8 The optimal testing resource allocation with the corresponding cost considering 4 levels of fault severity in each module

Module	m_1	m_2	m_3	m_4
1	313	243	112	57
2	332	87	100	138
3	135	170	48	166
4	145	215	85	61
5	301	199	171	33
Total	1226	914	516	455
Module	W	M	Reliability	Cost of removing faults
1	1118.349	725	0.9986	6815
2	1501.736	657	0.9012	6790
3	616.024	519	0.9185	6415
4	951.62	506	0.9980	5370
5	496.271	704	0.9957	6720
Total	4684	3111		32,110

And from the total of 3234 faults originally present in the software 3111 faults have been removed with the resource limitation of 5000 units. Amongst these total fault removal 1226 out of 1226 (actually present, refer Table 7) 1-Level severity faults, 914 out of 915, 2-Level severity faults, 516 out of 551, 3-Level severity faults, and 455 out of 542, 4-Level faults were removed.

Similarly, another framed data case is given in Table 9 wherein we are taking into consideration that the software consists of five modules with five levels of fault

Table 9 Parameter estimates for resource allocation problem with 5 levels of fault severity in each module

Module	a_1	a_2	a_3	a_4	a_5	Total faults
1	313	243	112	58	75	801
2	332	87	100	210	47	786
3	135	170	82	178	98	663
4	145	215	85	62	101	608
5	301	200	172	34	54	761
Total	1226	915	551	542	375	3609
Module	b_1	b_2	b_3	b_4	b_5	
1	0.23	0.019	0.11	0.012	0.002	
2	0.25	0.072	0.07	0.003	0.013	
3	0.39	0.043	0.005	0.012	0.08	
4	0.34	0.034	0.17	0.017	0.007	
5	0.27	0.018	0.019	0.024	0.019	

severity. The total testing resources available is assumed to be 6000 units. The total cost of removing the different types of faults is 41000 units. Also, it is desired that the reliability of each module is at least 0.90.

Using the parameters given in the Table 9, several runs of GA were made with Population Size 50, several generations 110, Selection Method as Tournament without Replacement, Crossover Probability as 0.9, and mutation probability as 0.1; and the solution was stabilized with these GA parameters. It is taken that cost of removing 1-Level severe fault is 5 units, 2-Level severe fault is 10 units, 3-Level severe fault is 15 units, 4-Level severe fault is 20 units and 5-Level severe fault is 25 units.

The optimal testing time allocation to each type of fault in the module and hence total fault removed (in integral values) from each module and their corresponding cost of removing is shown in Table 10.

From Table 10, we have that, from the total cost available for removing faults of 41,000 units 39,785 is used in attaining this optimal allocation of resources among modules. And from the total of 3609 faults originally present in the software 3418 faults have been removed with the resource limitation of 6000 units. Amongst these total fault removal, 1226 out of 1226 (actually present, refer Table 7) 1-Level severity faults, 910 out of 915, 2-Level severity faults, 515 out of 551, 3-Level severity faults, 469 out of 542, 4-Level severity faults, and 298 out of 375, 5-Level severity faults were removed.

Table 10 The optimal testing resource allocation with the corresponding cost considering 5 levels of fault severity in each module

Module	m_1	m_2	m_3	m_4	m_5
1	313	243	112	58	40
2	332	87	100	146	46
3	135	170	56	172	98
4	145	215	85	61	75
5	301	195	162	32	39
Total	1226	910	515	469	298
Module	W	m	Reliability	Cost of removing faults	
1	2441.815	766	0.9563	7835	
2	1583.2	711	0.9162	8100	
3	706.061	631	0.9517	9105	
4	897.874	581	0.9556	7245	
5	322.136	729	0.9580	7500	
Total	5951.086	3418		39,785	

6 Theoretical and Managerial Implications

The economic implications of software reliability are linked to the limited amount of time and money available to the testing team. These annoying realities of restricted resource supply and deadlines for delivering the software on time appear to compress the process of testing. Therefore to bring out the program on schedule and within budget, and also to meet the requirements of the client, a company must properly organize its testing process. Indeed, software testing is a trade-off between budget, time money, and reliability. And the need to model the allocation of testing resources is needed for this trade-off. By formulating and solving resource allocation problems for modular software systems, this chapter has tried to make important contributions to software reliability, emphasizing the significance of the magnitude of faults varying from the easiest to the hardest levels.

7 Conclusion

In this chapter, we have discussed the problem of modular software at the unit testing stage. A resource allocation problem aiming at the maximization of the total fault removal from modular software subject to availability of resources and budget with an aspired level of reliability for each module is formulated and solved using a Genetic Algorithm. For modeling the fault removal process of each module testing resource-dependent SRGM is used. The SRGM studied incorporates a novel idea of faults being categorized under varied levels (1-Level being simplest severity level and k-Level being hardest severity level) of fault severity. From the parameter estimation and goodness of fit results on software failure real data it is shown that the model accuracy improves as we move from 1-Level to 2-Level to 3-Level and so on. In this, an important remark is that it may happen that after incorporating a certain level of severity of faults, in the modeling framework, the accuracy of SRGM may get stabilized, depending on the criticality of the code of the software. The parameter estimation carried on data set showed that the accuracy of the model improved both by increasing the level of severity and by incorporating testing resources in the modeling of SRGM. By taking two numerical framed data cases, optimal resource allocation problems are solved by Genetic Algorithm considering four and five levels of fault severity.

8 Limitations and Future Scope

The chapter is done under the assumption of independence of the failures of different modules. In the future dependence of the failures from different modules can also be studied. The proposed problems have some limitations as they do not incorporate warranty and maintenance costs. Such costs can be taken in the future for studying the allocation and release planning decisions. In the future, we can explore the possibility of including multi-dimensional software reliability growth modeling to take care of the effect of not only testing resources but also other testing factors like testing coverage, testing time/number of test cases on the fault removal process simultaneously.

References

Aggarwal AG, Kapur PK, Gurjeet K, Ravi K (2012) Genetic algorithm based optimal testing effort allocation problem for modular software. Bvicam's Int J Inf Technol 4(1):445–451

Aggarwal AG., Kaur G, Kapur PK (7–8 Feb 2011) Testing time and resource dependent two dimensional software reliability model for faults of different severity and related optimal allocation problem, published in the proceedings of international congress on productivity, quality, reliability, optimization, and modeling (ICPQROM 2011), vol 1: Theoretical papers, Allied Publishers Pvt. Ltd., New Delhi, 160–175

Aggarwal AG, Kaur G, Kapur PK (14–16 Dec 2010) Optimal testing resource allocation for modular software considering imperfect debugging and change point using genetic algorithm. In: Published in IEEE proceedings of 2nd international conference on reliability, Safety and Hazard

Aggarwal AG, Gandhi N, Verma V, Tandon A (2019) Multi-release software reliability growth assessment: an approach incorporating fault reduction factor and imperfect debugging. Int J Math Oper Res 15(4):446–463

Aggarwal AG, Kapur PK, Nidhi N (2018) A discrete SRGM for a multi-release software system with faults of different severity. Int J Oper Res, Inderscience Enterp Ltd 32(2):156–168

Goel AL, Okumoto K (1979) Time-dependent error detection rate model for software reliability and other performance measures. IEEE Trans Reliab R-28(3):206–211

Goldberg DE (1989) Genetic algorithms in search of optimization and machine learning. Addison-Wesley

Huang CY (2005) Performance analysis of software reliability growth models with testing-effort and change-point. J Syst Softw 76:181–194

Huang CY, Lo JH, Kuo SK, Lyu MR (2004) Optimal allocation of testing resources considering cost, reliability, and testing effort. In: Proceedings of the 10th IEEE pacific international symposium on dependable computing

Kapur PK, Aggarwal AG, Kanica K, Gurjeet K (2009) Optimal testing resource allocation for modular software considering cost, testing effort and reliability using genetic algorithm, international journal of reliability, quality, and safety. Engineering 16(6):495–508

Kapur PK, Pham H, Aggarwal AG, Kaur G (2012) Two dimensional multi-release software reliability modeling and optimal release planning. IEEE Trans Reliab 61(3):758–768

Kapur PK, Younes S, Agarwala S (1995) Generalized Erlang model with n types of faults. ASOR Bulletin 14(1):5–11

Kapur PK, Pham H, Gupta A, Jha PC (2011) Software reliability assessment with OR applications. Springer

Kapur PK, Garg RB (1990) A software reliability growth model under imperfect debugging. RAIRO 24, 295–305
Kapur PK, Aggarwal AG, Gurjeet K (2010a) Simultaneous allocation of testing time and resources for a modular software. Int J Syst Assur Eng Manag 1(4):351–361
Kapur PK, Aggarwal AG, Abhishek T (2010b) Two-dimensional software reliability growth model with faults of different severity. Commun Dependability Qual Manag 13(3):98–110
Kapur PK, Bardhan A, Yadavalli V (2007) On allocation of resources during testing phase of a modular software. Int J Syst Sci 38(6):493–499
Kapur PK, Garg RB, Kumar S (1999) Contributions to hardware and software reliability. World Scientific, Singapore
Kareer N, Grover PS, Kapur PK (1990) An S-shaped reliability growth model with two types of errors. Microelectron Reliab 30(6):1085–1090
Kaur G, Aggarwal AG, Kedia A (2017) A study of optimal testing resource allocation problem for modular software with change point. Ann Comput Sci Inf Syst 14:77–84
Musa JD (1975) A theory of software reliability and its application. IEEE Trans Softw Eng SE-1, 312–327
Musa JD, Okumoto K (1984) A logarithmic poisson execution time model for software reliability measurement. In: Proceedings of 7th international conference on software engineering, pp 230–238
Musa JD, Iannino A, Okumoto K (1987) Software reliability: measurement, prediction, applications. McGraw Hill
Obha M (1984a) Software reliability analysis models. IBM J Res Dev 28:428–443
Obha M (1984b) Inflection S-shaped software reliability growth model. In: Osaki S, Hatoyama Y (eds), Lecture notes in economics and mathematical systems. Springer
Obha M, Yamada S (1984) S-shaped software reliability growth model. In: Proceedings of the 4th international conference on reliability and maintainability, pp 430–436
Ohetera H, Yamada S (1990) Optimal allocation and control problems for software testing resources. IEEE Trans Reliab 39(2):171–176
Pham H (2006) System software reliability. Springer, Reliability Engineering Series
Sastry K (2007) Single and multiobjective genetic algorithm toolbox for matlab in C++, IlliGAL Report No. 2007017
Shatnawi O, Kapur PK (2008) A Generalized software fault classification model. WSEAS Trans Comput 2(9):1375–1384
Trachtenberg M (1990) A general theory of software-reliability modeling. IEEE Trans Reliab 39:92–96
Xie M, Yang B (2001) Optimal testing time allocation for modular systems. Int J Qual Reliab Manag 18(4):854–863
Yamada S, Osaki S (1987) Optimal software release policies with simultaneous cost and reliability requirements. Eur J Oper Res 31:46–51
Yamada S, Obha M, Osaki S (1983) S-shaped software reliability growth modeling for software error detection. IEEE Trans Reliab 32(5):475–484
Yamada S, Ohtera H, Narihisa H (1986) Software reliability growth models with testing-effort. IEEE Trans Reliab R-35, 19–23
Yamada S, Osaki S, Narihisa H (1985) Software reliability growth models with two types of errors. Oper Res (RAIRO) 19(1):87–104
Zhu M, Pham H (2018) A two-phase software reliability modeling involving software fault dependency and Imperfect fault removal. Comput Lang Syst Struct 53:27–42

Integration of FAHP and COPRAS-G for Software Component Selection

Prarna Mehta, Abhishek Tandon, and Himanshu Sharma

Abstract Software Quality is directly proportional to firms' effectiveness. Thus, improving the quality of the software with respect to the clients' requirement has always been a chasing research field for many researchers. In today's world, software is being developed using independent components due to its ability to be reused in different software systems' architecture proficiently. This approach not only improvises the reliability of the software system but also accelerated development phase, reduced system failure risk and in monetary terms has been resourceful approach for a software firm. On the other hand, accommodating a wrong component in a software system can jeopardise it leading to complications to the firm. A structural decision-making mechanism is required for selection Commercial-Off the Shelf (COTS) components based on multiple criteria. In this chapter, Multiple Criteria Decision Making (MCDM) techniques have been implemented to determine the critical weights of the criteria using Fuzzy Analytical hierarchy process (FAHP) and then the COTS component alternatives are evaluated using Complex Proportional Assessment of alternatives with grey Relations (COPRAS-G). The novelty in the proposed algorithm is the application of COPRAS-G which is used to analyse the alternatives of COTS components maximising and minimising respective criteria.

Keywords FAHP · COTS · COPRAS-G · MCDM · Software reliability

1 Introduction

In the today's growing world, technology has been improving every minute of the day. Thus, there is an increase in the rate of software development, implying complex software systems. In order to keep up with the pace and evade complexities, it is

P. Mehta · H. Sharma (✉)
Department of Operational Research, University of Delhi, Delhi, India

A. Tandon
Shaheed Sukhdev College of Business Studies, University of Delhi, Delhi, India
e-mail: abhishektandon@sscbsdu.ac.in

A. G. Aggarwal et al. (eds.), *Optimization Models in Software Reliability*, Springer Series in Reliability Engineering,
https://doi.org/10.1007/978-3-030-78919-0_12

practical as well as economical to reuse software components at the development stage of a system. A component based software system is a process of designing software by accommodating reusable components, which breaks down the software complexity and keeps up with the demand of new software systems. As stated by Comella-Dorda et al. (2002), Commercial off the Shelf (COTS) components was publicly available for retail purpose which did not require any internal modifications. Garg et al. (2017) have defined COTS component as products that is ready to use, easy implementation and assimilate with other components in the software system. On the other hand, Gupta et al. (2012) outlined COTS component available in the market could be procured, rented or authorised to the global public. Imoize et al. (2019) have highlighted upon the significance of reusing software components to build new software.

Implementing existing components has been proved beneficial in the past with respect to reliability, budget, time and effort (Cortellessa et al. 2008). The objective of software developers to reuse COTS component is to reduce or shorten the time involved in development of software. In other words, software component once created can be reused in multiple problems independently by simply reassembling them to obtain the resulting system. Components integrate with other components in a given system through interfaces. There exist multiple independent COTS components in the market provided by numerable vendors. Siddiqui and Tyagi (2016) have emphasised how system's reliability is affected by each component. This puts weightage to the problem of selecting right components in accordance to the demand of a software system. Thus evaluation of COTS components becomes a significant prerequisite in order to achieve the full benefits.

Choosing a misfit COTS components may disrupt an organisation in both positive and negative way. Thus it is a tedious process of assessing and making a choice on the ground of some attributes. One has to carefully build a set of criteria that briefly supports preference for an alternative over another. In the real world, decision makers have to examine a number of qualitative as well as quantitative criteria for each COTS component. With the help of these criteria, one can rank and establish optimal alternative from an alternative set of components for a given system. Thus, COTS components selection problem can be considered as Multiple Criteria Decision Making (MCDM) problem. In this approach, different algorithms are adapted to select and rank components in accordance to some criteria scores. MCDM facilitates in selecting an optimal instance based on manifold and confounding set of criteria that exemplifies a COTS component. Garg (2017) has thoroughly discussed different COTS component selection and evaluation algorithms rendering pros and cons of each algorithm. Bali and Madan (2015) have chronologically given an outline of the development in COTS selection paradigms and also given a future perception. In this chapter, a MCDM framework has been proposed in order to select the best COTS components by considering contradicting objectives.

Many a times, decision-making is a complex procedure due to unknown information and conflicts in preferences is a by-product. A decision maker usually subjectively quantifies a set of criteria. In such a situation, one can desegregate intervals and linguistic terms to define complex preferences and it is called Fuzzy numbers. It improvises MCDM technique where human opinion and perception are ambiguous and uncertain. Thus, Fuzzy logic becomes a rational approach to combat such situations. One of the most extensively used MCDM techniques is Analytical hierarchy process (AHP) (Aghdaie et al. 2013) that does involve advance mathematical tools. It disintegrates a complex problem into a hierarchy starting from the objective to attain, simplifying it into criteria affecting it, to considering the number of alternatives. The main goal is to converts the thinking process of a decision maker into a structural form. AHP is simpler tool since it does not depend on numerical data but concentrates on the weightage of the criteria. However, the AHP's detriment is that it is inept in handling uncertain and vague human perceptions.

Many optimization models are constructed by combining AHP with other techniques like Fuzzy logic, mathematical programming, TOPSIS, DEA, etc. to deal with the fuzziness (Aruldoss et al. 2013). In this chapter, Fuzzy Analytical hierarchy process (FAHP) is deployed to derive weights of the criteria by analysing qualitative perceptions of decision makers. One cannot rely on criteria solely for correct result i.e. they might be misleading in certain cases since, the opinion of decision maker changes with time. Thus, prioritisation of criteria with the help of FAHP is relatively an important step to be taken under consideration (Thapar and Sarangal 2020).

After the weighting the criteria, it is important to prioritise the different components available in the market in accordance to the calculated weights. Complex Proportional Assessment of alternatives with grey Relations (COPRAS-G) assists in defining an experts' outlook in terms of an interval rather than a scalar value and hence ranks the alternatives in accordance to their utility level from highest to lowest. The criterion value is based on grey theory interval enriching real time decision-making process. The objective of this chapter is to integrate FAHP and COPRAS-G to assess COTS components by prioritising and assigning weights of importance to criteria that defines, rather characterises a COTS component.

To summarize, the objective of the chapter are as follows,

1. Selection of COTS components with the help of MCDM techniques.
2. Assigning weights to the selection criteria with the help of FAHP.
3. Assimilation of FAHP and COPRAS-G to prioritise COTS components.

There is no evidence in the literature that this hybrid methodology i.e., amalgamation of FAHP and COPRAS-G, has been implemented to analyse COTS component. In the past researches, this integrated algorithm has been deployed in other fields, for instance (Aghdaie et al. 2013) have adopted this approach for market segmentation and stating the model being efficient. This robust procedure was also promoted by Mobin et al. (2015) and applied on selection of suppliers.

The chapter is further segregated as Sect. 2 gives a comprehensive literature review, Sect. 3 talks about the methodology implemented, a numerical solved and results discussed in Sect. 4 and lastly in Sect. 5 the chapter is wrapped up by conclusion.

2 Literature Review

In the last two decades, researchers have been probing the benefits of COTS components, promoting its usage and amplifying the system's efficiency with the help of it. There is a bank of rich literature review available about the state-of-art of COTS components encapsulating the different selection techniques. Schneider and Han (2004) has drawn attention to the work that has been done in the past and has reflected upon problems that can be dealt as a research problem in the future. Wanyama and Far (2005) discusses various functionalities of the ideal COTS component selection and ways of their implementation. Mittal and Bhatia (2013) have reviewed the literature, distinguishing different criteria and assessing the quality of COTS model in accordance to ISO 9126 model. Vale et al. (2016) have given a through mapping of the work done so far in this field that gives a better understanding.

Amalgamations of a couple of methodologies have been proposed by numerable researchers to assess COTS components. These integrations have been unique in its own way in terms of the objective of the problem, or methodologies or with the end result. But all of these exclusive techniques diverge to the sole aim of proficient software system based on COTS components. For example, a robust technique called PRISM was developed by Lichota et al. (1997) where the software components were evaluated through multiple phases defining a generic software architecture. Another methodology, namely CISD, was suggested by Tran and Liu (1997) which was based on waterfall like approach. Grau et al. (2004) developed a COTS component selection system, namely, DesCOTS that considers all quality aspects for the selection process.

Zachariah and Rattihalli (2007) have implemented mathematical programming techniques like branch and bound algorithm and Goal programming technique for COTS selection. Neubauer and Stummer (2007) put forward a decision support system where, all possible solutions for COTS selection are suggested to the decision makers that interactively examine for the best solution. Sheng and Wang (2008) has implemented gap analysis for selection of COTS component by filling the gap between attributes and software system demands (Garg et al. 2017) has classified COTS selection as MCDM problem and evaluated using fuzzy based matrix. The set of criteria chosen to select COTS components were vendor capabilities, Business issues and Cost, which were further divided into multiple sub-criteria. Ernst et al. (2019) have given importance to speedy decision-making and confidence for selection of COTS component. Their aim is to make the process of selection and evaluation a continuous process, where components are examined in accordance to different software system demand.

COTS selection problem has been classified as an optimization process with the objective to optimize the quality with respect to multiple constraints in accordance to the suggested problem. Jung and Choi (1999) have not only optimised cost and quality under budget constraint but has also considered the compatibility between selected COTS component. In their study, the compatibility between the components was stated in terms of either- or condition. This gap was observed by Tang et al. (2011) and fulfilled by introducing a new optimising model. Shen et al. (2006) have proposed an optimization model where the objective to maximise the quality of the system subject to fuzzy budget constraint. Cortellessa et al. (2008) have discussed a novel non-linear optimisation model considering the framework of building or buying COTS components for a given delivery time and reliability constraints. Jha et al. (2011) has extended the optimization model (reliability is maximised while cost is minimised) by considering the compatibility between different COTS components under the built or buy decision framework.

Kontio (1995) has proposed a paradigm to evaluate off-the-shelf components called OTSO based on cost and software attributes. It is a combination of two algorithms, AHP and weighted sum methodology. Gupta et al. (2012) has developed a fuzzy model for optimised selection of COTS component using AHP and fuzzy mathematical programming. Garg et al. (2016) suggested the use of fuzzy distance-based technique to select and rank COTS components. The results were validated with that of AHP.

COPRAS has been majorly used to optimise problems in construction technology (Malinauskas and Kalibatas 2005; Ustinovichius et al. 2007; Zavadskas and Antucheviciene 2007). This methodology was also applied to select road design (Zavadskas et al. 2007b), sustainable development (Viteikiene 2006; Viteikiene and Zavadskas 2007; Zavadskas et al. 2007a), environmental problems (Kaklauskas and Zavadskas 2007; Kalibatas et al. 2007). Complex problems like selection of social media forums was handled using COPRAS-G (Tavana et al. 2013). Implementation of this technique has not been observed in the field of software reliability.

3 Methodology

In this chapter, we propose a problem of COTS component selection to build a software system. A COTS component is deployed in the system, that functions independently to other COTS components. It is observed from past literature that deploying multiple COTS components in a software system reduces complexities as well as enhances the reliability of the system. Jadhav and Sonar (2009) have discussed some steps to follow for selection of software packages. These steps can be implemented in this chapter and is given by the Fig. 1.

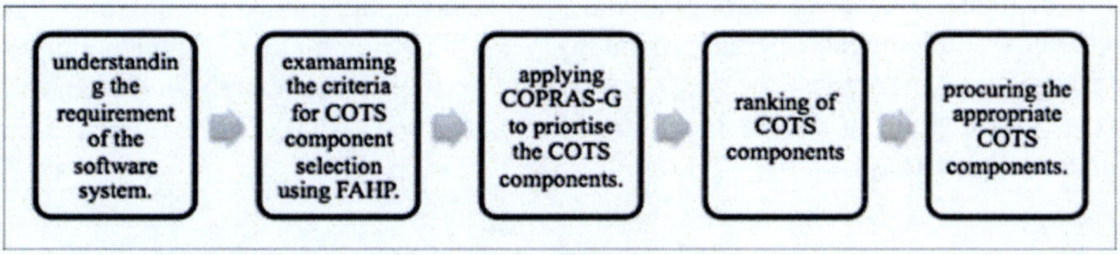

Fig. 1 A flowchart depicting the implemented methodology

Understanding the requirements of a software project and in accordance of the project scrutinizing the existing COTS components in the market has become a prerequisite. It is vital to comprehend the details and need of software project before starting the procedure of COTS components selection. Hence a pool of COTS components relevant to the problem is drawn in order to be evaluated. After obtaining the relevant COTS components, they are examined, scored and ranked based on some criteria and techniques.

In this chapter, we work on the COTS component evaluation and ranking technique. Since selection of components is done on the basis of multiple criteria, the problem is hence characterised as MCDM. It is a concept to assess and rank different alternatives. In the field of decision-making, MCDM is the most adopted approach and abundant algorithms are available in the literature. This chapter discusses about Fuzzy based MCDM technique that models considering human uncertainty and perception towards a given situation. Hence, COTS components are evaluated on the basis of certain set of criteria. These criteria are assigned some weights obtained with the help of FAHP. In the next step, these weights obtained from FAHP for respective criteria are used to obtain priority value for the COTS components under scrutiny. These components are then ranked in accordance to the priority function. The highest ranked COTS component is procured since it is a well suited unit for the given software project.

3.1 FAHP

Before the process of evaluation, selection and ranking of different alternatives, a set of criteria is decided upon. Thus, the decision-making team uses qualitative terminology to describe the weightage of each criterion. These linguistic terms are mapped over numerical data using fuzzy numbers for easy computation. Fuzzy numbers are implemented when there is uncertainty, vagueness, and incompleteness in the data. The term 'fuzziness' was first defined by Zadeh (1996) using fuzzy set theory (FST). While (Dubois and Prade 1979) extended mathematical operations over fuzzy numbers. A fuzzy number expands a real number and has properties analogous to number theory. Fuzzy sets are determined using optimal universe of discourse and membership functions. A membership function characterises fuzzy set by denoting each elements' degree of belongingness. The value of the membership degree lies between 0 and 1 for each element in the fuzzy set. There are many available membership functions for a given fuzzy set viz. triangular, trapezoidal, parabolic etc. In this chapter, Triangular fuzzy number (TFN) is implemented to describe the membership function, which is given by,

$$\mathrm{f_Z(x)} = \begin{cases} \frac{\mathrm{x-l}}{\mathrm{m-l}}, l \le x \le m \\ \frac{\mathrm{u-x}}{\mathrm{u-m}}, m \le x \le u \\ 0, otherwise \end{cases}$$

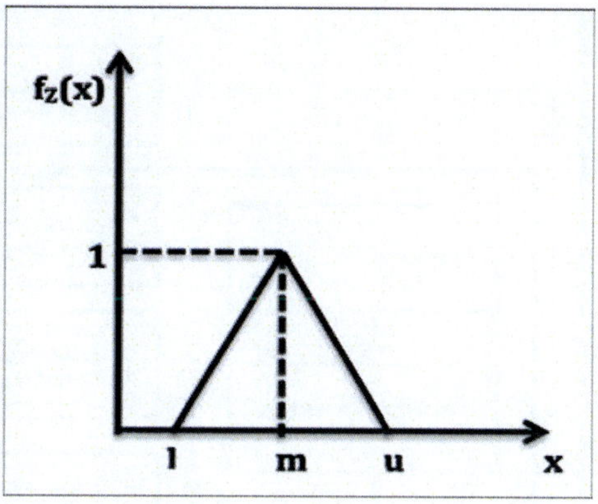

Fig. 2 A graph depicting a triangular fuzzy number

where, Z is a triangular fuzzy number, l, m, u are real numbers. TFN has the simplest form of membership function that trades linguistics terms with triplet numbers, thus helping in analysing human judgement. TFN membership function is depicted by Fig. 2.

From the past literature, it has been observed that AHP is the most widely used MCDM technique in decision-making. It was developed by Saaty (1977) that hierarchically structures a complex problem using weights, thus simplifying for decision maker to make decisive assessment of the problem. These weights assigned to each unit or criteria in the hierarchy imply its relative importance in achieving the objective. AHP is a methodical MCDM process that numerically signifies each criteria belonging to a set of criteria in accordance to a given problem. Thus, a set of criteria has to be selected which is then examined for its significance using AHP. After a problem is synthesised into a structural form, weights of each criterion at each level of the hierarchical are computed by making successive pairwise comparisons. A numerical scale was defined that quantifies the intensity of importance of a particular criterion (Saaty 1977). This scale ranges from 1/9 to 9 denoting different levels of importance. The result of the pairwise comparison at each level is encapsulated in a matrix, thus computing maximum eigenvector (λ_{max}) of each matrix. With the help of λ_{max} and Random Index (R.I.), consistency ratio is computed and which is given as,

$$\text{C.I.} = (\lambda_{max} - n)/(n - 1) \tag{1}$$

$$\text{C.R.} = \frac{\text{C.I.}}{\text{R.I.}} \tag{2}$$

Thus, looking at the consistency ratio one can state that AHP developed for a given problem will yield meaningful result or not. But this ratio computed on the basis of decision matrix, is not completely reliable and it is a time consuming process to calculate whenever the decision matrix is altered. In order to achieve C.R. less than 0.1, the decision matrix is modified, which gives a flawed result.

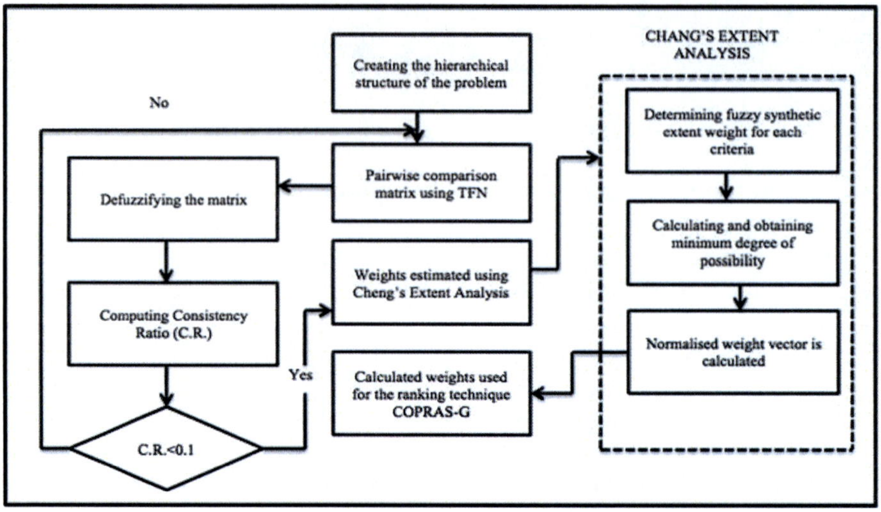

Fig. 3 An algorithm defining FAHP using Chang's extent analysis

A conventional AHP works successfully when the data is known, certain and crisp. But in real world problems, it is not necessary that the data is always crisp and definite. The decision makers usually have vague judgements, which is difficult to quantify and analyse. In such cases, fuzzy AHP is applied due to its ability to handle the fuzziness in the data. In this chapter, triangular fuzzy number is utilized that denotes the level of relationship between a pair of criteria. Chang's extent analysis is implemented to obtain weights of each criteria (Chang 1996) illustrated by Fig. 3.

3.2 *COPRAS-G*

COPRAS was developed by Zavadskas et al. (1994) where multiple contradicting objective functions are optimized given a set of constraints or criteria. These set of criteria scrutinizes the existing alternatives and rank them from best option to the worst for a given software system. Many MCDM problems are based on real world problems, thus systems with crisp value is considered to be theoretical concept. In COPRAS-G, alternatives are evaluated and ranked in accordance to the value of their respective utility function. The alternatives are characterized by various criteria, which are quantitatively described using grey theory. Figure 4 depicts the criteria used in the study. Here, the values are expressed in respect to some interval using grey relational grade (Julong 1989). In Grey theory, black systems and white systems are considered as ideal situations, in other words former signifies no information system while latter signifies complete information system. Practically, in daily life ideal situations does not arise. In such cases, grey is considered as intermediate

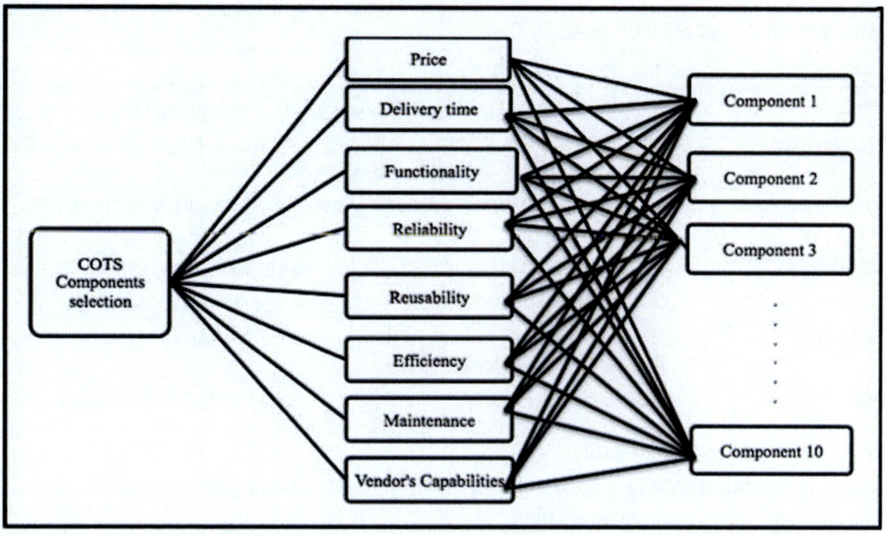

Fig. 4 A hierarchy depicting the criteria used to rank and prioritise the COTS components

zone that denotes partly known and unknown information. Liou et al. (2016) have applied a hybrid version of COPRAS-G to resolve complications in green supply chain problems. Madhuri et al. (2010) implemented this MCDM technique to select an appropriate website for a given service.

4 Numerical Illustration

From the literature review, it is evident that evaluation of COTS components is an imperative step at the development phase. Henceforth, in this section, a numerical has been performed to illustrate the proposed methodology. After thoroughly scrutinizing many research papers available in the literature and with the help of professional guidance selection of COTS components were based on criteria, namely, price, delivery time, functionality, reliability, reusability, efficiency, maintenance and vendor capabilities. Table 1 defines the set of criteria used in the proposed methodology. The objective is to minimise cost of producing and maintaining and to shorten the release time.

After establishing a set of criteria that characterises the COTS component, a decision matrix is constructed, given by the Table 2, with the help of TFN concept and by comparing the above selected criteria pairwise. TFN is used to quantify the ambiguousness present in the data. The decision matrix is then checked for its Consistency Ratio (C.R.) to be less than 0.10 with the help of AHP (Wind and Saaty 1980). For the given decision matrix, the C.R. is 0.0427.

Table 1 Description of set of criteria

Criteria	Description
Price	Includes purchasing cost, maintenance cost, and training cost
Delivery time	Takes into account the time taken by the v3ndor to release the product into the market or directly to the customer
Functionality	Describes the accuracy, suitability, interoperability and security of the coponent
Reliability	Determines the fault tolerance, recoverability and maturity of component
Reusability	Deals with level of ease in understanding, operating and learning
Efficiency	Measures the performance of the component in terms of response tme and resources utilized
Maintenance	Illustrates on the amount of effort required to test for faults, diagnose, and modify the component. It also takes into account the risk attached to the modification
Vendor's Capabilities	Implies installation, training, guidance, maintaining, credentials and financial stability

The goal of FAHP is to prioritise the criteria by calculating weights associated to each. The TFN values of criteria is defuzzied to crisp values using Changs' Extent Analysis and weightage of each criteria in characterising COTS component is computed, given by Table 3.

From the above table, it was observed that Price has the highest weightage whereas Functionality has the least. In the next step, 15 components are examined by experts with respect to the selected criteria while implementing COPRAS-G. This technique results in selecting the best alternative among a number of alternatives for given constraints. The set of criteria are classified on the basis of goal of maximisation or minimisation. In this problem, Price and Delivery time are to be minimised while the remaining criteria are to be maximised.

The initial decision matrix is generated with the help of expert opinion and is given by Table 4. In the table, the values of each criteria with respect to components are fuzzy in nature, where L denotes the lower limit and U denotes the upper limit for the respective criteria for each component. For a maximising criterion, the larger value is considered to be the best value and vice-versa (Ecer 2014).

The initial matrix is normalised and the resultant is depicted by Table 5. The formula for normalizing is as follows,

$$V = \frac{\hat{u}_{ij}}{\sum l_{ij} + \hat{u}_{ij}} \tag{3}$$

where, V defines the initial decision matrix, $\hat{u}_{ij}$ is the upper value of the ith criteria and j^{th} component and l_{ij} is the lower value of the ith criteria and j^{th} component. The critical weights obtained from FAHP given by Table 3 is used to compute weighted decision matrix with the help of the Eq. 4. Table 6 depicts the resultant of the weighted

Table 2 Decision matrix

Criteria		Price	Delivery time	Functionality	Reliability	Reusability	Efficiency	Maintenance	Vendor capabilities
Price	l	1	1/2	2/5	1	1/2	1/2	2/3	1/2
	m	1	2/3	1/2	1	2/3	2/3	1	2/3
	u	1	1	2/3	1	1	1	2	1
Delivery time	l	1	1	1/2	1/2	2/3	2/3	1	1
	m	1 1/2	1	2/3	2/3	1	1	1	1
	u	2	1	1	1	2	2	1	1
Functionality	l	1 1/2	1	1	1/2	1	1	2/3	1 1/2
	m	2	1 1/2	1	1	1	1 1/2	1	2
	u	2 1/2	2	1	1 1/2	1	2	2	2 1/2
Reliability	l	1	1	2/3	1	1/2	1	1/2	2/5
	m	1	1 1/2	1	1	2/3	1	2/3	1/2
	u	1	2	2	1	1	1	1	2/3
Reusability	l	1	1/2	1	1	1	1/2	1 1/2	2/3
	m	1 1/2	1	1	1 1/2	1	1	2	1
	u	2	1 1/2	1	2	1	1 1/2	2 1/2	2
Efficiency	l	1	1/2	1/2	1	2/3	1	1 1/2	2/3
	m	1 1/2	1	2/3	1	1	1	2	1
	u	2	1 1/2	1	1	2	1	2 1/2	2
Maintenance	l	1/2	1	1/2	1	2/5	2/5	1	1/2
	m	1	1	1	1 1/2	1/2	1/2	1	2/3
	u	1 1/2	1	1 1/2	2	2/3	2/3	1	1

(continued)

Table 2 (continued)

Criteria		Price	Delivery time	Functionality	Reliability	Reusability	Efficiency	Maintenance	Vendor capabilities
Vendor capabilities	l	1	1	2/5	1 1/2	1/2	1/2	1	1
	m	1 1/2	1	1/2	2	1	1	1 1/2	1
	u	2	1	2/3	2 1/2	1 1/2	1 1/2	2	1

Table 3 Weights obtained using FAHP

Criteria	Critical weights
Price	0.1604
Delivery time	0.1259
Functionality	0.088
Reliability	0.1412
Reusability	0.1028
Efficiency	0.1136
Maintenance	0.1503
Vendor's Capabilities	0.1177

normalised decision matrix.

$$V_{weighted} = V \times w_i \tag{4}$$

where, w_i is the weights of the criteria obtained from FAHP and $V_{weighted}$ is the weighted normalised decision matrix.

Next, sum of the criterion that has to be maximised (P) and the sum of criteria that has to be minimised (R) are calculated for each alternative in order to obtain the efficiency and rank the component from the best to the worst,

Where, $P_i = \frac{\sum_{i=1}^{2} l_{ij} + \widehat{\mathrm{u_{ij}}})}{2}$ and $R_i = \frac{\sum_{i=3}^{8} l_{ij} + \widehat{\mathrm{u_{ij}}})}{2}$; j = 1, …, 15

The criterion that has to be minimised are price and delivery time, while criteria that has to be maximised are functionality, reliability, reusability, efficiency, maintenance and vendor capabilities. The significance of each component is given by Q, Table 7 depicts the rank of the components using utility function (UF_i).

$$Q_i = P_i + \frac{\sum_{i=1}^{15} R_i}{R_i \sum_{i=1}^{15} \frac{1}{R_i}} \tag{5}$$

$$UF_i = \frac{Q_i}{Q_{max}} \times 100\% \tag{6}$$

Higher the value of utility function, better is the rank of the alternative. From the Table 7, one can verify that component 7 is the best alternative with respect to selected criteria with the value of utility function being 100% whereas component 12 being the worst with utility function value of 81.0343%.

Table 4 The initial decision matrix

Objective	Min		Min		Max		Max		Max		Max		Max		Max	
Criteria	Price		Delivery time		Functionality		Reliability		Reusability		Efficiency		Maintenance		Vendor capabilities	
Components	L	U	L	U	L	U	L	U	L	U	L	U	L	U	L	U
1	40	60	40	60	80	90	70	80	20	30	60	70	80	90	60	70
2	50	60	70	80	50	60	60	70	80	90	40	50	70	80	90	95
3	50	60	70	80	50	60	70	80	60	70	60	70	30	40	60	70
4	80	90	70	80	60	70	60	95	70	80	80	90	50	70	50	80
5	60	70	60	90	90	95	70	90	50	70	80	90	80	90	40	60
6	70	95	80	95	50	65	85	95	75	95	70	90	70	90	80	100
7	65	95	75	90	70	90	75	95	70	85	80	90	75	95	75	100
8	60	80	60	75	70	90	65	80	65	75	65	70	70	85	65	95
9	85	100	80	100	50	60	80	100	75	100	70	100	70	90	75	95
10	65	95	85	85	75	90	80	90	60	75	80	95	75	95	65	90
11	70	90	70	100	75	95	85	90	65	85	90	70	65	85	75	90
12	60	90	55	75	60	75	50	65	45	65	55	60	50	55	70	90
13	50	75	55	70	70	85	55	70	40	60	50	60	45	50	70	85
14	55	80	60	80	65	80	65	85	60	75	65	80	60	70	75	90
15	60	75	55	65	80	100	55	80	50	70	70	85	50	65	55	65
Column sum	920	1215	985	1225	995	1205	1025	1265	885	1125	1015	1170	940	1150	1005	1275
Total sum	1067.5		1105		1100		1145		1005		1092.5		1045		1140	

Table 5 Normalised decision matrix

Objective	Min		Min		Max		Max		Max		Max		Max		Max	
Criteria	Price		Delivery time		Functionality		Reliability		Reusability		Efficiency		Maintenance		Vendor capabilities	
Components	L	U	L	U	L	U	L	U	L	U	L	U	L	U	L	U
1	0.0375	0.0562	0.0362	0.0543	0.0727	0.0818	0.0611	0.0699	0.0199	0.0299	0.0549	0.0641	0.0766	0.0861	0.0526	0.0614
2	0.0468	0.0562	0.0633	0.0724	0.0455	0.0545	0.0524	0.0611	0.0796	0.0896	0.0366	0.0458	0.0670	0.0766	0.0789	0.0833
3	0.0468	0.0562	0.0633	0.0724	0.0455	0.0545	0.0611	0.0699	0.0597	0.0697	0.0549	0.0641	0.0287	0.0383	0.0526	0.0614
4	0.0749	0.0843	0.0633	0.0724	0.0545	0.0636	0.0524	0.0830	0.0697	0.0796	0.0732	0.0824	0.0478	0.0670	0.0439	0.0702
5	0.0562	0.0656	0.0543	0.0814	0.0818	0.0864	0.0611	0.0786	0.0498	0.0697	0.0732	0.0824	0.0766	0.0861	0.0351	0.0526
6	0.0656	0.0890	0.0724	0.0860	0.0455	0.0591	0.0742	0.0830	0.0746	0.0945	0.0641	0.0824	0.0670	0.0861	0.0702	0.0877
7	0.0609	0.0890	0.0679	0.0814	0.0636	0.0818	0.0655	0.0830	0.0697	0.0846	0.0732	0.0824	0.0718	0.0909	0.0658	0.0877
8	0.0562	0.0749	0.0543	0.0679	0.0636	0.0818	0.0568	0.0699	0.0647	0.0746	0.0595	0.0641	0.0670	0.0813	0.0570	0.0833
9	0.0796	0.0937	0.0724	0.0905	0.0455	0.0545	0.0699	0.0873	0.0746	0.0995	0.0641	0.0915	0.0670	0.0861	0.0658	0.0833
10	0.0609	0.0890	0.0769	0.0769	0.0682	0.0818	0.0699	0.0786	0.0597	0.0746	0.0732	0.0870	0.0718	0.0909	0.0570	0.0789
11	0.0656	0.0843	0.0633	0.0905	0.0682	0.0864	0.0742	0.0786	0.0647	0.0846	0.0824	0.0641	0.0622	0.0813	0.0658	0.0789
12	0.0562	0.0843	0.0498	0.0679	0.0545	0.0682	0.0437	0.0568	0.0448	0.0647	0.0503	0.0549	0.0478	0.0526	0.0614	0.0789
13	0.0468	0.0703	0.0498	0.0633	0.0636	0.0773	0.0480	0.0611	0.0398	0.0597	0.0458	0.0549	0.0431	0.0478	0.0614	0.0746
14	0.0515	0.0749	0.0543	0.0724	0.0591	0.0727	0.0568	0.0742	0.0597	0.0746	0.0595	0.0732	0.0574	0.0670	0.0658	0.0789
15	0.0562	0.0703	0.0498	0.0588	0.0727	0.0909	0.0480	0.0699	0.0498	0.0697	0.0641	0.0778	0.0478	0.0622	0.0482	0.0570

Table 6 Weighted normalised matrix

Objective	Min		Min		Max		Max		Max		Max		Max		Max	
Criteria	Price		Delivery time		Functionality		Reliability		Reusability		Efficiency		Maintenance		Vendor capabilities	
Components	L	U	L	U	L	U	L	U	L	U	L	U	L	U	L	U
1	0.0060	0.0090	0.0046	0.0068	0.0064	0.0072	0.0086	0.0099	0.0020	0.0031	0.0062	0.0073	0.0115	0.0129	0.0062	0.0072
2	0.0075	0.0090	0.0080	0.0091	0.0040	0.0048	0.0074	0.0086	0.0082	0.0092	0.0042	0.0052	0.0101	0.0115	0.0093	0.0098
3	0.0075	0.0090	0.0080	0.0091	0.0040	0.0048	0.0086	0.0099	0.0061	0.0072	0.0062	0.0073	0.0043	0.0058	0.0062	0.0072
4	0.0120	0.0135	0.0080	0.0091	0.0048	0.0056	0.0074	0.0117	0.0072	0.0082	0.0083	0.0094	0.0072	0.0101	0.0052	0.0083
5	0.0090	0.0105	0.0068	0.0103	0.0072	0.0076	0.0086	0.0111	0.0051	0.0072	0.0083	0.0094	0.0115	0.0129	0.0041	0.0062
6	0.0105	0.0143	0.0091	0.0108	0.0040	0.0052	0.0105	0.0117	0.0077	0.0097	0.0073	0.0094	0.0101	0.0129	0.0083	0.0103
7	0.0098	0.0143	0.0085	0.0103	0.0056	0.0072	0.0092	0.0117	0.0072	0.0087	0.0083	0.0094	0.0108	0.0137	0.0077	0.0103
8	0.0090	0.0120	0.0068	0.0085	0.0056	0.0072	0.0080	0.0099	0.0066	0.0077	0.0068	0.0073	0.0101	0.0122	0.0067	0.0098
9	0.0128	0.0150	0.0091	0.0114	0.0040	0.0048	0.0099	0.0123	0.0077	0.0102	0.0073	0.0104	0.0101	0.0129	0.0077	0.0098
10	0.0098	0.0143	0.0097	0.0097	0.0060	0.0072	0.0099	0.0111	0.0061	0.0077	0.0083	0.0099	0.0108	0.0137	0.0067	0.0093
11	0.0105	0.0135	0.0080	0.0114	0.0060	0.0076	0.0105	0.0111	0.0066	0.0087	0.0094	0.0073	0.0093	0.0122	0.0077	0.0093
12	0.0090	0.0135	0.0063	0.0085	0.0048	0.0060	0.0062	0.0080	0.0046	0.0066	0.0057	0.0062	0.0072	0.0079	0.0072	0.0093
13	0.0075	0.0113	0.0063	0.0080	0.0056	0.0068	0.0068	0.0086	0.0041	0.0061	0.0052	0.0062	0.0065	0.0072	0.0072	0.0088
14	0.0083	0.0120	0.0068	0.0091	0.0052	0.0064	0.0080	0.0105	0.0061	0.0077	0.0068	0.0083	0.0086	0.0101	0.0077	0.0093
15	0.0090	0.0113	0.0063	0.0074	0.0064	0.0080	0.0068	0.0099	0.0051	0.0072	0.0073	0.0088	0.0072	0.0093	0.0057	0.0067

Table 7 Components ranked according to the utility function

Components	P	R	1/R	Q	Utility	Rank
1	0.0443	0.0132	75.7019	0.0664	96.8830	4
2	0.0461	0.0168	59.4905	0.0635	92.6421	9
3	0.0388	0.0168	59.4905	0.0562	81.9535	14
4	0.0466	0.0213	46.9106	0.0603	87.9865	12
5	0.0496	0.0183	54.6090	0.0656	95.6732	7
6	0.0535	0.0224	44.7113	0.0666	97.1209	3
7	0.0549	0.0214	46.6845	0.0685	100.0000	1
8	0.0489	0.0182	54.9186	0.0650	94.7793	8
9	0.0536	0.0242	41.4025	0.0657	95.7984	6
10	0.0533	0.0217	46.0719	0.0668	97.4112	2
11	0.0529	0.0217	46.0719	0.0663	96.7889	5
12	0.0399	0.0187	53.5469	0.0555	81.0343	15
13	0.0396	0.0165	60.5615	0.0573	83.5373	13
14	0.0474	0.0181	55.1939	0.0635	92.6097	10
15	0.0442	0.0170	58.8977	0.0614	89.5561	11
Sum		0.0865	296.2024	0.0685		

5 Conclusion

In this chapter, an integrated and novel methodology has been applied to a manifold criteria problem of selection of apt COTS component for the development of software system. A set of relatively important features defining components such as price, delivery time, functionality, reusability, reliability, efficiency, maintenance and vendor capability are selected to evaluate components. The importance of these criteria is computed using FAHP where it was observed that price has a great impact on selection of components in comparison to other criteria. Multiple alternatives of COTS components were considered for ranking of the best alternative with respect to the set of features. COPRAS-G was implemented to compute the efficiency and rank the components with the help of scores of utility function. After reviewing research paper, it is evident that COPRAS-G has not been implemented to examine COTS component. With the help of FAHP, one can quantify linguistic data where as COPRAS-G takes into consideration contradicting objectives and results in a better solution.

6 Future Scope

This study can be extended by, comparing various MCDM techniques with COPRAS-G. COPRAS-G can be evaluated on larger datasets for generalised results.

References

Aghdaie MH, Zolfani SH, Zavadskas EK (2013) Market segment evaluation and selection based on application of fuzzy AHP and COPRAS-G methods. J Bus Econ Manag 14(1):213–233
Aruldoss M, Lakshmi TM, Venkatesan VP (2013) A survey on multi criteria decision making methods and its applications. Am J Inf Syst 1(1):31–43
Bali V, Madan S (2015) COTS evaluation & selection process in design of component based software system: an overview and future direction. Glob J Comput Sci Technol
Chang D-Y (1996) Applications of the extent analysis method on fuzzy AHP. Eur J Oper Res 95(3):649–655
Comella-Dorda S, Dean JC, Morris E, Oberndorf PA (2002) process for COTS software product evaluation. International conference on COTS-based software systems. Springer, pp 86–96
Cortellessa V, Marinelli F, Potena P (2008) An optimization framework for "build-or-buy" decisions in software architecture. Comput Oper Res 35(10):3090–3106
Dubois D, Prade H (1979) Fuzzy real algebra: some results. Fuzzy Sets Syst 2(4):327–348
Ecer F (2014) A hybrid banking websites quality evaluation model using AHP and COPRAS-G: a Turkey case. Technol Econ Dev Econ 20(4):758–782
Ernst N, Kazman R, Bianco P (2019) Component comparison, evaluation, and selection: a continuous approach. In: 2019 IEEE international conference on software architecture companion (ICSA-C). IEEE, pp 87–90
Garg R (2017) A systematic review of COTS evaluation and selection approaches. Accounting 3(4):227–236
Garg R, Sharma R, Sharma K (2016) Ranking and selection of commercial off-the-shelf using fuzzy distance based approach. Decis Sci Lett 5(2):201–210
Garg R, Sharma R, Sharma K (2017) MCDM based evaluation and ranking of commercial off-the-shelf using fuzzy based matrix method. Decis Sci Lett 6(2):117–136
Grau G, Carvallo JP, Franch X, Quer C (2004) DesCOTS: a software system for selecting COTS components. In: Proceedings: 30th euromicro conference, 2004. IEEE, pp 118–126
Gupta P, Mehlawat MK, Verma S (2012) COTS selection using fuzzy interactive approach. Optim Lett 6(2):273–289
Imoize AL, Idowu D, Bolaji T (2019) A brief overview of software reuse and metrics in software engineering. World Sci News 122:56–70
Jadhav AS, Sonar RM (2009) Evaluating and selecting software packages: a review. Inf Softw Technol 51(3):555–563
Jha P, Arora R, Kumar UD (2011) An Optimization framework for "build-or-buy" strategy for component selection in a fault tolerant modular software system under recovery block scheme. Rat Math 21(1):91–105
Julong D (1989) Introduction to grey system theory. J Grey Syst 1(1):1–24
Jung H-W, Choi B (1999) Optimization models for quality and cost of modular software systems. Eur J Oper Res 112(3):613–619
Kaklauskas A, Zavadskas EK (2007) Decision support system for innovation with a special emphasis on pollution. Int J Environ Pollut 30(3–4):518–528
Kalibatas D, Krutinis M, Viteikiene M (2007) Multi-objective evaluation of microclimate in dwelling. Technol Econ Dev Econ 13(1):24–31

Kontio J (1995) OTSO: a systematic process for reusable software component selection
Lichota RW, Vesprini RL, Swanson B (1997) PRISM product examination process for component based development. In: Proceedings fifth international symposium on assessment of software tools and technologies. IEEE, pp 61–69
Liou JJ, Tamošaitienė J, Zavadskas EK, Tzeng G-H (2016) New hybrid COPRAS-G MADM Model for improving and selecting suppliers in green supply chain management. Int J Prod Res 54(1):114–134
Madhuri BC, Chandulal A, Padmaja M (2010) Selection of best web site by applying COPRAS-G method. Int J Comput Sci Inf Technol 1(2):138–146
Malinauskas P, Kalibatas D (2005) The selection of rational constructional technology processes variants using COPRAS method. Technol Econ Dev Econ 11(3):197–205
Mittal S, Bhatia PK (2013) Software component quality models from ISO 9126 perspective: a review. IJMRS's Int J Eng Sci 2(2)
Mobin M, Roshani A, Saeedpoor M, Mozaffari MM (2015) Integrating FAHP with COPRAS-G method for supplier selection (case study: an Iranian manufacturing company). In: Proceedings of the American society for engineering management
Neubauer T, Stummer C (2007) Interactive decision support for multiobjective COTS selection. In: 2007 40th annual hawaii international conference on system sciences (HICSS'07). IEEE, p 283b
Saaty TL (1977) A scaling method for priorities in hierarchical structures. J Math Psychol 15(3):234–281
Schneider J-G, Han J (2004) Components–the past, the present, and the future. In: Workshop on component-oriented programming
Shen X, Chen Y, Xing L (2006) Fuzzy optimization models for quality and cost of software systems based on COTS. In: Proceedings of the sixth international symposium on operations research and its applications (ISORA'06). Xinjiang, China, pp 312–318
Sheng J, Wang B (2008) Evaluating COTS components using gap analysis. In: 2008 The 9th international conference for young computer scientists. IEEE, pp 1248–1253
Siddiqui Z, Tyagi K (2016) Application of fuzzy-MOORA method: ranking of components for reliability estimation of component-based software systems. Decis Sci Lett 5(1):169–188
Tang J, Mu L-F, Kwong C, Luo X (2011) An optimization model for software component selection under multiple applications development. Eur J Oper Res 212(2):301–311
Tavana M, Momeni E, Rezaeiniya N, Mirhedayatian SM, Rezaeiniya H (2013) A novel hybrid social media platform selection model using fuzzy ANP and COPRAS-G. Expert Syst Appl 40(14):5694–5702
Thapar SS, Sarangal H (2020) Quantifying reusability of software components using hybrid fuzzy analytical hierarchy process (FAHP)-Metrics approach. Appl Soft Comput 88:105997
Tran V, Liu D-B (1997) A procurement-centric model for engineering component-based software systems. In: Proceedings fifth international symposium on assessment of software tools and technologies. IEEE, pp 70–79
Ustinovichius L, Zavadkas E, Podvezko V (2007) Application of a quantitative multiple criteria decision making (MCDM-1) approach to the analysis of investments in construction. Control Cybern 36(1):251
Vale T, Crnkovic I, De Almeida ES, Neto PAdMS, Cavalcanti YC, de Lemos Meira SR (2016) Twenty-eight years of component-based software engineering. J Syst Softw 111:128–148
Viteikiene M (2006) Sustainable residential areas evaluation. Technol Econ Dev Econ 12(2):152–160
Viteikiene M, Zavadskas EK (2007) Evaluating the sustainability of Vilnius city residential areas. J Civ Eng Manag 13(2):149–155
Wanyama T, Far BH (2005) Towards providing decision support for COTS selection. In: Canadian conference on electrical and computer engineering, 2005. IEEE, pp 908–911
Wind Y, Saaty TL (1980) Marketing applications of the analytic hierarchy process. Manage Sci 26(7):641–658

Zachariah B, Rattihalli R (2007) A multicriteria optimization model for quality of modular software systems. Asia-Pacific J Oper Res 24(06):797–811

Zadeh LA (1996) Fuzzy sets. In: Fuzzy sets, fuzzy logic, and fuzzy systems: selected papers by Lotfi A Zadeh. World Sci 394–432

Zavadskas EK, Antucheviciene J (2007) Multiple criteria evaluation of rural building's regeneration alternatives. Build Environ 42(1):436–451

Zavadskas EK, Kaklauskas A, Kaklauskiene J (2007a) Modelling and forecasting of a rational and sustainable development of Vilnius: emphasis on pollution. Int J Environ Pollut 30(3–4):485–500

Zavadskas EK, Kaklauskas A, Peldschus F, Turskis Z (2007b) Multi-attribute assessment of road design solutions by using the COPRAS method. Balt J Road & Bridg Eng 2(4)

Zavadskas EK, Kaklauskas A, Sarka V (1994) The new method of multicriteria complex proportional assessment of projects. Technol Econ Dev Econ 1(3):131–139

Estimation and Testing Procedures for the Reliability Functions of Exponentiated Generalized Family of Distributions and a Characterization Based on Records

Taruna Kumari and Anupam Pathak

Abstract In this chapter, characterization based on record values for a family of distributions namely; exponentiated generalized family of distributions is provided. Two measures of reliability are considered, namely; $R(t) = P(X > t)$ and $P = P(X > Y)$. Point as well as interval estimation procedures are developed for unknown parameter(s), $R(t)$ and P, based on records. Two types of point estimators are considered, namely; (i) uniformly minimum variance unbiased estimators and (ii) maximum likelihood estimators. Testing procedures are also developed for the hypotheses related to various parametric functions. A comparative study of different methods of estimation is done through simulation studies. Real data example is used to illustrate the results.

Keywords Exponentiated generalized family of distributions · Characterization · Point estimation · Confidence interval · Records · Monte-Carlo simulation

1 Introduction

In this chapter we develop the concept of reliability, introduce the exponentiated generalized (EG) family of distributions and it's characterization. We develop point estimation procedures based on records. As far as point estimation is concerned, we derive uniformly minimum variance unbiased estimators (UMVUES) and maximum likelihood estimators (MLES). A new technique of obtaining these estimators is developed, in which first of all the estimators of powers of parameter are obtained. These estimators are used to obtain the estimators of sampled probability density function (*pdf*) at a specified point, which are subsequently used to obtain the estimators of $R(t)$ and P. The estimators of P are derived for the cases, when X and Y belong to the same as well as different families of distributions. Therefore, in the proposed

T. Kumari
Department of Statistics, University of Delhi, Delhi 110007, India

A. Pathak (✉)
Department of Statistics, Ramjas College, University of Delhi, Delhi 110007, India

A. G. Aggarwal et al. (eds.), *Optimization Models in Software Reliability*,
Springer Series in Reliability Engineering,
https://doi.org/10.1007/978-3-030-78919-0_13

method, the estimators of parameter(s), $R(t)$ and P are interrelated, where as in literature, authors have handled these three estimation problems separately. Moreover, in our method, one does not requires expressions for $R(t)$ and P to estimate them. Further, confidence intervals for parameters, $R(t)$ and P are constructed and testing procedures are also developed for various hypotheses. Finally, we present numerical findings along with real data analysis and conclusions are made on our results.

1.1 Reliability and Reliability Function

Reliability means 'the probability of a device (or item or organism) performing it's defined purpose adequately for a specified period of time, under the operating conditions encountered'. Longer life is identified with greater reliability. Reliability is a popular concept that has been celebrated for years as a commendable attribute of a person or a product. Reliability technology has a potentially wide range of application areas, such as safety/risk analysis, environmental protection, quality/reliability management and verification, optimization of maintenance and operation, engineering design etc.

The reliability function $R(t)$ is defined as the probability of failure-free operation until time t. Thus, if the random variable (rv) X denotes the lifetime of an item, then $R(t) = P(X > t)$. Another measure of reliability under stress-strength set-up is the probability $P = P(X > Y)$, which represents the reliability of an item of random strength X subject to random stress Y. A lot of work has been done in the literature for the point estimation and testing of $R(t)$ and P. For example one may refer to Kelley et al. (1976), Sathe and Shah (1981) and Chao (1982). Constantine et al. (1986) derived UMVUE and MLE of P for gamma distribution. Awad and Gharraf (1986) estimated P for Burr distribution. For estimation of $R(t)$ corresponding to Maxwell distribution, one may refer to Tyagi and Bhattacharya (1989). Chaturvedi and Tomer (2002) derived UMVUE of $R(t)$ and P for negative binomial distribution. For exponentiated Weibull, half logistic, generalized Lomax and Rayleigh distributions, the inferential procedures are available in Chaturvedi and Pathak (2012), Chaturvedi et al. (2016), Pathak and Chaturvedi (2013, 2014), respectively. Inferences have been drawn for $R(t)$ and P for some families of lifetime distributions by Chaturvedi and Pathak (2014), Chaturvedi and Kumari (2015, 2017, 2019), Chaturvedi and Rani (1997), Chaturvedi and Singh (2008), Chaturvedi and Tomer (2003).

1.2 Record and Record Values

Record values are found in many situations of daily life as well as in many statistical applications. Often, we are interested in observing new records and in recording them, for example Olympic records or world records in sport. This theory is largely based on the theory of order statistics and is especially closely related to extreme order statistics. Record values and the associated statistics are of particular interest in the areas of climatology, sports, traffic, medicine, economics etc. A large number of record data saved for a long time motivated the development of several mathematical models reflecting the corresponding record processes and forecasting the future record results. Chandler (1952) defined the model of record statistics as a model for successive extremes in a sequence of independent and identically distributed (iid) random variables (rvs). Ahsanullah (1995) and Arnold et al. (1998) are two useful references which have clarified the notion of records and their properties. Several inferential procedures for the parameters of different distributions, based on record data, have been developed by Arashi and Emadi (2008), Balakrishnan et al. (1995), Belaghi et al. (2015), Habibi et al. (2006), Kumari et al. (2019), Nagaraja (1988a), Razmkhah et al. (2012) and others.

1.3 Characterizations of Distributions

A characterization is a certain distributional or statistical property of a statistic or statistics that uniquely determines the associated stochastic model. In recent years considerable attention has been paid to the problem of characterizing the *pdf* of a rv based on conditional expectations in general, and in particular, on it's mean residual life function. Characterizations of distributions through conditional expectation have been considered by Franco and Ruiz (1996), Khan and Alzaid (2004), Kumari and Pathak (2014a, b, c), Nagaraja (1988b), Raqab (2002) and Ahsanullah et al. (2013). In this chapter we have presented the characterization of EG family of distributions through conditional expectation for record values.

1.4 The EG Family of Distributions

The exponentiated distributions are quite different from their baseline functions and need special investigations. Adding a parameter to baseline cumulative distribution function (*cdf*) *G(x)*, by exponentiation, produces a *cdf F(x)*, say, this is richer and more flexible to modelling data. For example, $F(x) = [G(x)]^{\alpha}$ is flexible enough to accommodate both monotone as well as non-monotone hazard rates. In particular, if *G(x)* is exponential such that $G(x) = \left(1 - e^{-\theta x}\right)$, then the *pdf* $g(x) = \theta e^{-\theta x}$ is monotone decreasing on the positive half of the real line. However, $F(x) = \left[1 - e^{-\theta x}\right]^{\alpha}$ has *pdf* $f(x) = \alpha\theta e^{-\theta x}\left[1 - e^{-\theta x}\right]^{\alpha-1}$ which is unimodal on $\left[0, \infty\right)$ with mode at $X = (\ln\alpha)/\theta$. Furthermore, while the exponential distribution *G(x)* has constant hazard rate θ, it can be shown that the exponentiated exponential *F(x)* has increasing hazard rate if $\alpha > 1$, constant hazard rate if $\alpha = 1$ and decreasing hazard rate if $\alpha < 1$.

In order to investigate and construct more flexible families of distributions, Cordeiro et al. (2013) proposed the EG family of distributions with *pdf* $f(x; \alpha, \beta)$ and the *cdf* $F(x; \alpha, \beta)$ given by

$$f(x; \alpha, \beta) = \alpha\beta g(x)\{1 - G(x)\}^{\beta-1}\left[1 - \{1 - G(x)\}^{\beta}\right]^{\alpha-1} \tag{1}$$

and

$$F(x; \alpha, \beta) = \left[1 - \{1 - G(x)\}^{\beta}\right]^{\alpha}, \tag{2}$$

respectively, for x belonging to the support of G and $\alpha, \beta > 0$, where $g(\cdot)$ denotes the *pdf* corresponding to $G(\cdot)$. α and β are the shape parameters of the EG family of distributions. Cordeiro et al. (2013) derived simple representation for EG family of distributions. They studied general properties of the particular members of EG family of distributions and obtained MLES of it's parameters.

From above, the reliability function $R(t)$ at a specified time $t(>0)$ and the hazard function $h(x; \alpha, \beta)$ are given by

$$R(t) = 1 - \left[1 - \{1 - G(t)\}^{\beta}\right]^{\alpha} \tag{3}$$

and

$$h(x; \alpha, \beta) = \frac{\alpha\beta g(x)\{1 - G(x)\}^{\beta-1}\left[1 - \{1 - G(x)\}^{\beta}\right]^{\alpha-1}}{1 - \left[1 - \{1 - G(x)\}^{\beta}\right]^{\alpha}}, \tag{4}$$

respectively.

1.5 Notations and Definition

Let $X_1, X_2, \ldots, X_n \ldots$ be an infinite sequence of iid rvs from an absolutely continuous distribution function (*df*) $F(\cdot)$ and *pdf* $f(\cdot)$. An observation X_j is called a lower record value if it's value is less than that of all previous observations, i.e., X_j is a lower record if $X_j < X_i$ for all $i<j$. The record time sequence $\{T_n, n \geq 1\}$ is defined in the following manner:

$$T_1 = 1, \text{ with probability } 1,$$

and for $n \geq 2$,

$$T_n = \min\{j : j > T_{n-1}, X_j < X_{T_{n-1}}\}.$$

The lower record value sequence $R_1,\ R_2,\ \ldots,\ R_n$ is defined as:

$$R_n = X_{T_n},\ n = 1,\ 2,\ \ldots$$

The *pdf* of R_n and the joint *pdf* of $R_n,\ R_m\ (R_n < R_m)$, such that $n > m$ {see (Arnold et al. 1998)} are respectively, given by

$$f_{R_n}(r_n) = \frac{1}{\Gamma(n)}[-\ln F(r_n)]^{n-1} f(r_n);\ r_n > 0 \tag{5}$$

and

$$f_{R_n,\ R_m}(r_n,\ r_m) = \frac{[-\ln F(r_m)]^{m-1}}{\lceil m \ \lceil (n-m)}\left[-\ln\left(\frac{F(r_n)}{F(r_m)}\right)\right]^{n-m-1} \frac{f(r_n) f(r_m)}{F(r_m)};\ r_n,\ r_m > 0. \tag{6}$$

Suppose, we observe the first n lower record values $R_1 = r_1,\ R_2 = r_2,\ \ldots,\ R_n = r_n$ from the *df F(·)* and the *pdf f(·)*. Then the joint *pdf* of the first n lower record values {see (Arnold et al. 1998)} is given by

$$f_{R_1,\ R_2,\ \ldots,\ \ldots,\ R_n}(r_1,\ r_2,\ \ldots,\ r_n)$$
$$= f(r_n)\prod_{i=1}^{n-1} h(r_i), \text{ where, } r_1 > r_2 > \ldots > r_n \text{ and } h(r_i) = \frac{f(r_i)}{F(r_i)} \tag{7}$$

2 A Characterization of EG Family of Distribution

This section presents a new characterization of EG family of distributions based on lower record values.

Theorem 1 *Let X be an absolutely continuous (with respect to Lebesgue measure) rv with cdf* $F(x; \alpha, \beta)$. *Assume that* $F(0; \alpha, \beta) = 0$ *and* $F(\infty; \alpha, \beta) = 1$. *Then X has a EG family of distributions given at* (1) *if and only if*

$$(\alpha + k)E\left[\left[1 - \{1 - G(r_{n+1})\}^{\beta}\right]^k \middle| r_m = x\right] = \alpha E\left[\left[1 - \{1 - G(r_n)\}^{\beta}\right]^k \middle| r_m = x\right].$$

Proof Using (5) and (7), the conditional *pdf* of $f_{R_{n+1}|R_m}(r_{n+1}|r_m),\ R_{n+1} < R_m\ (n + 1 > m)$ is given by.

$$f_{R_{n+1}|R_m}(r_{n+1}|r_m) = \frac{1}{(n-m)!}\left[-\ln\frac{F(r_{n+1})}{F(r_m)}\right]^{n-m} \frac{f(r_{n+1})}{F(r_m)}. \tag{8}$$

Then,

$$\mu_{n+1|m} = E\Big[\big[1-\{1-G(r_{n+1})\}^{\beta}\big]^{k}\Big|r_m = x\Big]$$

$$= \frac{1}{(n-m)!F(x;\alpha,\beta)} \int_0^x \big[1-\{1-G(y)\}^{\beta}\big]^{k}\Big[-\ln\frac{F(y;\alpha,\beta)}{F(x;\alpha,\beta)}\Big]^{n-m} f(y;\alpha,\beta)dy$$

$$= \frac{\alpha\beta}{(n-m)!\big[1-\{1-G(x)\}^{\beta}\big]^{\alpha}} \int_0^x \big[1-\{1-G(y)\}^{\beta}\big]^{\alpha+k-1}$$

$$\Big[-\alpha\ln\Big(\frac{1-\{1-G(y)\}^{\beta}}{1-\{1-G(x)\}^{\beta}}\Big)\Big]^{n-m} \cdot g(y)\{1-G(y)\}^{\beta-1}dy.$$

Using the transformation $-\alpha\ln\Big(\frac{1-\{1-G(y)\}^{\beta}}{1-\{1-G(x)\}^{\beta}}\Big) = z$, we get

$$\mu_{n+1|m} = \frac{\big[1-\{1-G(x)\}^{\beta}\big]^{k}}{(n-m)!} \int_0^{\infty} z^{n-m}\exp\Big\{-\Big(\frac{\alpha+k}{\alpha}\Big)z\Big\}dz$$

$$= \Big(\frac{\alpha}{\alpha+k}\Big)^{n-m+1}\big[1-\{1-G(x)\}^{\beta}\big]^{k}.$$

Thus,

$$(\alpha+k)E\Big[\big[1-\{1-G(r_{n+1})\}^{\beta}\big]^{k}\Big|r_m = x\Big] = \alpha\big[1-\{1-G(x)\}^{\beta}\big]^{k}\Big(\frac{\alpha}{\alpha+k}\Big)^{n-m}. \tag{9}$$

Replacing $n+1$ by n in (9), we get

$$(\alpha+k)E\Big[\big[1-\{1-G(r_{n})\}^{\beta}\big]^{k}\Big|r_m = x\Big] = \alpha\big[1-\{1-G(x)\}^{\beta}\big]^{k}\Big(\frac{\alpha}{\alpha+k}\Big)^{n-m-1}. \tag{10}$$

From (9) and (10), we get.

$$(\alpha+k)E\Big[\big[1-\{1-G(r_{n+1})\}^{\beta}\big]^{k}\Big|r_m = x\Big] = \alpha E\Big[\big[1-\{1-G(r_{n})\}^{\beta}\big]^{k}\Big|r_m = x\Big].$$

Now to prove the sufficiency

$$(\alpha+k)E\Big[\big[1-\{1-G(r_{n+1})\}^{\beta}\big]^{k}\Big|r_m = x\Big] = \alpha E\Big[\big[1-\{1-G(r_{n})\}^{\beta}\big]^{k}\Big|r_m = x\Big]$$

then,

$$\frac{(\alpha+k)}{(n-m)!}\int_0^x \left[1-\{1-G(y)\}^{\beta}\right]^k \left[-\ln\frac{F(y;\alpha,\beta)}{F(x;\alpha,\beta)}\right]^{n-m} \frac{f(y;\alpha,\beta)}{F(x;\alpha,\beta)}dy$$

$$=\frac{\alpha}{(n-m-1)!}\int_0^x \left[1-\{1-G(y)\}^{\beta}\right]^k \left[-\ln\frac{F(y;\alpha,\beta)}{F(x;\alpha,\beta)}\right]^{n-m-1} \frac{f(y;\alpha,\beta)}{F(x;\alpha,\beta)}dy.$$

Canceling $F(x;\alpha,\beta)$ from both sides, we get

$$\frac{(\alpha+k)}{(n-m)!}\int_0^x \left[1-\{1-G(y)\}^{\beta}\right]^k \left[-\ln\frac{F(y;\alpha,\beta)}{F(x;\alpha,\beta)}\right]^{n-m} f(y;\alpha,\beta)dy$$

$$=\frac{\alpha}{(n-m-1)!}\int_0^x \left[1-\{1-G(y)\}^{\beta}\right]^k \left[-\ln\frac{F(y;\alpha,\beta)}{F(x;\alpha,\beta)}\right]^{n-m-1} f(y;\alpha,\beta)dy.$$

Differentiating both sides of the above equation with respect to x and simplifying for *(n-m)* times, we get

$$(\alpha+k)\frac{f(x;\alpha,\beta)}{F(x;\alpha,\beta)}\int_0^x \left[1-\{1-G(y)\}^{\beta}\right]^k f(y;\alpha,\beta)dy$$
$$=\alpha\left[1-\{1-G(x)\}^{\beta}\right]^k f(x;\alpha,\beta),$$

or,

$$(\alpha+k)\int_0^x \left[1-\{1-G(y)\}^{\beta}\right]^k f(y;\alpha,\beta)dy=\alpha F(x;\alpha,\beta)\left[1-\{1-G(x)\}^{\beta}\right]^k.$$

Differentiating the above equation with respect to x, we get

$$(\alpha+k)\left[1-\{1-G(x)\}^{\beta}\right]^k f(x;\alpha,\beta)$$
$$=\alpha f(x;\alpha,\beta)\left[1-\{1-G(x)\}^{\beta}\right]^k$$
$$+\alpha\beta F(x;\alpha,\beta)k\left[1-\{1-G(x)\}^{\beta}\right]^{k-1} g(x)\{1-G(x)\}^{\beta-1},$$

which on simplification yields

$$\frac{f(x;\alpha,\beta)}{F(x;\alpha,\beta)}=\frac{\alpha\beta g(x)\{1-G(x)\}^{\beta-1}}{\left[1-\{1-G(x)\}^{\beta}\right]}.$$

3 Point Estimation Procedures

Let $R_1,\ R_2,\ \ldots,\ R_n$ are n lower records from (1).

Lemma 1 *Let* $S=-\ln\left[1-\{1-G(r_n)\}^{\beta}\right]$. *Then, S is complete and sufficient for the distribution given at* (1). *Moreover, the pdf of S is given by*

$$f(s;\ \alpha)=\frac{\alpha^n s^{n-1}}{\Gamma(n)}\exp(-\alpha s);\ \alpha,\ s>0.$$

Proof From (4), the joint *pdf* of $R_1,\ R_2,\ \ldots,\ R_n$ is.

$$f^*(r_1,\ r_2,\ \ldots,\ r_n;\ \alpha,\ \beta)=(\alpha\beta)^n\prod_{i=1}^{n}\frac{g(r_i)\{1-G(r_i)\}^{\beta-1}}{\left[1-\{1-G(r_i)\}^{\beta}\right]}\exp(-\alpha S). \quad (11)$$

It follows from (11) and factorization theorem {see (Rohatgi and Saleh 2012), p. 361)} that S is a sufficient statistics for α.

From (1), the *pdf* of R_n is given by

$$f_{R_n}(r_n;\ \alpha,\ \beta)=\frac{\alpha\beta}{\Gamma(n)}g(r_n)\{1-G(r_n)\}^{\beta-1}\left[1-\{1-G(r_n)\}^{\beta}\right]^{\alpha-1}$$
$$\left(-\alpha\ln\left[1-\{1-G(r_n)\}^{\beta}\right]\right)^{n-1}. \quad (12)$$

The distribution of S follows from (12) and standard transformation technique of rvs.

Since the distribution of S belongs to one-parameter exponential family of distributions for known β, it is also complete {see (Rohatgi and Saleh 2012), p. 367)}.

3.1 UMVUES of α, R(t) *and* P, *When* β *is Known*

The following theorem provides UMVUE of powers of α.

Theorem 2 *For* $q\in(-\infty,\ \infty),\ q\neq 0$, *the UMVUE of* α^q *is given by*

$$\hat{\alpha}^q=\begin{cases}\frac{\Gamma(n)}{\Gamma(n-q)}S^{-q};\ n>q,\\ 0;\ otherwise.\end{cases}$$

Proof From Lemma 1,

$$E\left[\frac{\Gamma(n)}{\Gamma(n-q)}S^{-q}\right] = \alpha^q;\ n > q.$$

Hence, the theorem follows from Lehmann-Scheffè theorem {see (Rohatgi and Saleh 2012), p. 367)}.

The following lemma provides UMVUE of the sampled *pdf* (1) at a specified point 'x'.

Lemma 2 *The UMVUE of the sampled pdf (1) at a specified point 'x' is given by*

$$\hat{f}(x;\alpha,\beta) = \begin{cases} \dfrac{(n-1)\beta\{1-G(x)\}^{\beta-1}g(x)}{S\left[1-\{1-G(x)\}^{\beta}\right]}\left(1+S^{-1}\ln\left[1-\{1--G(x)\}^{\beta}\right]\right)^{n-2}; \\ \qquad -S < \ln\left[1-\{1-G(x)\}^{\beta}\right], \\ 0;\ otherwise. \end{cases}$$

Proof We can write (1) as

$$f(x;\alpha,\beta) = \frac{\beta\{1-G(x)\}^{\beta-1}g(x)}{\left[1-\{1-G(x)\}^{\beta}\right]}\sum_{i=0}^{\infty}\frac{1}{i!}\left(\ln\left[1-\{1-G(x)\}^{\beta}\right]\right)^{i}\alpha^{i+1}. \quad (13)$$

Using Lemma 1 of Chaturvedi and Tomer (2002), Theorem 2 and (13), the UMVUE of $f(x;\alpha,\beta)$ at a specified point 'x' is given by

$$\hat{f}(x;\alpha,\beta) = \frac{\beta\{1-G(x)\}^{\beta-1}g(x)}{\left[1-\{1-G(x)\}^{\beta}\right]}\sum_{i=0}^{\infty}\frac{1}{i!}\left(\ln\left[1-\{1-G(x)\}^{\beta}\right]\right)^{i}\hat{\alpha}^{i+1}$$

$$= \frac{(n-1)\beta\{1-G(x)\}^{\beta-1}g(x)}{S\left[1-\{1-G(x)\}^{\beta}\right]}\sum_{i=0}^{n-2}\binom{n-2}{i}\left(S^{-1}\ln\left[1-\{1-G(x)\}^{\beta}\right]\right)^{i}$$

and the lemma follows.

In the following theorem, we obtain the UMVUE of the reliability function *R(t)*.

Theorem 3 *The UMVUE of the reliability function R(t) is given by*

$$\hat{R}(t) = \begin{cases} 1-\left(1+S^{-1}\ln\left[1-\{1-G(t)\}^{\beta}\right]\right)^{n-1}; \\ \qquad -S < \ln\left[1-\{1-G(t)\}^{\beta}\right], \\ 1;\ otherwise. \end{cases}$$

Proof Since $g(x;s) = f(x;\alpha)f(s;\alpha)$ is a continuous function of *(X, S)* on the rectangle $[t, \infty) \times [0, \infty)$, the conditions of Fubini's theorem {see (Bilodeau et al. 2010), p. 207)} are satisfied for the change of order of integration. Let us consider the expected value of the integral $\int_t^\infty \hat{f}(x; \alpha, \beta)dx$ with respect to s, i.e.,

$$\int_0^\infty \left\{ \int_t^\infty \hat{f}(x; \alpha, \beta)dx \right\} f(s; \alpha)ds = \int_t^\infty \left[E_S\left(\hat{f}(x; \alpha, \beta) \right) \right] dx$$

$$= \int_t^\infty f(x; \alpha, \beta)dx = R(t). \quad (14)$$

We conclude from (14) that the UMVUE of *R(t)* can be obtained simply by integrating $\hat{f}(x; \alpha, \beta)$ from t to ∞. Thus, from Lemma 2,

$$\hat{R}(t) = \frac{(n-1)\beta}{S} \int_t^\infty \frac{\{1 - G(x)\}^{\beta-1} g(x)}{\left[1 - \{1 - G(x)\}^\beta\right]} \left(1 + S^{-1} \ln\left[1 - \{1 - G(x)\}^\beta\right]\right)^{n-2} dx$$

$$= (n-1) \int_{S^{-1}\ln\left[1-\{1-G(t)\}^\beta\right]}^{0} (1+v)^{n-2} dv; \quad -S < \ln\left[1 - \{1 - G(t)\}^\beta\right]$$

Hence, the theorem follows.

In order to obtain the UMVUE of *P*, let *X* and *Y* be two independent rvs following the classes of distributions $f(x; \alpha_1, \beta_1)$ and $f(y; \alpha_2, \beta_2)$, respectively, where

$$f(x; \alpha_1, \beta_1) = \alpha_1 \beta_1 g(x) \{1 - G(x)\}^{\beta_1 - 1} \left[1 - \{1 - G(x)\}^{\beta_1}\right]^{\alpha_1 - 1}; \; x > 0, \; \alpha_1, \beta_1 > 0$$

and

$$f(y; \alpha_2, \beta_2) = \alpha_2 \beta_2 h(y) \{1 - H(y)\}^{\beta_2 - 1} \left[1 - \{1 - H(y)\}^{\beta_2}\right]^{\alpha_2 - 1}; \; y > 0, \; \alpha_2, \beta_2 > 0.$$

Let $\{R_n\}$ and $\{R_m^*\}$ be the record value sequences for *X's* and *Y's*, respectively and define $S = -\ln\left[1 - \{1 - G(r_n)\}^{\beta_1}\right]$ and $T = -\ln\left[1 - \{1 - H(r_m^*)\}^{\beta_2}\right]$.

The following theorem provides the UMVE of *P*, when *X* and *Y* belong to different family of distributions.

Theorem 4 *The UMVUE of P is given by*

$$\hat{P} = \begin{cases} 1-(m-1)\int\limits_0^c \left(1+S^{-1}\ln\left[1-\left\{1-G\left[H^{-1}\left\{1-\left(1-e^{Tv}\right)^{\beta_2^{-1}}\right\}\right]\right\}^{\beta_1}\right]\right)^{n-1} \\ (1-v)^{m-2}dv;\ G^{-1}\left\{1-\left(1-e^{-S}\right)^{\beta_1^{-1}}\right\} > H^{-1}\left\{1-\left(1-e^{-T}\right)^{\beta_2^{-1}}\right\}, \\ 1-(m-1)\int\limits_0^1 \left(1+S^{-1}\ln\left[1-\left\{1-G\left[H^{-1}\left\{1-\left(1-e^{Tv}\right)^{\beta_2^{-1}}\right\}\right]\right\}^{\beta_1}\right]\right)^{n-1} \\ (1-v)^{m-2}dv;\ G^{-1}\left\{1-\left(1-e^{-S}\right)^{\beta_1^{-1}}\right\} < H^{-1}\left\{1-\left(1-e^{-T}\right)^{\beta_2^{-1}}\right\}, \end{cases}$$

where $c = -T^{-1}\ln\left[1-\left\{1-H\left[G^{-1}\left\{1-(1-e^{-S})^{\beta_1^{-1}}\right\}\right]\right\}^{\beta_2}\right]$.

Proof It follows from Lemma 2 that the UMVUES of $f(x;\alpha_1,\beta_1)$ and $f(y;\alpha_2,\beta_2)$ at a specified point 'x' and 'y' are.

$$\hat{f}(x;\alpha_1,\beta_1) = \begin{cases} \dfrac{(n-1)\beta_1\{1-G(x)\}^{\beta_1-1}g(x)}{S\left[1-\{1-G(x)\}^{\beta_1}\right]}\left(1+S^{-1}\ln\left[1-\{1-G(x)\}^{\beta_1}\right]\right)^{n-2}; \\ \qquad -S < \ln\left[1-\{1-G(x)\}^{\beta_1}\right], \\ 0;\ otherwise \end{cases}$$

and

$$\hat{f}(y;\alpha_2,\beta_2) = \begin{cases} \dfrac{(m-1)\beta_2\{1-H(y)\}^{\beta_2-1}h(y)}{T\left[1-\{1-H(y)\}^{\beta_2}\right]}\left(1+T^{-1}\ln\left[1-\{1-H(y)\}^{\beta_2}\right]\right)^{m-2}; \\ \qquad -T < \ln\left[1-\{1-H(y)\}^{\beta_2}\right], \\ 0;\ otherwise, \end{cases}$$

respectively.

From the arguments similar to those adopted in the proof of Theorem 3,

$$\hat{P} = \int\limits_{y=0}^{\infty}\int\limits_{x=y}^{\infty} \hat{f}(x;\alpha_1,\beta_1)\hat{f}(y;\alpha_2,\beta_2)dxdy$$

$$\hat{P} = \int\limits_{y=0}^{\infty} \hat{R}(y;\alpha_1,\beta_1)\hat{f}(y;\alpha_2,\beta_2)dy;\ -S < \ln\left[1-\{1-G(y)\}^{\beta_1}\right],$$
$$-T < \ln\left[1-\{1-H(y)\}^{\beta_2}\right]$$

$$\hat{P} = 1 - (m-1)\beta_2 \int_{c'}^{\infty} \left(1 + S^{-1}\ln\left[1 - \{1 - G(y)\}^{\beta_1}\right]\right)^{n-1}$$

$$\cdot \frac{\{1 - H(y)\}^{\beta_2 - 1} h(y)}{T\left[1 - \{1 - H(y)\}^{\beta_2}\right]} \left(1 + T^{-1}\ln\left[1 - \{1 - H(y)\}^{\beta_2}\right]\right)^{m-2} dy,$$

where $c' = \max\left[G^{-1}\left\{1 - \left(1 - e^{-S}\right)^{\beta_1^{-1}}\right\},\ H^{-1}\left\{1 - \left(1 - e^{-T}\right)^{\beta_2^{-1}}\right\}\right]$.

The theorem now follows on considering the two cases and putting $v = -T^{-1}\ln\left[1 - \{1 - H(y)\}^{\beta_2}\right]$.

In the following theorem, we obtain the UMVUE of P, when X and Y belong to same family of distributions.

Theorem 5 *When $G(\cdot) \overset{d}{=} H(\cdot)$ and $\beta_1 = \beta_2 = \beta$, say, the UMVUE of P is*

$$\hat{P} = \begin{cases} 1 - (m-1)\sum_{i=0}^{m-2}(-1)^i\binom{m-2}{i}B(i+1,\ n)\left(\frac{S}{T}\right)^{i+1}; & S < T, \\ 1 - (m-1)\sum_{j=0}^{n-1}(-1)^j\binom{n-1}{j}B(j+1,\ m-1)\left(\frac{T}{S}\right)^{j}; & S > T. \end{cases}$$

Proof Taking $G(\cdot) \overset{d}{=} H(\cdot)$ and $\beta_1 = \beta_2 = \beta$, in Theorem 4, for $S < T$, we get

$$\hat{P} = 1 - (m-1)\int_0^{S/T}\left(1 - \frac{T}{S}v\right)^{n-1}(1-v)^{m-2}dv$$

$$= 1 - (m-1)\int_0^1 (1-w)^{n-1}\left(1 - \frac{S}{T}w\right)^{m-2}\left(\frac{S}{T}\right)dw$$

$$= 1 - (m-1)\sum_{i=0}^{m-2}(-1)^i\binom{m-2}{i}\left(\frac{S}{T}\right)^{i+1}\int_0^1 w^i(1-w)^{n-1}dw$$

and the first assertion follows. Similarly, we can prove the second assertion.

3.2 *MLES of* R(t) *and* P*, When All Parameters Are Unknown*

Let $R_1,\ R_2, \ldots,\ R_n$ be the first n lower records from (1). Let us denote by $\Theta = (\alpha,\ \beta)$, where α and β are unknown. The log-likelihood function for observing Θ, based on the first n lower records is given by

$$\ln L(\Theta|r_1, r_2, \ldots, r_n) = n\ln\alpha + n\ln\beta + \alpha\ln\left[1 - \{1 - G(r_n)\}^{\beta}\right]$$
$$+ \sum_{i=1}^{n}\ln g(r_i) + (\beta - 1)\sum_{i=1}^{n}\ln\{1 - G(r_i)\} - \sum_{i=1}^{n}\ln\left[1 - \{1 - G(r_i)\}^{\beta}\right]. \tag{15}$$

Differentiating (15) with respect to all unknown parameters and equating these differential coefficients to zero, we get

$$\frac{n}{\alpha} + \ln\left[1 - \{1 - G(r_n)\}^{\beta}\right] = 0 \tag{16}$$

and

$$\frac{n}{\beta} - \frac{\alpha\{1 - G(r_n)\}^{\beta}\ln\{1 - G(r_n)\}}{\left[1 - \{1 - G(r_n)\}^{\beta}\right]} + \sum_{i=1}^{n}\ln\{1 - G(r_i)\}$$
$$+ \sum_{i=1}^{n}\frac{\{1 - (r_i)\}^{\beta}\ln\{1 - G(r_i)\}}{\left[1 - \{1 - G(r_i)\}^{\beta}\right]} = 0. \tag{17}$$

Solving (16) and (17) simultaneously, let $\tilde{\alpha}$ and $\tilde{\beta}$ are the MLES of α and β, respectively. Moreover, from (16), we get

$$\tilde{\alpha} = \frac{-n}{\ln\left[1 - \{1 - G(r_n)\}^{\tilde{\beta}}\right]}. \tag{18}$$

Corollary 1 *When β is known, the MLE of α is given by*

$$\tilde{\alpha} = \frac{-n}{\ln\left[1 - \{1 - G(r_n)\}^{\beta}\right]} = \frac{n}{S}.$$

Proposition 1 *The MLE of α given in Corollary 1 has an inverse Gamma (IG) distribution with parameter n and $n\alpha$, i.e., $\tilde{\alpha} \sim IG(n, n\alpha)$.*

Proof Proposition directly follows from Lemma 1 and Corollary 1.

In the following lemma, we provide the MLE of the sampled *pdf* at a specified point 'x'.

Lemma 3 *The MLE of $f(x;\ \alpha,\ \beta)$ at a specified point 'x' is*

$$\tilde{f}(x; \alpha, \beta) = \tilde{\alpha}\tilde{\beta}g(x)\{1 - G(x)\}^{\tilde{\beta}-1}\left[1 - \{1 - G(x)\}^{\tilde{\beta}}\right]^{\tilde{\alpha}-1}.$$

Proof The proof follows from (1) and one-to-one property of the MLES.

The following theorem and corollary follows from the invariance property of MLES.

Theorem 6 *The MLE of R(t) is given by*

$$\tilde{R}(t) = 1 - \left[1 - \{1 - G(t)\}^{\tilde{\beta}}\right]^{\tilde{\alpha}}.$$

Corollary 2 *When β is known the MLE of R(t) is given by*

$$\tilde{R}(t) = 1 - \left[1 - \{1 - G(t)\}^{\beta}\right]^{\tilde{\alpha}}.$$

The following theorem provides the MLE of P, when X and Y belong to different family of distributions.

Theorem 7 *The MLE of P, when X and Y belongs to different family of distributions is*

$$\tilde{P} = \tilde{\alpha}_2 \int_0^1 \left(1 - \left[1 - \left\{1 - G\left[H^{-1}\left(1 - u^{\tilde{\beta}_2^{-1}}\right)\right]\right\}^{\tilde{\beta}_1}\right]^{\tilde{\alpha}_1}\right)(1 - u)^{\tilde{\alpha}_2 - 1} du,$$

where $\left(\tilde{\alpha}_1, \tilde{\beta}_1\right)$ and $\left(\tilde{\alpha}_2, \tilde{\beta}_2\right)$ are the MLES of (α_1, β_1) and (α_2, β_2) based on first n and m lower records $(r_1, r_2, \ldots, r_n)$ and $\left(r_1^*, r_2^*, \ldots, r_m^*\right)$, respectively.

Proof The proof of the above theorem is similar to the proof of Theorem 4.

The following theorem provides the MLE of P, when X and Y belong to same family of distributions.

Corollary 3 *The MLE of P, when X and Y belong to same family of distributions and $\beta_1 \neq \beta_2$ are known*

$$\tilde{P} = \tilde{\alpha}_2 \int_0^1 \left(1 - \left[1 - \left\{1 - G\left[H^{-1}\left(1 - u^{\beta_2^{-1}}\right)\right]\right\}^{\beta_1}\right]^{\tilde{\alpha}_1}\right)(1 - u)^{\tilde{\alpha}_2 - 1} du,$$

where $(\tilde{\alpha}_1, \tilde{\alpha}_2)$ are the MLES of (α_1, α_2) based on first n and m lower records $(r_1, r_2, \ldots, r_n)$ and $\left(r_1^*, r_2^*, \ldots, r_m^*\right)$, respectively.

Theorem 8 *When $G(\cdot) \stackrel{d}{=} H(\cdot)$ and $\beta_1 = \beta_2 = \beta$, say, the MLE of P is given by $\widetilde{P} = \frac{\widetilde{\alpha}_1}{\widetilde{\alpha}_1 + \widetilde{\alpha}_2}$.*

Proposition 2 *The pdf of MLE of P, when $G(\cdot) \stackrel{d}{=} H(\cdot)$ and $\beta_1 = \beta_2 = \beta$, say, is given by*

$$f(\widetilde{P}) = \frac{1}{B\left(\frac{n}{2}, \frac{m}{2}\right)} \frac{C^{\frac{m}{2}}(\widetilde{P})^{\frac{m}{2}-1}(1 - \widetilde{P})^{\frac{n}{2}-1}}{(1 + \widetilde{P}(C - 1))^{\frac{n+m}{2}}}; 0 < \widetilde{P} < 1, \text{ where } C = \frac{m\alpha_2}{n\alpha_1}.$$

Proof From Proposition 1, $\tilde{\alpha}_1 \sim IG(n, n\alpha_1)$ and $\tilde{\alpha}_2 \sim IG(m, m\alpha_2)$. It is clear that $W_1 = \frac{n\alpha_1}{\tilde{\alpha}_1} \sim Gamma(2n)$ and $W_2 = \frac{m\alpha_2}{\tilde{\alpha}_2} \sim Gamma(2m)$. Using Theorem 8, $\tilde{P}$ can be rewritten as $\tilde{P} = \frac{1}{1+CW}$, where $W = \frac{W_1}{W_2} = \frac{n\alpha_1}{m\alpha_2}\frac{\tilde{\alpha}_2}{\tilde{\alpha}_1} \sim Beta_2\left(\frac{n}{2}, \frac{m}{2}\right)$ and $C = \frac{m\alpha_2}{n\alpha_1}$.

The result follows on using standard transformation technique of rvs.

Remark From Proposition 1, $E(\tilde{\alpha}) = \frac{n}{n-1}\alpha;\ n > 1$ and $V(\tilde{\alpha}) = \frac{n^2}{(n-1)^2(n-2)}\alpha^2;\ n > 2$. It is clear that $\frac{n-1}{n}\tilde{\alpha}$ is an unbiased estimator of α. Moreover, as $n \to \infty$, $E(\tilde{\alpha}) \to \alpha$ and $V(\tilde{\alpha}) \to 0$. Hence $\tilde{\alpha}$ is a consistent estimator of α.

4 Confidence Interval for α, β, $R(t)$ and P

The Fisher information matrix of $\Theta = (\alpha, \beta)$ is

$$I(\Theta) = -E\begin{pmatrix} \frac{\partial^2 \ln L}{\partial\alpha^2} & \frac{\partial^2 \ln L}{\partial\alpha\partial\beta} \\ \frac{\partial^2 \ln L}{\partial\alpha\partial\beta} & \frac{\partial^2 \ln L}{\partial\beta^2} \end{pmatrix}, \text{ where}$$

$$\frac{\partial^2 \ln L}{\partial\alpha^2} = \frac{-n}{\alpha^2},$$

$$\frac{\partial^2 \ln L}{\partial\alpha\partial\beta} = \frac{-\{1-G(r_n)\}^{\beta}\ln\{1-G(r_n)\}}{\left[1-\{1-G(r_n)\}^{\beta}\right]}$$

and

$$\frac{\partial^2 \ln L}{\partial\beta^2} = \frac{-n}{\beta^2} - \alpha\left[\frac{\ln\{1-G(r_n)\}}{\{1-G(r_n)\}^{-\beta}-1}\right]^2\{1-G(r_n)\}^{-\beta} + \sum_{i=1}^{n}\left[\frac{\ln\{1-G(r_i)\}}{\{1-G(r_i)\}^{-\beta}-1}\right]^2\{1-G(r_i)\}^{-\beta}.$$

It is very difficult to obtain the expectation of the above expression, so we use observe Fisher information matrix which can be obtained by removing the expectation sign. The asymptotic variance–covariance matrix of the MLES is the inverse of $I(\Theta)$, which will enable us to find out the variances of $\tilde{\alpha}$ and $\tilde{\beta}$, and thereafter we can construct confidence intervals for α and β, respectively.

Assuming asymptotic normality of the MLES, confidence intervals for α and β are given by

$$\left(\tilde{\alpha} - z_{\nu/2}SD(\tilde{\alpha}),\ \tilde{\alpha} + z_{\nu/2}SD(\tilde{\alpha})\right) \text{ and } \left(\tilde{\beta} - z_{\nu/2}SD(\tilde{\beta}),\ \tilde{\beta} + z_{\nu/2}SD(\tilde{\beta})\right), \tag{19}$$

respectively, where $SD(\cdot) = \sqrt{Var(\cdot)}$ and $z_{\nu/2}$ is the $100(1-\nu/2)$ percentile point of standard normal distribution.

Using above confidence intervals, one can easily obtain the $100(1-\nu)\%$ asymptotic confidence interval for *R(t)* as

$$\left(\begin{array}{c} 1-\left[1-\{1--G(t)\}^{\tilde{\beta}-z_{\nu/2}SD(\tilde{\beta})}\right]^{\tilde{\alpha}-z_{\nu/2}SD(\tilde{\alpha})}, \\ 1-\left[1-\{1-G(t)\}^{\tilde{\beta}+z_{\nu/2}SD(\tilde{\beta})}\right]^{\tilde{\alpha}+z_{\nu/2}SD(\tilde{\alpha})} \end{array}\right). \tag{20}$$

Now, we develop confidence interval for the ratio of shape parameters, i.e., $\tau = \frac{\alpha_1}{\alpha_2}$, when β_1 and β_2 are known. Using the fact that $2\alpha_1 S \sim \chi^2_{2n}$ and $2\alpha_2 T \sim \chi^2_{2m}$, we have $F = \frac{2\alpha_1 S/2n}{2\alpha_2 T/2m} \sim F_{2n,\,2m}$. Therefore, F is a pivotal quantity for τ, and a $100(1-\nu)\%$ confidence interval for τ is given as

$$\left(kF_{2n,\,2m}(1-\nu/2),\; kF_{2n,\,2m}(\nu/2)\right), \tag{21}$$

where $k = \frac{n\ln\left[1-\{1-H(r^*_m)\}^{\beta_2}\right]}{m\ln\left[1-\{1-G(r_n)\}^{\beta_1}\right]}$ and $F_{2n,\,2m}(\delta)$ is the δth percentile of F-distribution with $(2n, 2\,m)$ degrees of freedom.

Next, we develop confidence interval for common shape parameter α. Here, we suppose that the shape parameters of two family of distributions are equal, i.e., $\alpha_1 = \alpha_2 = \alpha$, when β_1 and β_2 are known. Moreover, $X \sim f(x;\alpha,\beta_1)$ and $Y \sim f(y;\alpha,\beta_2)$. Therefore, the joint distribution of record values from these distributions can be written as

$$\begin{aligned} &L(\Theta|r_1, r_2, \ldots, r_n, r^*_1, r^*_2, \ldots, r^*_m) \\ &= \alpha^{n+m}\beta_1^n\beta_2^m\left[1-\{1-G(r_n)\}^{\beta_1}\right]^{\alpha}\prod_{i=1}^{n}\frac{g(r_i)\{1-G(r_i)\}^{\beta_1-1}}{\left[1-\{1-G(r_i)\}^{\beta_1}\right]} \\ &\left[1-\left\{1-H(r^*_m)\right\}^{\beta_2}\right]^{\alpha}\prod_{j=1}^{m}\frac{h(r^*_j)\left\{1-H(r^*_j)\right\}^{\beta_2-1}}{\left[1-\left\{1-H(r^*_j)\right\}^{\beta_2}\right]}, \end{aligned}$$

where $\Theta = (\alpha,\ \beta_1,\ \beta_2)$.

It can be easily shown that the MLE of the common shape parameter α is given by

$$\tilde{\alpha} = \frac{-(n+m)}{\ln\left[1-\{1-G(r_n)\}^{\beta_1}\right]+\ln\left[1-\left\{1-H(r^*_m)\right\}^{\beta_2}\right]} = \frac{(n+m)}{S+T}.$$

Using the fact that $2\alpha(S+T) \sim \chi^2_{2n+2m}$, The 100(1−ν)% confidence interval for α is given by

$$\left(\frac{\chi^2_{2n+2m}(1-\nu/2)}{2h'}, \frac{\chi^2_{2n+2m}(\nu/2)}{2h'}\right), \text{ where } h' = S + T \tag{22}$$

In order to obtain the confidence interval for P, we consider the case when X and Y belong to same family of distribution and $\beta_1 = \beta_2 = \beta$ is known. From Proposition 2, it is clear that

$$Z = \frac{(W_1/2n)}{(W_2/2m)} \sim F_{2n,\,2m} \; and \; P = \frac{1}{1 + \frac{1}{Z}\frac{\tilde{\alpha}_2}{\tilde{\alpha}_1}} \Rightarrow Z = \frac{P}{1-P}\frac{\tilde{\alpha}_2}{\tilde{\alpha}_1} \sim F_{2n,\,2m}.$$

Using above, 100(1 − ν)% confidence interval for P is given by

$$\left(\left\{\frac{\tilde{\alpha}_2}{\tilde{\alpha}_1 F_{2n,\,2m}(1-\nu/2)} + 1\right\}^{-1}, \left\{\frac{\tilde{\alpha}_2}{\tilde{\alpha}_1 F_{2n,\,2m}(\nu/2)} + 1\right\}^{-1}\right), \tag{23}$$

where $F_{2n,\,2m}(\nu/2)$ is the upper 100 $(1-\nu/2)$ percentile point of F-distribution with $(2n,\, 2\, m)$ degrees of freedom.

5 Testing Procedures for Various Hypotheses

Here, we consider the case when β is known. An important hypothesis in life-testing experiments is $H_o : \alpha = \alpha_o$ against $H_1 : \alpha \neq \alpha_o$. It follows from (11) that, under H_o

$$\underset{\Theta_o}{Sup}\, L(\Theta|r_1,\, r_2,\, \ldots,\, r_n) = \alpha_o^n \beta^n \prod_{i=1}^{n} \frac{g(r_i)\{1-G(r_i)\}^{\beta-1}}{[1-\{1-G(r_i)\}^{\beta}]} \exp(-\alpha_o S);$$

$$\Theta_o = \{\alpha : \alpha = \alpha_o\}$$

and

$$\underset{\Theta}{Sup}\, L(\Theta|r_1,\, r_2,\, \ldots,\, r_n) = \left(\frac{n}{S}\right)^n \beta^n \prod_{i=1}^{n} \frac{g(r_i)\{1-G(r_i)\}^{\beta-1}}{[1-\{1-G(r_i)\}^{\beta}]} \exp(-n);$$

$$\Theta = \{\alpha : \alpha > 0\}.$$

Therefore, the likelihood ratio (LR) is given by

$$\frac{\underset{\Theta_o}{Sup}\, L(\Theta|r_1,\ r_2,\ ...,\ r_n)}{\underset{\Theta}{Sup}\, L(\Theta|r_1,\ r_2,\ ...,\ r_n)} = \left(\frac{\alpha_o S}{n}\right)^n \exp(-\alpha_o S + n). \tag{24}$$

We note that the first term on the right hand side of (24) is monotonically increasing and the second term is monotonically decreasing function of S. Denoting by $\chi^2_{2n}(\cdot)$, the chi-square statistic with $2n$ degrees of freedom and using the fact that $2\alpha_o S \sim \chi^2_{2n}$, the critical region is given by

$$\{0 < S < k_o\}U\left\{k_o^{'} < S < \infty\right\},$$

where k_o and $k_o^{'}$ are obtained such that $k_o = (2\alpha_o)^{-1}\chi^2_{2n}(1 - \nu/2)$ and $k_o^{'} = (2\alpha_o)^{-1}\chi^2_{2n}(\nu/2)$.

Another important hypothesis in life-testing experiments is $H_o : \alpha \leq \alpha_o$ against $H_1 : \alpha > \alpha_o$. It follows from (11) that, for $\alpha_1 < \alpha_2$,

$$\frac{L(\Theta_1|r_1,\ r_2,\ ...,\ r_n)}{L(\Theta_2|r_1,\ r_2,\ ...,\ r_n)} = \frac{\alpha_1^n}{\alpha_2^n}\exp((\alpha_2 - \alpha_1)S);$$
$$\Theta_1 = \{\alpha : \alpha = \alpha_1\},\ \Theta_2 = \{\alpha : \alpha = \alpha_2\}. \tag{25}$$

It follows from (25) that $f(x; \alpha, \beta)$ has monotone likelihood ratio in S. Thus, the uniformly most powerful critical region (UMPCR) for testing H_o against H_1 is given by

$$\phi = \begin{cases} 1, & if\ S < k_o^{''}, \\ 0, & otherwise, \end{cases} \quad \text{where } k_o^{''} = (2\alpha_o)^{-1}\chi^2_{2n}(1 - \nu).$$

6 Simulation Study and Real Data Analysis

6.1 A Particular Case and Algorithm

In this section, we consider exponentiated exponential distribution as a particular case of EG family of distributions, which is given at (1).

The rv X is said to follow the exponentiated exponential distribution, if it's *pdf* and *cdf* are given by

$$f(x; \alpha, \beta) = \alpha\beta e^{-\beta x}\left(1 - e^{-\beta x}\right)^{\alpha-1};\ x,\ \alpha,\ \beta > 0 \tag{26}$$

and

$$F(x; \alpha, \beta) = \left(1 - e^{-\beta x}\right)^{\alpha};\ x,\ \alpha,\ \beta > 0. \tag{27}$$

In order to investigate the behavior of the proposed estimators, a simulation study using Monte-Carlo method is carried out. Since all considered inference procedures depends only on the smallest record, therefore algorithm of drawing smallest record is of great importance.

The simulation procedure involves the following steps:

1. Choose the value of n, α and β.
2. Using (26), one can easily show that $-\alpha \ln\left(1 - e^{-\beta x}\right)$ follows *Exp(1)*. Such a rv X admits the representation.
   $X \overset{d}{=} \frac{-1}{\beta} \ln\left(1 - \exp\left(-\frac{X^*}{\alpha}\right)\right)$, where X* is an *Exp(1)* variate.
   Consequently, for $n = 1, 2, 3, 4, \ldots$, we have

$$R_n \overset{d}{=} \frac{-1}{\beta} \ln\left(1 - \exp\left(\frac{-1}{\alpha} \sum_{i=1}^{n} X_i^*\right)\right),$$

where R_n is the nth lower record {see Arnold et al. (1998)} (28)

3. Use (28) to generate $k = 1000$ independent random samples of lower records each of size n.
4. Compute k values of S corresponding to above k lower records.
5. Compute k estimates of α, $R(t)$ and P corresponding to k values of S and then obtain their averages along with mean square error (MSE), respectively.

6.2 Simulation Studies

In order to check the performance of the estimators of α (when β is known), we have simulated 1000 random samples of lower records using (26) and (28), each of size n = 4(4)16 with different pair of values of $(\alpha, \beta) = (1.5, 1.5), (1.5, 2.0), (2.0, 1.5)$ and $(2.0, 2.0)$. With the help of Theorem 2 and Corollary 1, average maximum likelihood (ML) and uniformly minimum variance (UMVU) estimates of α along with their mean square errors (MSES) are obtained. These results are reported in Table 1. Under the same set-up, using Theorem 3 and Corollary 2, we have computed average ML and UMVU estimates of $R(t)$ along with their MSES. For $t = 0.3(0.3)1.5$, results are presented in Table 2.

In order to investigate the performance of estimators of P obtained in Theorem 5 and 8 (when $\beta_1 = \beta_2 = \beta$ are known), we have simulated 1000 random samples of lower records using (26) and (28), from each of the populations X and Y with sizes $(n, m) = (4, 4), (8, 8), (12, 12)$ and $(16, 16)$ with $\alpha_1 = 1.5(0.5)2.0$, $\alpha_2 = 1.0(1.0)5.0$ and $\beta_1 = \beta_2 = \beta = 1.5(0.5)2.0$. For different combinations of parametric values, we have computed actual values of P, average ML and UMVU estimates of P along with their MSES. Obtained results are presented in Table 3. In order to investigate the performance of estimators of P obtained in Theorem 4 and Corollary 3 (when

Table 1 Estimates of with their corresponding MSES in parenthesis

$n \rightarrow$		4		8		12		16	
$\alpha \downarrow$	$\beta \downarrow$	$\tilde{\alpha}$	$\hat{\alpha}$	$\tilde{\alpha}$	$\hat{\alpha}$	$\tilde{\alpha}$	$\hat{\alpha}$	$\tilde{\alpha}$	$\hat{\alpha}$
1.5	1.5	2.0131 (7.9022)	1.5099 (4.297)	1.7051 (0.5172)	1.492 (0.3638)	1.6395 (0.3027)	1.5029 (0.238)	1.5948 (0.1885)	1.4951 (0.1578)
	2.0	1.9752 (1.9581)	1.4814 (0.9747)	1.7177 (0.5306)	1.503 (0.37)	1.6425 (0.296)	1.5056 (0.2317)	1.6007 (0.1887)	1.5006 (0.1569)
2.0	1.5	2.692 (4.3832)	2.019 (2.1965)	2.2718 (0.9581)	1.9878 (0.6772)	2.1723 (0.5052)	1.9913 (0.3996)	2.1362 (0.3625)	2.0027 (0.3023)
	2.0	2.643 (3.6515)	1.9823 (1.8217)	2.2774 (0.9594)	1.9927 (0.6757)	2.1789 (0.506)	1.9974 (0.3982)	2.1407 (0.3523)	2.0069 (0.2923)

Table 2 Estimates of $R(t)$ with their corresponding MSES in parenthesis

$n \rightarrow$				4		8		12		16	
$\alpha \downarrow$	$\beta \downarrow$	t	$R(t)$	$\tilde{R}(t)$	$\hat{R}(t)$	$\tilde{R}(t)$	$\hat{R}(t)$	$\tilde{R}(t)$	$\hat{R}(t)$	$\tilde{R}(t)$	$\hat{R}(t)$
1.5	1.5	0.3	0.7819	0.7961 (0.019)	0.7841 (0.029)	0.7879 (0.011)	0.7801 (0.014)	0.7882 (0.008)	0.7828 (0.009)	0.7362 (0.006)	0.7819 (0.007)
		0.6	0.5429	0.587 (0.032)	0.5408 (0.038)	0.5703 (0.017)	0.5451 (0.017)	0.559 (0.011)	0.5417 (0.011)	0.5569 (0.008)	0.5438 (0.008)
		0.9	0.3624	0.4162 (0.028)	0.3612 (0.027)	0.3897 (0.013)	0.3618 (0.012)	0.3819 (0.008)	0.3632 (0.007)	0.3761 (0.006)	0.3621 (0.005)
		1.2	0.2374	0.283 (0.019)	0.2347 (0.015)	0.261 (0.008)	0.2375 (0.007)	0.2522 (0.004)	0.2368 (0.004)	0.2492 (0.003)	0.2377 (0.003)
		1.5	0.1539	0.1916 (0.012)	0.154 (0.008)	0.1731 (0.004)	0.1554 (0.003)	0.1651 (0.002)	0.1537 (0.002)	0.1628 (0.002)	0.1543 (0.001)
	2.0	0.3	0.6969	0.7226 (0.025)	0.6958 (0.035)	0.7138 (0.014)	0.6985 (0.017)	0.7074 (0.01)	0.6965 (0.011)	0.7057 (0.008)	0.6973 (0.008)
		0.6	0.4158	0.4702 (0.03)	0.4159 (0.031)	0.4456 (0.015)	0.4174 (0.014)	0.4353 (0.009)	0.4163 (0.009)	0.4297 (0.007)	0.4154 (0.007)
		0.9	0.2374	0.2857 (0.02)	0.2371 (0.016)	0.2607 (0.008)	0.2372 (0.006)	0.2532 (0.005)	0.2377 (0.004)	0.2491(0.003)	0.2377 (0.003)
		1.2	0.1329	0.166 (0.009)	0.1322 (0.006)	0.148 (0.003)	0.1324 (0.002)	0.1425 (0.002)	0.1324 (0.001)	0.1406 (0.001)	0.1331 (0.001)
		1.5	0.0737	0.094 (0.003)	0.0729 (0.002)	0.0828 (0.001)	0.0734 (0.001)	0.0798 (0.001)	0.0737 (0)	0.0783 (0)	0.0738 (0)
2.0	1.5	0.3	0.8687	0.8644 (0.013)	0.8669 (0.019)	0.8671 (0.007)	0.8683 (0.009)	0.8684 (0.005)	0.8692 (0.006)	0.868 (0.004)	0.8685 (0.004)

(continued)

Table 2 (continued)

$n \rightarrow$				4		8		12		16	
$\alpha \downarrow$	$\beta \downarrow$	t	$R(t)$	$\tilde{R}(t)$	$\hat{R}(t)$	$\tilde{R}(t)$	$\hat{R}(t)$	$\tilde{R}(t)$	$\hat{R}(t)$	$\tilde{R}(t)$	$\hat{R}(t)$
		0.6	0.6478	0.6819 (0.028)	0.648 (0.037)	0.6664 (0.016)	0.6471 (0.018)	0.6606 (0.011)	0.6472 (0.012)	0.6572 (0.008)	0.6469 (0.009)
		0.9	0.4513	0.5016 (0.032)	0.4487 (0.034)	0.4786 (0.015)	0.4508 (0.015)	0.4705 (0.01)	0.4517 (0.01)	0.4644 (0.008)	0.4502 (0.007)
		1.2	0.3033	0.3552 (0.025)	0.3021 (0.022)	0.3301 (0.011)	0.3037 (0.01)	0.3206 (0.006)	0.303 (0.006)	0.3159 (0.005)	0.3028 (0.004)
		1.5	0.1997	0.2433 (0.016)	0.1991 (0.012)	0.2208 (0.006)	0.1998 (0.005)	0.2137 (0.004)	0.1999 (0.003)	0.2102 (0.002)	0.2 (0.002)
	2.0	0.3	0.7964	0.8031 (0.018)	0.7925 (0.027)	0.8017 (0.011)	0.7954 (0.013)	0.8007 (0.008)	0.7961 (0.009)	0.7984 (0.006)	0.7947 (0.007)
		0.6	0.5117	0.5591 (0.032)	0.5101 (0.037)	0.5381 (0.016)	0.5116 (0.017)	0.5287 (0.011)	0.5106 (0.011)	0.5248 (0.008)	0.5112 (0.008)
		0.9	0.3033	0.3584 (0.026)	0.3051 (0.022)	0.3297 (0.011)	0.3032 (0.01)	0.3214 (0.007)	0.3038 (0.006)	0.3165 (0.005)	0.3034 (0.004)
		1.2	0.1732	0.2131 (0.013)	0.1724 (0.009)	0.1924 (0.005)	0.1733 (0.004)	0.1853 (0.003)	0.1728 (0.002)	0.1822 (0.002)	0.173 (0.002)
		1.5	0.0971	0.124 (0.006)	0.0973 (0.004)	0.1094 (0.002)	0.0973 (0.001)	0.1051 (0.001)	0.0973 (0.001)	0.1031 (0.001)	0.0974 (0.001)

Table 3 Estimates of P with their corresponding MSES in parenthesis

$(n, m) \rightarrow$				(4, 4)		(8, 8)		(12, 12)		(16, 16)	
$\alpha_1 \downarrow$	$\beta \downarrow$	$\alpha_2 \downarrow$	P	$\tilde{P}$	$\hat{P}$	$\tilde{P}$	$\hat{P}$	$\tilde{P}$	$\hat{P}$	$\tilde{P}$	$\hat{P}$
1.5	1.5	1.0	0.6	0.5868 (0.026)	0.5974 (0.032)	0.592 (0.014)	0.5975 (0.016)	0.5973 (0.009)	0.6012 (0.01)	0.5964 (0.007)	0.5993 (0.008)
		2.0	0.4286	0.4401 (0.027)	0.4327 (0.034)	0.431 (0.014)	0.4268 (0.016)	0.4303 (0.01)	0.4275 (0.01)	0.4294 (0.007)	0.4272 (0.008)
		3.0	0.3333	0.3533 (0.024)	0.3366 (0.028)	0.3422 (0.012)	0.3334 (0.014)	0.3383 (0.008)	0.3323 (0.009)	0.3382 (0.006)	0.3337 (0.006)
		4.0	0.2727	0.2931 (0.021)	0.2719 (0.024)	0.2826 (0.01)	0.2716 (0.011)	0.2802 (0.007)	0.2729 (0.007)	0.2782 (0.005)	0.2726 (0.005)
		5.0	0.2308	0.2541 (0.019)	0.2311 (0.02)	0.2436 (0.009)	0.2318 (0.009)	0.2386 (0.006)	0.2306 (0.006)	0.2372 (0.004)	0.2312 (0.004)
	2.0	1.0	0.6	0.5903 (0.027)	0.6011 (0.033)	0.5955 (0.014)	0.6013 (0.015)	0.5952 (0.01)	0.5991 (0.01)	0.5968 (0.007)	0.5997 (0.007)
		2.0	0.4286	0.4353 (0.027)	0.4274 (0.034)	0.4308 (0.014)	0.4266 (0.016)	0.4291 (0.009)	0.4262 (0.01)	0.4317 (0.007)	0.4296 (0.008)
		3.0	0.3333	0.3471 (0.024)	0.33 (0.029)	0.3421 (0.012)	0.3333 (0.013)	0.3391 (0.008)	0.3331 (0.009)	0.337 (0.006)	0.3325 (0.006)
		4.0	0.2727	0.2942 (0.021)	0.2729 (0.023)	0.2827 (0.01)	0.2717 (0.011)	0.28 (0.007)	0.2726 (0.007)	0.2781 (0.005)	0.2725 (0.005)
		5.0	0.2308	0.2535 (0.018)	0.2303 (0.019)	0.2439 (0.009)	0.2322 (0.009)	0.238 (0.006)	0.2301 (0.006)	0.2367 (0.004)	0.2307 (0.004)
2.0	1.5	1.0	0.6667	0.6505 (0.024)	0.6677 (0.028)	0.659 (0.012)	0.6679 (0.013)	0.6609 (0.008)	0.6669 (0.009)	0.6616 (0.006)	0.6661 (0.006)

(continued)

Table 3 (continued)

$(n, m) \rightarrow$				(4, 4)		(8, 8)		(12, 12)		(16, 16)	
$\alpha_1 \downarrow$	$\beta \downarrow$	$\alpha_2 \downarrow$	P	$\tilde{P}$	$\hat{P}$	$\tilde{P}$	$\hat{P}$	$\tilde{P}$	$\hat{P}$	$\tilde{P}$	$\hat{P}$
		2.0	0.5	0.5011 (0.028)	0.5011 (0.036)	0.5019 (0.015)	0.502 (0.017)	0.4996 (0.01)	0.4996 (0.011)	0.499 (0.007)	0.4989 (0.008)
		3.0	0.4	0.4128 (0.027)	0.4023 (0.034)	0.4043 (0.014)	0.3986 (0.015)	0.4057 (0.009)	0.4019 (0.01)	0.4027 (0.007)	0.3998 (0.007)
		4.0	0.3333	0.3478 (0.024)	0.3305 (0.028)	0.3408 (0.012)	0.3318 (0.013)	0.3396 (0.008)	0.3336 (0.009)	0.3372 (0.006)	0.3327 (0.006)
		5.0	0.2857	0.308 (0.022)	0.2877 (0.025)	0.2964 (0.011)	0.2859 (0.011)	0.2935 (0.007)	0.2864 (0.007)	0.2899 (0.005)	0.2845 (0.005)
	2.0	1.0	0.6667	0.6478 (0.024)	0.6646 (0.029)	0.6585 (0.012)	0.6674 (0.013)	0.6606 (0.008)	0.6666 (0.009)	0.6635 (0.006)	0.668 (0.006)
		2.0	0.5	0.5013 (0.027)	0.5016 (0.035)	0.5016 (0.015)	0.5017 (0.017)	0.4998 (0.01)	0.4998 (0.011)	0.5004 (0.008)	0.5004 (0.008)
		3.0	0.4	0.4138 (0.026)	0.4035 (0.033)	0.406 (0.014)	0.4003 (0.015)	0.4044 (0.009)	0.4006 (0.01)	0.402 (0.007)	0.3991 (0.007)
		4.0	0.3333	0.3529 (0.024)	0.3361 (0.029)	0.3429 (0.012)	0.334 (0.013)	0.3395 (0.008)	0.3335 (0.008)	0.3385 (0.006)	0.334 (0.006)
		5.0	0.2857	0.3048 (0.021)	0.2842 (0.024)	0.2963 (0.011)	0.2857 (0.011)	0.292 (0.007)	0.2848 (0.007)	0.2912 (0.005)	0.2858 (0.006)

$\beta_1 \neq \beta_2$ are known), we have generated 1000 random samples of lower records using (26) and (28), from each of the populations *X* and *Y* with sizes $(n, m) = (4, 4), (8, 8), (12, 12)$ and $(12, 16)$ with different combinations of $\alpha_1 = 1.5(0.5)2.0$, $\alpha_2 = 1.0(1.0)5.0$, $\beta_1 = 1.5(0.5)2.0$ and $\beta_2 = 2.0(1.0)3.0$. For different combinations of parametric values, we have computed actual values of *P*, average ML and UMVU estimates of *P* along with their MSES. Obtained results are presented in Table 4.

To check the performance of the estimators of $R(t)$ (when β is known) graphically, we have simulated 1000 random samples of lower records using (26) and (28), each of size $n = 4(4)16$ with $\alpha = 1.5$ and $\beta = 2.0(1.0)4.0$. Using Theorem 3 and Corollary 2, average ML and UMVU estimates of $R(t)$ along with their MSES are obtained. For different values of *t*, MSES of ML and UMVU estimators of $R(t)$ are plotted against *t*, in Fig. 1. Under the same set-up, MSES of ML and UMVU estimators of $R(t)$ are plotted against their estimates in Fig. 2.

To check the performance of the estimators, for different set of values of α and β, we have again simulated 1000 random samples of lower records using (26) and (28), each of size $n = 4(4)16$ with $\alpha = 1.5(1.0)3.5$ and $\beta = 1.5$. Using Theorem 3 and Corollary 2, average ML and UMVU estimates of $R(t)$ along with their MSES are obtained. For different values of *t*, MSES of ML and UMVU estimators of $R(t)$ are plotted against *t* in Fig. 3. Under the same set-up, MSES of ML and UMVU estimators of $R(t)$ are plotted against their estimates in Fig. 4.

Furthermore, to investigate the performance of estimators of *P* obtained in Theorems 5 and 8 (when $\beta_1 = \beta_2 = \beta$ are known) graphically, we have generated 1000 random samples of lower records using (26) and (28), from each of the populations *X* and *Y* with sizes $(n, m) = (4, 4), (8, 8), (12, 12)$ and $(16, 16)$ with different combinations of $\alpha_1 = 1.5(1.0)3.5$ and $\beta_1 = \beta_2 = \beta = 1.5$. For different combinations of above parametric values and for $\alpha_2 = 1.0(1.0)6.0$, the computed MSES of ML and UMVU estimators of *P* against their estimates are plotted in Fig. 5. Under the same set-up, MSE of ML and UMVU estimators of *P*, which are obtained by using Theorem 4 and Corollary 3 (when $\beta_1 \neq \beta_2$ are known), for $\alpha_2 = 1.0(1.0)6.0$ with different combinations of $\alpha_1 = 2.0(0.5)3.0$, $\beta_1 = 2.5$ and $\beta_2 = 1.0(0.5)2.0$ are plotted against their estimates in Fig. 6.

In order to obtain the ML estimates of α and β (when both the parameters are unknown), we have simulated 1000 random samples of lower records using (26) and (28), each of size $n = 4(4)16$ with $(\alpha, \beta) = (1.5, 1.5), (1.5, 2.0), (2.0, 1.5)$ and $(2.0, 2.0)$. For each sample of lower records, the ML estimates of α and β are obtained by using (15), (18) and using the *optimize*() function with *maximum* = *T*, which is available in *R 3.2.4*. Average estimates along with their corresponding MSES are reported in Table 5. Under the same set-up, actual values and average estimates of $R(t)$ along with their corresponding MSES for $t = 0.30$ and 0.60, are presented in Table 6. Moreover, by using Theorem 7 and the same technique discussed earlier, we have constructed Table 7 for various values of α_1, α_2, β_1 and β_2. For different values of *(n, m,)* Table 7 gives the actual values, average estimates of *P* along with their corresponding MSES.

In Table 8 (Table 9), 95% confidence interval for *P* along with length of the interval based on ML estimates of α_1 and α_2, when $\beta_1 = \beta_2 = \beta$ are known (when

Table 4 Estimates of P with their corresponding MSES in parenthesis

$(n, m) \rightarrow$					(4, 4)		(8, 8)		(12, 12)		(12, 16)	
$\alpha_1 \downarrow$	$\beta_2 \downarrow$	$\beta_1 \downarrow$	$\alpha_2 \downarrow$	P	$\tilde{P}$	$\hat{P}$	$\tilde{P}$	$\hat{P}$	$\tilde{P}$	$\hat{P}$	$\tilde{P}$	$\hat{P}$
1.5	2.0	1.5	1.0	0.6761	0.6659 (0.024)	0.6775 (0.029)	0.6698 (0.012)	0.6757 (0.013)	0.6704 (0.008)	0.6743 (0.009)	0.6746 (0.007)	0.6759 (0.008)
			2.0	0.5245	0.5304 (0.028)	0.5251 (0.035)	0.5276 (0.015)	0.5246 (0.017)	0.5267 (0.01)	0.5247 (0.011)	0.5289 (0.009)	0.5246 (0.01)
			3.0	0.4345	0.4524 (0.028)	0.4376 (0.034)	0.4407 (0.014)	0.4325 (0.016)	0.4385 (0.009)	0.4329 (0.01)	0.4426 (0.009)	0.4352 (0.009)
			4.0	0.3741	0.3968 (0.026)	0.376 (0.031)	0.3862 (0.013)	0.3753 (0.014)	0.382 (0.009)	0.3746 (0.009)	0.3834 (0.008)	0.3744 (0.008)
			5.0	0.3303	0.3534 (0.024)	0.3293 (0.027)	0.3431 (0.012)	0.3305 (0.012)	0.3392 (0.008)	0.3307 (0.008)	0.3406 (0.007)	0.3308 (0.007)
		2.0	1.0	0.6	0.5916 (0.026)	0.6027 (0.032)	0.5951 (0.014)	0.6008 (0.015)	0.5967 (0.009)	0.6005 (0.01)	0.5993 (0.008)	0.6002 (0.009)
			2.0	0.4286	0.4357 (0.027)	0.4278 (0.034)	0.4332 (0.014)	0.429 (0.016)	0.4325 (0.01)	0.4297 (0.011)	0.4334 (0.008)	0.4285 (0.009)
			3.0	0.3333	0.3512 (0.024)	0.3343 (0.029)	0.342 (0.012)	0.3331 (0.013)	0.3398 (0.008)	0.3338 (0.009)	0.3394 (0.007)	0.3318 (0.008)
			4.0	0.2727	0.2922 (0.021)	0.2708 (0.023)	0.2844 (0.01)	0.2734 (0.011)	0.279 (0.007)	0.2717 (0.007)	0.2813 (0.006)	0.2728 (0.006)
			5.0	0.2308	0.2566 (0.019)	0.2337 (0.02)	0.2431 (0.009)	0.2313 (0.009)	0.2388 (0.006)	0.2309 (0.006)	0.2392 (0.005)	0.2304 (0.005)
2.0	3.0	1.5	1.0	0.8333	0.816 (0.014)	0.8326 (0.016)	0.8245 (0.007)	0.8331 (0.007)	0.8271 (0.005)	0.8328 (0.005)	0.8282 (0.004)	0.832 (0.004)

(continued)

Table 4 (continued)

$(n, m) \rightarrow$					(4, 4)		(8, 8)		(12, 12)		(12, 16)	
$\alpha_1 \downarrow$	$\beta_2 \downarrow$	$\beta_1 \downarrow$	$\alpha_2 \downarrow$	P	$\tilde{P}$	$\hat{P}$	$\tilde{P}$	$\hat{P}$	$\tilde{P}$	$\hat{P}$	$\tilde{P}$	$\hat{P}$
			2.0	0.7333	0.7223 (0.021)	0.73 (0.027)	0.7292 (0.011)	0.7332 (0.012)	0.732 (0.008)	0.7347 (0.008)	0.7327 (0.007)	0.7333 (0.008)
			3.0	0.6643	0.6639 (0.025)	0.6634 (0.032)	0.6625 (0.013)	0.6618 (0.015)	0.6645 (0.009)	0.664 (0.01)	0.6662 (0.009)	0.6637 (0.009)
			4.0	0.6127	0.6193 (0.027)	0.6126 (0.035)	0.6181 (0.014)	0.6142 (0.016)	0.6163 (0.01)	0.6135 (0.011)	0.6165 (0.009)	0.6117 (0.01)
			5.0	0.5722	0.5817 (0.028)	0.5696 (0.036)	0.5794 (0.015)	0.5727 (0.017)	0.5777 (0.01)	0.573 (0.011)	0.5791 (0.01)	0.5727 (0.01)
		2.0	1.0	0.7714	0.7523 (0.019)	0.769 (0.021)	0.7622 (0.009)	0.7709 (0.01)	0.7661 (0.006)	0.7719 (0.006)	0.7689 (0.005)	0.7723 (0.006)
			2.0	0.6429	0.6382 (0.025)	0.6429 (0.031)	0.6407 (0.013)	0.643 (0.015)	0.6436 (0.009)	0.6451 (0.01)	0.6435 (0.008)	0.6425 (0.009)
			3.0	0.5583	0.5651 (0.028)	0.5598 (0.036)	0.5617 (0.015)	0.5586 (0.017)	0.5606 (0.01)	0.5585 (0.011)	0.5624 (0.009)	0.558 (0.01)
			4.0	0.4977	0.5113 (0.028)	0.4991 (0.035)	0.5034 (0.015)	0.4966 (0.017)	0.5025 (0.01)	0.4979 (0.011)	0.503 (0.009)	0.4964 (0.01)
			5.0	0.4515	0.4684 (0.027)	0.4509 (0.033)	0.4595 (0.015)	0.4502 (0.016)	0.4592 (0.01)	0.4528 (0.011)	0.4598 (0.009)	0.4517 (0.009)

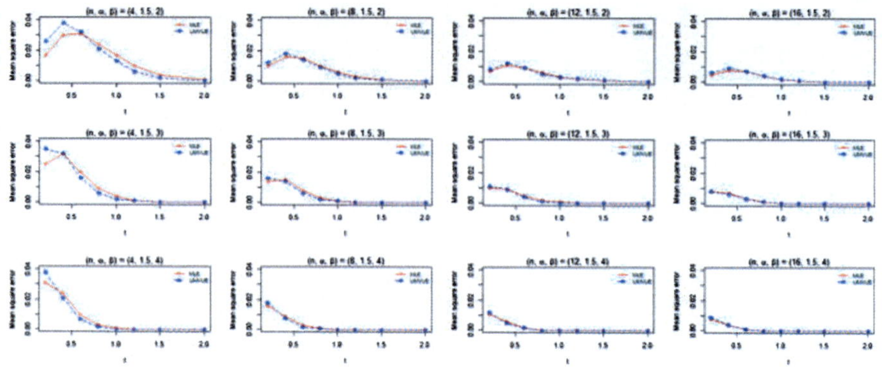

Fig. 1 Plot of MSES of estimators of $R(t)$ against different values of t

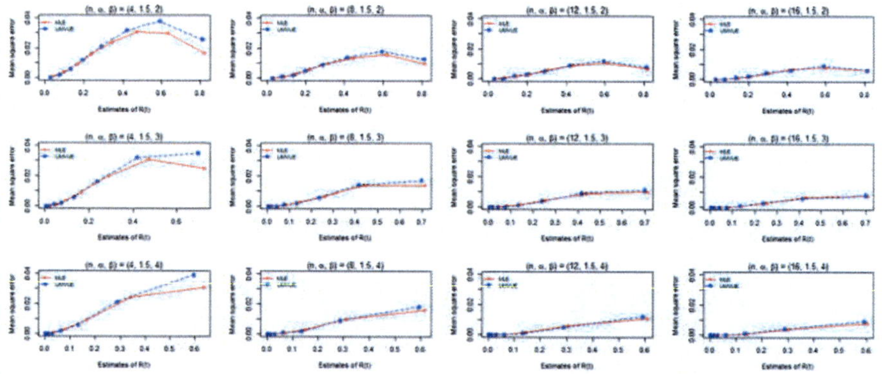

Fig. 2 Plot of estimates of $R(t)$ against their corresponding MSES

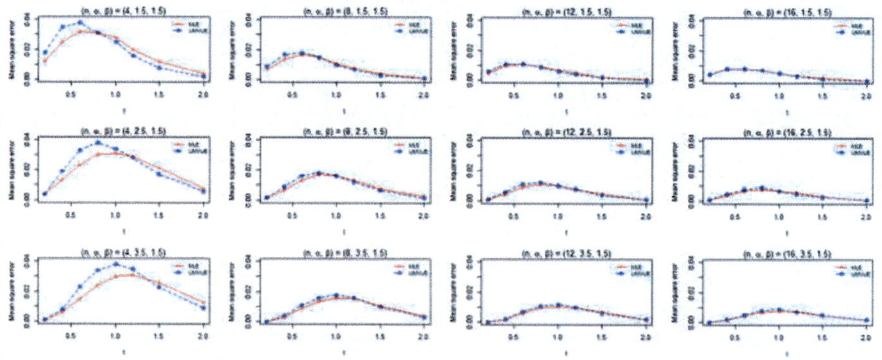

Fig. 3 Plot of MSES of estimators of $R(t)$ against different values of t

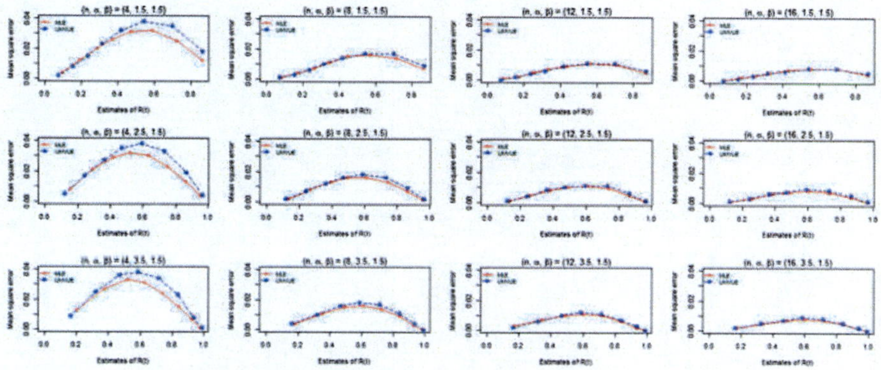

Fig. 4 Plot of estimates of *R(t)* against their corresponding MSES

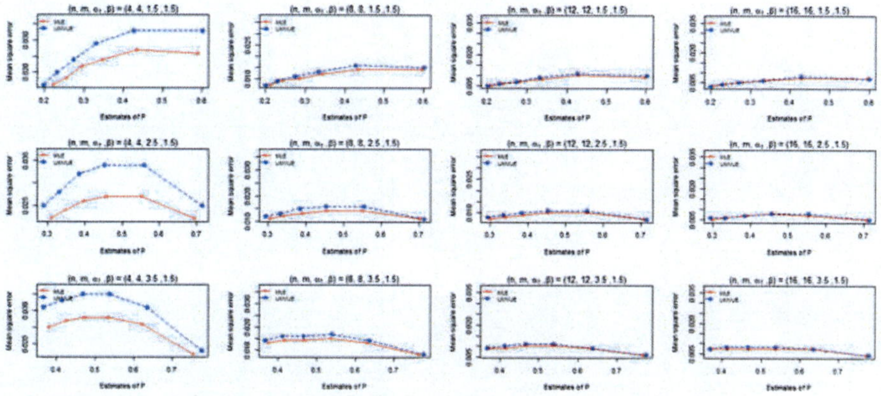

Fig. 5 Plot of estimates of *P* against their MSE of corresponding estimators

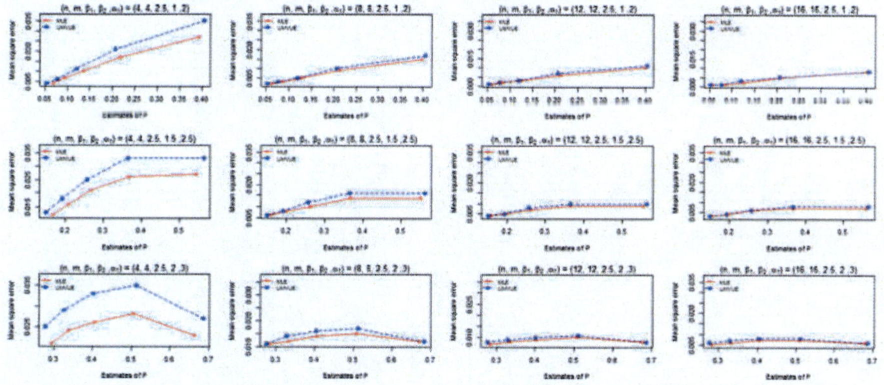

Fig. 6 Plot of estimates of *P* against their corresponding MSES

Table 5 ML estimates of α and β, when both the parameter are unknown

$n \rightarrow$		4		8		12		16	
$\alpha \downarrow$	$\beta \downarrow$	$\tilde{\alpha}$	$\tilde{\beta}$	$\tilde{\alpha}$	$\tilde{\beta}$	$\tilde{\alpha}$	$\tilde{\beta}$	$\tilde{\alpha}$	$\tilde{\beta}$
1.5	1.5	2.7091 (9.6266)	2.3021 (1.4127)	1.8628 (1.0387)	2.1181 (1.1849)	1.6773 (0.3451)	2.0677 (1.0974)	1.6444 (0.2331)	2.0885 (1.1038)
1.5	2.0	2.4369 (7.6553)	2.5267 (0.8125)	1.7496 (0.6076)	2.3721 (0.7828)	1.6667 (0.3741)	2.3522 (0.7454)	1.6337 (0.2101)	2.3211 (0.7082)
2.0	1.5	3.6071 (18.5206)	2.2629 (1.3055)	2.5205 (1.8752)	2.1041 (1.1266)	2.307 (0.7852)	2.0343 (1.0117)	2.1805 (0.4463)	1.994 (0.9875)
2.0	2.0	3.1793 (8.7642)	2.5539 (0.7909)	2.4407 (1.5319)	2.3915 (0.6985)	2.2537 (0.6312)	2.3398 (0.6858)	2.164 (0.3909)	2.3358 (0.6785)

Table 6 ML estimates of *R(t)*, when both the parameter are unknown

$n \rightarrow$				4	8	12	16
$\alpha \downarrow$	$\beta \downarrow$	$t \downarrow$	$R(t) \downarrow$	$\tilde{R}(t)$	$\tilde{R}(t)$	$\tilde{R}(t)$	$\tilde{R}(t)$
1.5	1.5	0.3	0.7819	0.7362 (0.0347)	0.7211 (0.0304)	0.7196 (0.028)	0.7184 (0.027)
		0.6	0.5429	0.491 (0.0568)	0.4586 (0.0523)	0.4601 (0.0521)	0.4553 (0.0505)
	2.0	0.3	0.6969	0.6818 (0.0354)	0.673 (0.0246)	0.6773 (0.0214)	0.6703 (0.0215)
		0.6	0.4158	0.4153 (0.0409)	0.3835 (0.0317)	0.4001 (0.0335)	0.4054 (0.0353)
2.0	1.5	0.3	0.8687	0.8237 (0.0252)	0.818 (0.02)	0.8112 (0.0197)	0.8116 (0.018)
		0.6	0.6478	0.5742 (0.057)	0.5619 (0.0515)	0.5661 (0.0514)	0.5652 (0.0505)
	2.0	0.3	0.7964	0.7765 (0.0255)	0.7634 (0.0204)	0.7578 (0.0181)	0.7601 (0.0173)
		0.6	0.5117	0.5044 (0.0425)	0.477 (0.0371)	0.4795 (0.0353)	0.4768 (0.0359)

Table 7 ML estimates of *P*, when both the parameter are unknown

(n, m)→					(4, 4)	(8, 8)	(12, 12)	(16, 16)
$\alpha_1 \downarrow$	$\beta_2 \downarrow$	$\beta_1 \downarrow$	$\alpha_2 \downarrow$	$P \downarrow$	$\tilde{P}$	$\tilde{P}$	$\tilde{P}$	$\tilde{P}$
1.5	2.0	1.5	1.0	0.6761	0.6545 (0.0631)	0.6504 (0.057)	0.6589 (0.0552)	0.6498 (0.0533)
			2.0	0.5245	0.5028 (0.076)	0.4848 (0.0758)	0.4899 (0.0689)	0.4971 (0.0685)
		2.0	1.0	0.6	0.6073 (0.0602)	0.6036 (0.0572)	0.6056 (0.0566)	0.6102 (0.056)
			2.0	0.4286	0.4259 (0.0703)	0.4434 (0.0632)	0.4268 (0.0627)	0.423 (0.0626)
2.0	3.0	1.5	1.0	0.8333	0.7799 (0.0436)	0.778 (0.0416)	0.7764 (0.0415)	0.7812 (0.0372)
			2.0	0.7333	0.6699 (0.0659)	0.6668 (0.0622)	0.6653 (0.0605)	0.6764 (0.0574)
		2.0	1.0	0.7714	0.7366 (0.0472)	0.7312 (0.0457)	0.7225 (0.0425)	0.7328 (0.0426)
			2.0	0.6429	0.5972 (0.0624)	0.5991 (0.0613)	0.6042 (0.0612)	0.5976 (0.0584)

$\beta_1 = \beta_2 = \beta$ are unknown), using (23) for various value of $\alpha_1, \alpha_2, \beta$ and *(n, m)* are provided.

In Table 10, we have constructed 95% confidence interval for $\tau = \frac{\alpha_1}{\alpha_2}$ along with length of the interval, using (21) for various value of $\alpha_1, \alpha_2, \beta_1, \beta_2$ and *(n, m)*. Whereas, in Table 11, we have constructed 95% confidence interval for $\alpha_1 = \alpha_2 = \alpha$ along with length of the interval, using (22) for various value of α, β_1, β_2 and *(n, m)*.

For the theory developed in Sect. 5, for testing the hypothesis $H_o : \alpha = 2 (= \alpha_o)$ against $H_1 : \alpha \neq 2 (= \alpha_o)$, we have generated the following random sample of lower record values of size $n = 10$ from (26) with $\alpha = 2$ and $\beta = 1.5$.

Table 8 95% confidence interval for P based on ML estimates of α_1 and α_2 when $\beta_1 = \beta_2 = \beta$ is known

$(n, m)\rightarrow$				(4, 4)	(8, 8)	(12, 12)	(12, 16)
$\beta\downarrow$	$\alpha_1\downarrow$	$\alpha_2\downarrow$	$P\downarrow$	[95% confidence interval] Length of the interval			
1.0	1.5	1.5	0.5	[0.1844, 0.8163] 0.6319	[0.2647, 0.7329] 0.4683	[0.308, 0.6962] 0.3882	[0.313,4 0.678] 0.3646
	1.5	2.0	0.4286	[0.1408, 0.763] 0.6223	[0.2112, 0.6712] 0.46	[0.2509, 0.633] 0.3821	[0.259, 0.6172] 0.3582
	2.0	1.5	0.5714	[0.2308, 0.855] 0.6242	[0.3159, 0.7788] 0.4629	[0.3688, 0.7506] 0.3817	[0.3835, 0.7416] 0.3581
	2.0	2.0	0.5	[0.1843, 0.8162] 0.6319	[0.2644, 0.7327] 0.4683	[0.3135, 0.7017] 0.3881	[0.3232, 0.6878] 0.3646
2.0	1.5	1.5	0.5	[0.1866, 0.8184] 0.6319	[0.2644, 0.7327] 0.4683	[0.3049, 0.6932] 0.3882	[0.3174, 0.6821] 0.3647
	1.5	2.0	0.4286	[0.1538, 0.7813] 0.6275	[0.2146, 0.6757] 0.4611	[0.2557, 0.6389] 0.3832	[0.2629, 0.622] 0.3591
	2.0	1.5	0.5714	[0.2262, 0.8517] 0.6256	[0.3297, 0.7895] 0.4598	[0.3653, 0.7477] 0.3824	[0.3817, 0.7401] 0.3584
	2.0	2.0	0.5	[0.1841, 0.816] 0.6319	[0.269, 0.7372] 0.4683	[0.3078, 0.696] 0.3882	[0.3157, 0.6803] 0.3646

1.0845, 0.6753, 0.2772, 0.2312, 0.2058, 0.1409, 0.0758, 0.0272, 0.0184, 0.0164.

Now with the help of chi-square table at 5% level of significance (LOS), we obtained $k_o = 2.3977$ and $k_o^{'} = 8.5424$. Hence, in this case we do not reject H_o at 5% LOS as $S = 3.7173$.

Again, for testing $H_o : \alpha \leq 1 (= \alpha_o)$ against $H_1 : \alpha > 1(= \alpha_o)$, we have considered the same sample given above. Now at 5% LOS, we obtained $k_o^{''} = 15.7052$ and hence in this case we reject H_o as $S = 3.7173$.

6.3 *Real Data Example*

First data set:

To illustrate the developed approaches in previous sections, we consider the real data set which is also used in Lawless (2003), page 3. This data is from Nelson (1982),

Table 9 95% confidence interval for P based on ML estimates of α_1 and α_2 when $\beta_1 = \beta_2 = \beta$ is unknown

$(n, m)\rightarrow$				(4, 4)	(8, 8)	(12, 12)	(12, 16)
$\beta\downarrow$	$\alpha_1\downarrow$	$\alpha_2\downarrow$	$P\downarrow$	[95% confidence interval] Length of the interval			
1.0	1.5	1.5	0.5	[0.3599, 0.917] 0.5571	[0.2667, 0.735] 0.4683	[0.428, 0.7939] 0.366	[0.4048, 0.7361] 0.3312
	1.5	2.0	0.4286	[0.1593, 0.7883] 0.629	[0.2659, 0.7342] 0.4683	[0.2774, 0.6641] 0.3867	[0.2295, 0.5498] 0.3203
	2.0	1.5	0.5714	[0.1956, 0.827] 0.6314	[0.2383, 0.7046] 0.4663	[0.2601, 0.6441] 0.3841	[0.4218, 0.7494] 0.3276
	2.0	2.0	0.5	[0.103, 0.693] 0.59	[0.2009, 0.6571] 0.4563	[0.1688, 0.5111] 0.3424	[0.4319, 0.7571] 0.3252
2.0	1.5	1.5	0.5	[0.3141, 0.9] 0.5859	[0.2441, 0.7112] 0.4671	[0.3539, 0.7383] 0.3844	[0.285, 0.6203] 0.3354
	1.5	2.0	0.4286	[0.0462, 0.4875] 0.4414	[0.1777, 0.6223] 0.4446	[0.2129, 0.5821] 0.3692	[0.2997, 0.6369] 0.3373
	2.0	1.5	0.5714	[0.4095, 0.9316] 0.5221	[0.378, 0.8225] 0.4445	[0.3553, 0.7394] 0.3841	[0.3628, 0.7001] 0.3373
	2.0	2.0	0.5	[0.3264, 0.905] 0.5786	[0.4909, 0.8803] 0.3893	[0.1658, 0.5057] 0.34	[0.3207, 0.6593] 0.3386

concerning the data on time to breakdown an insulating fluid between electrodes at a voltage of 36 kV (minutes). The times to breakdown at voltage 36 kV are.

$$X = \begin{pmatrix} 1.97, 0.59, 2.58, 1.69, 2.71, 25.50, 0.35, 0.99, 3.99, 3.67, \\ 2.07, 0.96, 5.35, 2.90, 13.77 \end{pmatrix}.$$

Second data set:

We consider the data set regarding the time in minutes for the first goal scoring during the final stages matches of the European Champions League for two consecutive years 2011–2012 and 2012–2013, for the return matches (the data are available online at http://www.it.soccerway.com). We have considered only the matches with at least one goal scored and we have divided all times by 90, i.e., the total of minutes of a soccer match, in order to obtain data belonging to the unit interval. The obtained observed data are the following:

Table 10 95% confidence interval for α_1/α_2

(n, m)→					(4, 4)	(8, 8)	(12, 12)	(12, 16)
β_1 ↓	β_2 ↓	α_1 ↓	α_2 ↓	α_1/α_2 ↓	[95% confidence interval] Length of the interval			
1.0	2.0	1.5	1.5	1	[0.2965, 5.828] 5.5315	[0.4161, 3.1728] 2.7567	[0.4807, 2.4756] 1.9949	[0.4965, 2.2905] 1.794
		1.5	2.0	0.75	[0.2269, 4.459] 4.2321	[0.3101, 2.3642] 2.0542	[0.3601, 1.8543] 1.4942	[0.3724, 1.7182] 1.3458
		2.0	1.5	1.3333	[0.4077, 8.0129] 7.6052	[0.555, 4.2316] 3.6767	[0.6404, 3.2977] 2.6573	[0.6695, 3.0885] 2.419
		2.0	2.0	1	[0.3013, 5.9224] 5.6211	[0.4183, 3.1892] 2.771	[0.4829, 2.4867] 2.0038	[0.4968, 2.2921] 1.7953
2.0	1.0	1.5	1.5	1	[0.2936, 5.7709] 5.4772	[0.4095, 3.1227] 2.7131	[0.4868, 2.5067] 2.0199	[0.498, 2.2973] 1.7994
		1.5	2.0	0.75	[0.2259, 4.4391] 4.2132	[0.3126, 2.3839] 2.0712	[0.3638, 1.8732] 1.5095	[0.375, 1.73] 1.355
		2.0	1.5	1.3333	[0.3969, 7.8014] 7.4044	[0.5515, 4.2055] 3.654	[0.6445, 3.3187] 2.6743	[0.662, 3.0542] 2.3922
		2.0	2.0	1	[0.3005, 5.9059] 5.6054	[0.3005, 5.9059] 5.6054	[0.4782, 2.4624] 1.9843	[0.4955, 2.2858] 1.7903

Table 11 95% confidence interval for $\alpha_1 = \alpha_2 = \alpha$

(n,m)→				(4, 4)	(8, 8)	(12, 12)	(12, 16)
β_1 ↓	β_2 ↓	α_1 ↓	α_2 ↓	[95% confidence interval] Length of the interval			
1	2	1.5	1.5	[0.6495, 2.7121] 2.0626	[0.8516, 2.3038] 1.4522	[0.9618, 2.1585] 1.1967	[0.9967, 2.1044] 1.1077
		2.0	2.0	[0.8594, 3.5889] 2.7294	[1.146, 3.1002] 1.9542	[1.2843, 2.8823] 1.598	[1.3261, 2.7998] 1.4737
2	1	1.5	1.5	[0.6479, 2.7055] 2.0576	[0.8575, 2.3197] 1.4622	[0.9637, 2.1629] 1.1992	[0.996, 2.103] 1.107
		2.0	2.0	[0.8641, 3.6083] 2.7442	[1.1461, 3.1005] 1.9544	[1.278, 2.8683] 1.5903	[1.3308, 2.8098] 1.479

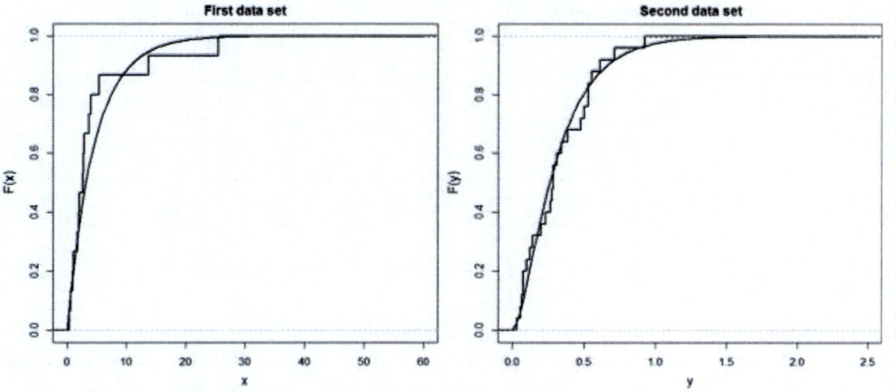

Fig. 7 Empirical and fitted cdf for X and Y data sets

$$Y = \begin{pmatrix} 0.267, 0.611, 0.344, 0.533, 0.033, 0.478, 0.200, \\ 0.056, 0.556, 0.711, 0.078, 0.533, 0.922, 0.067, 0.389, \\ 0.289, 0.233, 0.144, 0.100, 0.278, 0.500, 0.078, 0.289, 0.311, 0.122. \end{pmatrix}$$

We first apply the Kolmogorov–Smirnov (KS) test to check whether the distribution given at (26), fits the given X and Y-populations. Fitting the (26) distribution to the data x and y, we obtain the following ML estimates of $(\alpha_1,\ \beta_1)$ and $(\alpha_2,\ \beta_2)$ as.

$$(\tilde{\alpha}_1,\ \tilde{\beta}_1)_{complete\ data} = (0.9647,\ 0.2118)\ and = (1.7842,\ 4.3276).$$

According to the KS test, we do not reject the null hypothesis that both the data observed for X (KS = 0.2181; p-value = 0.4142) and the data observed for Y (KS = 0.10926; p-value = 0.9265) are drawn from (26). Figure 7, confirms the good fit of (26), for these two data sets.

From the whole sequence of data given in X and Y, we consider the observed lower record sets, given respectively by:

$$r = (1.97, 0.59, 0.35) \text{and } r* = (0.267, 0.033).$$

Using these lower record values we obtained $(\widetilde{\alpha}_1, \widetilde{\beta}_1) = (2.8412,\ 1.2215)$ and $(\widetilde{\alpha}_2, \widetilde{\beta}_2) = (1.1245,\ 5.6053)$. Now on using Theorem 7 and these results we get $\widetilde{P}$ = 0.9728 and the ML estimates of P obtained using the whole data set is equal to 0.9286, these results seem highly satisfactory.

Furthermore, to see the log likelihood profile of β_1 and β_2, we have converted the log likelihood obtained in (15) in single parameter β (with the help of expression given at (18)). Then for observed values of r and $r*$ record data sets, we have plotted the Log likelihood for (26) as a special case in Fig. 8. The vertical line in Fig. 8, corresponds to the ML estimates of β_1 and β_2, based on records. Thus Fig. 8, confirms

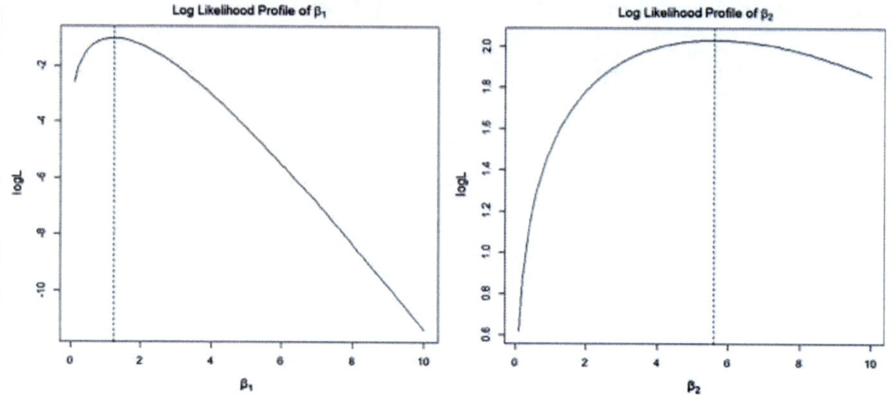

Fig. 8 Log likelihood profile of β_1 and β_2, for r and r^* record data set

the accuracy of ML estimates of β_1 and β_2. Hence the accuracy of the ML estimates of α_1 and α_2.

In Fig. 9, by using the expression obtained in Theorem 6 and the ML estimates of (α, β) based on records, we have plotted the ML estimates of $R(t)$ against different values of t, for the first and second data set.

Finally, we present the hypothesis testing for our real data sets. From the first data set, it's clear that, $r_n = 0.35$. If we assume, $\beta_1 = 1.2215$, is the known value of β_1. Then for testing the hypothesis $H_o : \alpha_1 = 2.8412 (= \alpha_{1o})$ against $H_1 : \alpha_1 \neq 2.8412 (= \alpha_{1o})$, we have the data set r of lower record values of size $n = 3$. Now with the help of chi-square table at 5% LOS, we obtained $k_o = 0.2178$ and $k_o' = 2.5428$. Hence, in this case we do not reject H_o at 5% LOS as $S = 1.0559$.

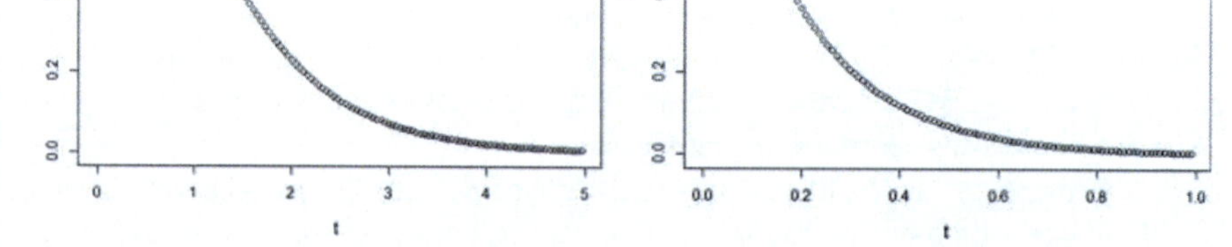

Fig. 9 Plot of ML estimates of $R(t)$ against different values of t for X and Y data sets

For testing $H_o : \alpha_1 \leq 3(= \alpha_{1o})$ against $H_1 : \alpha_1 > 3(= \alpha_{1o})$, we again considered the first data set of lower record values given above. Now at 5% LOS, we obtained $k_o^{''} = 0.2726$ and hence in this case also we do not reject H_o as $S = 1.0559$.

Similarly form the first second set, it's clear that, $r_m^* = 0.033$. If we assume, $\beta_2 = 5.6053$, is the known value of β_2. Then for testing the hypothesis $H_o : \alpha_2 = 1.1245(= \alpha_{2o})$ against $H_1 : \alpha_1 \neq 1.1245(= \alpha_{1o})$, we have the set r^* of lower record values of size $m = 2$. Now with the help of chi-square table at 5% LOS, we obtained $k_o = 0.2154$ and $k_o^{'} = 4.9548$. Hence, in this case we do not reject H_o at 5% LOS as $T = 1.7786$.

Again, for testing $H_o : \alpha_2 \leq 1.5(= \alpha_{2o})$ against $H_1 : \alpha_2 > 1.5(= \alpha_{2o})$, we again considered the second data set of lower record values given above. Now at 5% LOS, we obtained $k_o^{''} = 0.2369$ and hence in this case we do not reject H_o as $T = 1.7786$.

7 Conclusion

In this Chapter, we have developed the estimation procedures for the EG family of distributions based on lower record values. Considerations were given to both, point as well as interval estimations. Hypotheses were developed for various parametric functions. All comparisons are made on the basis of MSES. We used the software R (www.r-project.org) for the computations of functions in the various expressions.

1. From Table 1, we conclude that UMVUE of α gives better estimates than MLE of α, which is true for all values of N.
2. From Table 2 and Figs. (1, 2, 3 and 4) for smaller values of t (i.e., for higher values of *R(t)*) it can be seen that the performance of MLE of *R(t)* performs better than the performance of UMVUE of *R(t)*. Whereas, for higher values of t (i.e., for lower values of *R(t)*) they shows that the performance of UMVUE of *R(t)* performs better than the performance of MLE of *R(t)*.
3. From Tables 3 and 4, Figs. 5 and 6, it is observed that the MLE of *P* is more efficient than the UMVUE of *P*.
4. From Figs. (1, 2, 3, 4, 5 and 6), it can be seen that as *n* or (*n, m*) increases; estimates from both, MLE and UMVUE comes close to each other.
5. From Tables (8, 9, 10 and 11), it is observed that length of the confidence interval decreases as (*n, m*) increases.
6. Two real data-set are fitted to exponentiated exponential distribution and ML estimates of *P* are calculated on the basis of lower records.

8 Future Scope

In future this work can be extended for Bayesian analysis and different types of censoring schemes.

References

Ahsanullah M (1995) Record statistics. Nova Science Publishers, New York

Ahsanullah M, Shakil M, Golam Kibria BMG (2013) A characterization of power function distribution based on lower records. ProbStat Forum 06:68–72

Arnold BC, Balakrishnan N, Nagarajah H (1998) Records. Wiley, New York

Arashi M, Emadi M (2008) Evidential inference based on record data and inter-record times. Stat Pap 49(2):291–301

Awad AM, Gharraf MK (1986) Estimation of P(Y<X) in the Burr case: a comparative study. Commun Stat Simul Comput 15(2):389–403

Balakrishnan N, Ahsanullah M, Chan PS (1995) On the logistic record values and associated inference. J Appl Stat Sci 2(3):233–248

Belaghi RA, Arashi M, Tabatabaey SMM (2015) On the construction of preliminary test estimator based on record values for the Burr XII model. Commun Stat Theory Methods 44(1):1–23. https://doi.org/10.1080/03610926.2012.733473

Bilodeau GG, Thie PR, Keough GE (2010) An introduction to analysis. Jones and Bartlett, U.K. Learning International, U.K.

Chandler KN (1952) The distribution and frequency of record values. J Roy Stat Soc B 14(2):220–228

Chao A (1982) On comparing estimators of Pr{X > Y} in the exponential case. IEEE Trans Reliab 31(4):389–392

Chaturvedi A, Pathak A (2012) Estimation of the reliability function for exponentiated Weibull distribution. J Stat Appl 7(3/4):113

Chaturvedi A, Pathak A (2014) Estimating the reliability function for a family of exponentiated distributions. J Prob Stat. https://doi.org/10.1155/2014/563093

Chaturvedi A, Kang SB, Pathak A (2016) Estimation and testing procedures for the reliability functions of generalized half logistic distribution. J Korean Stat Soc 45(2):314–328

Chaturvedi A, Kumari T (2015) Estimation and testing procedures for the reliability functions of a family of lifetime distributions. InterStat. YEAR/2015/abstracts/1504001.php. http://interstat.statjournals.net/INDEX/Apr15.html

Chaturvedi A, Kumari T (2017) Estimation and testing procedures for the reliability functions of a general class of distributions. Commun Stat Theory Methods 46(22):11370–11382

Chaturvedi A, Kumari T (2019) Estimation and testing procedures of the reliability functions of generalized inverted scale family of distributions. Stat J Theor Appl Stat 53(1):148–176

Chaturvedi A, Rani U (1997) Estimation procedures for a family of density funtions representing various life-testing models. Metrika 46(1):213–219

Chaturvedi A, Singh KG (2008) A family of lifetime distributions and related estimation and testing procedures for the reliability function. J Appl Statist Sci 16(2):35–50

Chaturvedi A, Tomer SK (2002) Classical and Bayesian reliability estimation of the negative binomial distribution. J Appl Stat Sci 11(1):33–43

Chaturvedi A, Tomer SK (2003) UMVU estimation of the reliability function of the generalized life distributions. Stat Papers 44(3):301–313

Constantine K, Tse SK, Karson M (1986) Estimators of P(Y < X) in the gamma case. Commun Stat Simul Comput 15(2):365–388

Cordeiro GM, Ortega EM, da Cunha DC (2013) The exponentiated generalized class of distributions. J Data Sci 11(1):1–27

Franco M, Ruiz JM (1996) On characterization of continuous distributions by coditional expectation of record values. Sankhyā Indian J Stat Ser A 135–141

Habibi AR, Arghami NR, Ahmadi J (2006) Statistical evidence in experiments and in record value. Commun Stat Theory Methods 35(11):1971–1983

Kelley GD, Kelley JA, Schucany WR (1976) Efficient estimation of P(Y < X) in the exponential case. Technometrics 18(3):359–360

Khan AH, Alzaid AA (2004) Characterization of distributions through linear rgression of non-adjacent generalized order statistics. J Appl Stat Sci 13:123–136

Kumari T, Chaturvedi A, Pathak A (2019) Estimation and testing procedures for the reliability functions of Kumaraswamy-G distributions and a characterization based on records. J Stat Theory Practice 13:22. https://doi.org/10.1007/s42519-018-0014-7

Kumari T, Pathak A (2014a) Recurrence relations for single and product moments of generalized order statistics from generalized power Weibull distribution and is characterization. Int J Sci Eng Res 5(1):1914:1917

Kumari T, Pathak A (2014b) Relations for moments of generalized order statistics from Chen distribution and its characterization. J Comb Inf Syst Sci 39(1–4):49–56

Kumari T, Pathak A (2014c) Conditional expectation of certain distributions of dual generalized order statistics. Int J Math Anal 8(3):141–148

Lawless JF (2003) Statistical models and methods for lifetime data. Wiley, New York

Nagaraja HN (1988a) Record values and related statistics—a review. Commun Stat Theory Methods 17(7):2223–2238

Nagaraja HN (1988b) Some characterizations of continuous distributions based on regressions of adjacent order statistics and record values. Sankhyā Indian J Stat Ser A 70–73

Nelson W (1982) Applied life data analysis. Wiley, New York

Pathak A, Chaturvedi A (2013) Estimation of the reliability function for four-parameter exponentiated generalized lomax distribution. IJSER 5(1):1171–1180

Pathak A, Chaturvedi A (2014) Estimation of the reliability function for two-parameter exponentiated Rayleigh or Burr type X distribution. Stat Optim Inf Comput 2(4):305–322

Raqab MZ (2002) Inferences for generalized exponential distribution based on record statistics. J Stat Plan Inference 104(2):339–350

Razmkhah M, Morabbi H, Ahmadi J (2012) Comparing two sampling schemes based on entropy of record statistics. Stat Pap 53(1):95–106

Rohatgi VK, Saleh AKMdE (2012) An introduction to probability and statistics. Wiley, New York

Sathe YS, Shah SP (1981) On estimating P(X>Y) for the exponential distribution. Commun Stat Theory Methods 19:39–47

Tyagi RK, Bhattacharya SK (1989) A note on the MVU estimation of reliability for the Maxwell failure distribution. Estadistica 41(137):73–79

Modelling of Non-linear Multi-objective Programming and TOPSIS in Software Quality Assessment Under Picture Fuzzy Framework

Sameer Anand, Ritu Bibyan, and Aakash

Abstract The open-source software (OSS) market comprises of thousands of products and applications with different quality. The primary concern of the individuals and organizations is to evaluate the quality OSS products and packages. In this chapter, we have demonstrated a MCDM based model to assess the quality of OSS by taking performance and cost-based criteria related with OSS to analyze its quality on the basis of feedback gathered from the users and the experts. To avoid the uncertainty attached with the opinion of the expert, the Maximum-Entropy-Minimum-Variance-Ordered-Weighted-Aggregation (MEMV-OWA) operator has been incorporated. The criteria weights are calculated by solving a non-linear multi-objective programming problem proposed in the MEMV-OWA operator. The picture fuzzy set information has been used which is an addendum of intuitionistic fuzzy set, representing the human opinion more precisely. In this chapter, we have proposed a model named MEMV-OWA-PF-TOPSIS (Maximum Entropy Minimum Variance-Ordered Weighted Aggregation-Picture-Fuzzy-TOPSIS). A step by step procedure has been exhibited to show its implementation in real-life problems. A numerical illustration related to the software quality assessment of OSS on the basis of their performance and cost-based criteria has also been provided in this study.

Keywords Software quality assessment · Uncertainty · TOPSIS · MEMV-OWA · Picture fuzzy set

S. Anand
Shaheed Sukhdev College of Business Studies, University of Delhi, New Delhi, India
e-mail: sameeranand@sscbsdu.ac.in

R. Bibyan (✉) · Aakash
Department of Operational Research, University of Delhi, New Delhi, India

A. G. Aggarwal et al. (eds.), *Optimization Models in Software Reliability*,
Springer Series in Reliability Engineering,
https://doi.org/10.1007/978-3-030-78919-0_14

1 Introduction

An Open source software (OSS) is distributed with specific kind of license which permits end-users to access the source code legally. The programmers can modify the software in any way they want with the grant of the license. Many licenses exist in the software market, but, the software is open source if it can be purposed again, which implies that source code can be taken by anyone and can deliver their code or program. And also, It is accessible in the form of source code with no supplementary cost, means users can access the code and makes changes to it easily.

There are thousands of projects and millions of registered developers in the software market sector. Due to which the quality of the software is questioned and at times becomes a matter of fuss Karakoidas et al. (2007). The ample variety among the available OSS products offering similar functionalities makes it a challenging task to rank software for experts and customers it becomes challenging to choose which software to use. This decision-making problem has grabbed considerable attention in the software market and academia.

There are various MCDM methods and techniques which have been studied so far by researchers, but the essential elements remain the same in MCDM problems. In MCDM for each technique, a fixed or absolute number of trails are considered, at least one decision-maker and two criteria. These elements help in decision building by sorting, selecting, rating or ranking of trails. Therefore it is consequently said that MCDM is just not a collection of theories, technologies, methods but it is even a specific way to handle the decision-making problems. It has been implemented on an increasing number of domains from the last few decades. It has a feature that it can efficiently deal with conflicting criteria.

In this research work, we have proposed a model to rank the OSS based on its quality and cost by using expert opinion. Various criteria are taken into consideration which are relevant to OSS quality. The MEMV-OWA operator has used the uncertainty attached with the response of expert. This has been done by maximizing the information and minimizing the variation in their point of view or response. A non-linear multi-objective programming problem is constructed to extract the weights using MEMV-OWA technique.

We have also used the notion of Picture Fuzzy sets (PFS) introduced by Zadeh (1965), which is the extension of Intuitionistic fuzzy sets (IFS). It presents human opinions more precisely as compared to IFS as it considers options like acceptance, neutral, rejection and refusal/desist. So far, in the literature, no work has been done to handle the unknown criteria weights using PFS in the respective field of software. To fill this research gap, we have attempted to introduce an MCDM model with Picture Fuzzy settings. Therefore, our objective in this research study is to:

1. To identify the most important criteria for assessing the quality of software
2. To construct an effective software selection model by applying integrated MEMV-OWA-PF-TOPSIS.

The rest of the chapter is structured as follow: In the Sect. 2, we have briefly provided related work in the field of MCDM, Software Quality and Picture Fuzzy.

Later, a picture fuzzy sets based weighted distance-based similarity measure is used to rank software explained in Sect. 3 by step by step process. The study in this chapter intends to establish a complex MCDM problem using TOPSIS technique: a similarity index based technique. So far, in the literature, no work has been done to handle the unknown criteria weights by using information based on PFS in the field of software. We have presented an example with specific criteria to rank OSS in Sect. 4. At the end of the chapter in Sect. 5, the entire conclusion of the study with the results is discussed and future scope has also been provided.

2 Research Background

The concern for various researchers belonging to different areas has a common interest in the field of selection of software which results in many approaches. These approaches are generally a combination of different MCDM techniques, making it simple to deal with the complexity of the research problem. Reliability of software depends on various factors; thus, MCDM problem can easily be used to evaluate the software. The approaches which come under MCDM are: "Analytic Hierarchy Process (AHP)", "Fuzzy Analytic Hierarchy Process (FAHP)", "Compromise programming (CP)", "Preference Ranking Organization Method for Enrichment Evaluation (PROMETHEE)", "Technique for Order of Preference by Similarity to Ideal Solution (TOPSIS)", "Artificial Network Process (ANP)" (Coyle 2004; Karayalcin 1982; Charnes and Cooper 1961; Brans et al. 1986; Deng 1999; Chen 2000; Saaty 1996).

In general, the quality of a software depends on various factors, one of the factors is reliability which is considered to learn and understand the reason behind a software failure (Yadav and Khan 2012). There are traditional models which analysis qualities of software which can be used to evaluate a software (Boehm 1978; Grady 1992; McCall 1977; Jacobson et al. 1999; Dromey 1996; Linda and Shaw 1998). In this chapter, we have incorporated MEMV-OWA operator with picture fuzzy TOPSIS to handle the uncertainty in the point of view of an expert. The OWA operator was initially by Yager (1988). Later, an OWA operator for maximizing variance (MV-OWA) was introduced by Fullér and Majlender (2001).

The fuzzy set theory was first given by Zadeh (1965) and he defined it as a class of objects with a sequence of grades membership. This idea opened a new area of research for the researchers. Some of the researchers worked on its extension and the application of a fuzzy set. One of the significant and essential extensions of the fuzzy set is intuitionistic fuzzy sets, and it was given by Atanassov (1999). The theory was focused on the extension of definitions fuzzy set objects, new objects and their characteristics. Cuong and Kreinovich (2013) introduced Picture fuzzy set (PFS) which take into consideration human's opinion. It has more than two options like yes, no, neutral, and refusal. Later, Cường (2014) also proposed some operations based on PFS.

The PFSs has three essential components, namely the degree of belongingness, non-belongingness, and neutral. The property of PFS is that aggregation of these components must not exceed 1. In literature, the aggregation operators for PFSs and its application on MCDM is presented by Garg (2017). Some of the significant developments are: Singh (2015) brought the concept of finding the correlation coefficient of PFSs. A generalized picture distance measure was developed by Son (2016) and used it to picture fuzzy clustering. Van Viet and Van Hai (2017) introduced the system based on picture fuzzy set and named it as picture inference system (PIS) in. Hwang and Yoon (1981) bought a technique named TOPSIS, and used crisp values to handle MCDM problems. Later, it was extended by different makers to utilize it using an extension of fuzzy sets. Kuo (2017) modified the TOPSIS method with different ranking indexes. Tian et al. (2018) used TOPSIS by calculating weights using the best–worst method and to evaluate MCGDM problem with intuitionistic fuzzy information.

Wei (2016) used the concept of cross-entropy measure of PFSs in MCDM problems. Later in 2017, he utilized arithmetic and geometric operation of PFSs on MADM method. Wu and Wei (2017) used pictures of fuzzy aggregation operators in MADM problems like enterprise resource planning (ERP) selection. Furthermore, Ashraf et al. (2018) introduced the concept of cubic PFS which is also an extended form of PFSs. Dombi aggregation operators were introduced by Jana et al. (2019) for PFSs situations and applied them on MADM problems. Wang et al. (2018) ranked the characteristics of energy performance contracting projects (EPC) using MCDM technique and information from picture fuzzy. Peng (2017) introduced an operator called "Picture Fuzzy Ordered Weighted Geometric" Operator to deal with the MADM problems. He also introduced "Picture Fuzzy Induces Ordered Weighted Geometric" operator. Wang and Li (2018) used a hesitant picture fuzzy set and used it in MCDM.

Wang et al. (2019) worked on the MCDM problem by incorporating maximum deviation technique to calculate the weights. He developed a method to compare PLTSs (probabilistic linguistic term sets) based distance measure. Zhang et al. (2019) focused on "picture 2-tuple linguistic numbers" (P2TLNs) and created a score, action rules and accuracy functions, And later used them to solve MCDM problem established on the distance from the average solution.

Zaidan et al. (2015) presented a comparative study on Open Source of Electronic Medical Records by using MCDM techniques. To rank these software systems Weighted Sum Method (WSM), Weighted Product Method (WPM), Simple Additive Weighting (SAW), and TOPSIS were used. The aggregation of AHP and TOPIS is frequently used for ranking of different alternatives. One such research was done for selection of ETL (Extract, Transform and Load) software by Hanine et al. (2016) whereas Kara and Cheikhrouhou (2014) used the fuzzy AHP with TOPSIS for the ranking of alternatives. A hybrid approach was given by (Efe (2016)) using fuzzy AHP and Fuzzy TOPSIS. Yazgan et al. (2009) addressed the issue of prioritizing ERP software by creating Artificial Network Process (ANP) model. The results of the ANP model were used to train an artificial neural network (ANN), model. Lee et al. (2014) proposed an AHP application to solve the issue of evaluation of Open

Source Customer Relationship Management software. A model for software selection was formed with the combination of "Stepwise Weight Assessment Ratio Analysis" (SWARA) and the PROMETHEE by (Shukla et al. 2016). Wei et al. (2019) proposed the VIKOR method to evaluate multi-criteria group decision making problem having 2-TLNNS (2-tuple linguistic neutrosophic numbers).

3 Methodology

In this chapter, we have demonstrated a picture fuzzy-based TOPSIS model by incorporating non- linear multi-objective MEMV-OWA operator to rank the Open Source Software. The MEMV-OWA operator is adopted to evaluate the unknown criteria weights for the selection of software. We have also emerged the Picture Fuzzy information while constructing a TOPSIS model.

Preliminaries

Definition 1 The operator OWA with dimension "n" is a function $F : R_n \rightarrow R$ associated with weight vector say $W = (w_1, w_2, w_{3,} \ldots w_n)$ such that.

$$0 \leq w_i \leq 1;\ i = 1{,}2{,}\ldots{,}n$$
$$w_1 + w_2 + w_3 + \cdots + w_n = 1$$

Further, it has a property that 1

$$F(a_1 + a_2 + a_3 + \cdots + a_n) = \sum_{i}^{n} w_i b_i \tag{1}$$

In the assemblage of $(a_1, a_2, a_3, \ldots, a_n)$ arranged in descending order, we can say that b_i is one of the largest ith element.

Definition 2 Measure of Orness exemplify the location of an OWA operator, which is nearer to either orlike or andlike level. The weights of the OWA operator are near to one another concealed by a specific degree of orness. Therefore orness is defined as.

$$orness(W) = \sum_{i=1}^{n} \frac{n-i}{n-1} w_i = \alpha \in [0, 1] \tag{2}$$

(i) If α is closer to zero, then we can say that the interconnection between the various attributes has high andlike value. This implements that decision-maker is utmost noncommittal.

(ii) If α is closer to one, then we can say that the interconnection between the various attributes has high orlike value. This implements that decision-maker is utmost optimistic.

(iii) If $w_i = \frac{1}{n}$, then $\alpha = 0.5$, which implements that the decision-maker faces reasonable assessment.

Definition 3 For a specific level of orness, the variability in the weighting vector is determined by a measure of variance. The measure of variance is defined as:

$$D^2(W) = \frac{1}{n}\sum_{i=1}^{n}[w_i - E(w)]^2 = \frac{1}{n}\sum_{i=1}^{n}w_i^2 - \frac{1}{n^2} \tag{3}$$

While considering multiple attribute decision making, the variability in the weighting vector should be controlled to ignore the overestimation of a single attribute

Definition 4 The measure of entropy discovers that to what extent the information is exploited or utilized under an uncertain environment and conditions. The other term used for this is measure of dispersion and is given as:

$$Disp(W) = -\sum_{i=1}^{n} w_i \ln(w_i) \tag{4}$$

(i) If $w_i = 1$ and $w_j = 0$ $(j \neq i)$, then $Disp(W)$ is minimum i.e. 0. It implies that only a single attribute is studied during the course of aggregation.

(ii) If $w_i = \frac{1}{n}$, then $Disp(W)$ is maximum i.e. $Disp(W) = \ln(n)$. It implies that all attributes are studied during the course of aggregation.

Definition 5 The MEMV-OWA is a basically a bi-objective non -linear programming model which is constructed to evaluate the weighting vector. In this, the uncertain information based on the experience of decision makers is exploited by maximizing the entropy and on the other hand to ignore the over estimation of the preferences by decision-maker, we minimize variance of the weighting vector. The mathematical non-linear programming model is represented as:

Model:

$$Maximize: \sum_{i=1}^{n} w_i \ln(w_i)$$

$$Minimize: \frac{1}{n}\sum_{i=1}^{n}[w_i - E(w)]^2$$

$$subjectto: \sum_{i=1}^{n} \frac{n-i}{n-1} w_i = \alpha; 0 \leq \alpha \leq 1$$

$$\sum_{i=1}^{n} w_i = 1, 0 \leq w_i \leq 1, i = 1, 2, \ldots, n$$

Definition 6 A set whose all elements have degree of membership of belongingness in them is said to be Fuzzy Set. Suppose, $X = \{x_1, x_2, \ldots, x_n\}$ is a universal set, then fuzzy set Z defined on X is written as.

$$Z = \{(x, \mu_A(x)) | x \in X\}$$

where membership function is $\mu_Z(x) : X \to [0, 1]$ such that x ∈ X to the set Z.

Definition 7 An intuitionistic fuzzy set I on universal set X is defined as:

$I = \{(x, \mu_I(x), \vartheta_I(x)) | x \in X\}$.

where$\mu_I(x) \in [0, 1]$, is the membership degree and , $\vartheta_I(x) \in [0, 1]$ is the non-membership degree of x in I with the following condition:

$$0 \leq \mu_I(x) + \vartheta_I(x) \leq 1$$

For all $x \in X$, the lack of uncertainty is reflected by the amount $\pi_I(x) = 1 - (\mu_I(x) + \vartheta_I(x))$. This amount $\pi_I(x)$ is the degree of the hesitancy of $x \in X$ to the set I.

Definition 8 A picture fuzzy set (PFS) P on X is defined as:

$$P = \{(x, \mu_P(x), \eta_P(x), \vartheta_P(x)) | x \in X\}$$

where, $\mu_P(x)$ = positive membership degree, $\eta_P(x)$ = neutral membership degree,$\vartheta_P(x)$ = negative membership degree with the following condition:

$$0 \leq \mu_P(x) + \eta_P(x) + \vartheta_P(x) \leq 1$$

For all $x \in X$, the amount $\pi_P(x) = 1 - (\mu_P(x) + \eta_P(x) + \vartheta_P(x))$ is called the refusal degree of $x \in X$ to the set P.

Let M and S be two PFSs defined on X, then some operators are represented below:

(i) $M \subseteq S$ iff $\mu_M(x) \leq \mu_S(x), \eta_M(x) \leq \eta_S(x) and \vartheta_M(x) \geq \vartheta_S(x)$ for all $x \in X$.

(ii) $M = S \text{iff}\, M \subseteq S$ and $S \subseteq M$.
(iii) $M \cap S = \{x, \min(\mu_M(x), \mu_S(x)), \max(\eta_M(x), \eta_S(x)), \max(\vartheta_M(x)\vartheta_S(x)) | x \in X\}$
(iv) $M \cup S = \{x, \max(\mu_M(x), \mu_S(x)), \min(\eta_M(x), \eta_S(x)), \min(\vartheta_M(x)\vartheta_S(x)) | x \in X\}$
(v) $M^C = \{(x, \vartheta_S(x), \eta_S(x), \mu_S(x)) | x \in X\}$

Distance Measures

The picture fuzzy-based distance measures with picture fuzzy information are given as:

(a) Distance Measure between Two Picture Fuzzy Numbers

Let M and S be two PFSs defined on $X = \{x_{1,}x_{2,}x_{3,\ldots,}x_n\}$, then the distance between M and N is given by:

$$D_P(M, S) = \frac{1}{3n}\sum_{i=1}^{n}([|\mu_M(x_i) - \mu_S(x)| + |\eta_M(x_i) - \eta_S(x_i)| + |\vartheta_M(x_i) - \vartheta_S(x_i)| + \max[|\mu_M(x_i) - \mu_S(x_i)|, |\eta_M(x_i) - \eta_S(x_i)|, |\vartheta_M(x_i) - \vartheta_S(x_i)|]]) \tag{5}$$

The $D_P(M, S)$ is distance measure if it holds the following conditions:

(i) $0 \le D_P(M, S) \le 1$
(ii) $D_P(M, S) = 0$ iff $M = S$
(iii) $D_P(M, S) = D_P(S, M)$
(iv) For any $M, S, O \in PFSs(X)$, we have $D_P(M, O) \ge D_P(M, S)$ and $D_P(M, O) \ge D_P(S, O)$.

(b) Weighted Distance Measure

Let M and S be two PFSs defined on $X = \{x_{1,}x_{2,}x_{3,\ldots,}x_n\}$ and the weights of the "m" criteria be w_j holding condition that $\sum_{j=1}^{m} w_j = 1$. Then, the weighted distance measure is given by:

$$D_P^w(M, S) = \frac{1}{3n}\sum_{i=1}^{n} w_j([|\mu_M(x_i) - \mu_S(x)| + |\eta_M(x_i) - \eta_S(x_i)| + |\vartheta_M(x_i) - \vartheta_S(x_i)| + \max[|\mu_M(x_i) - \mu_S(x_i)|, |\eta_M(x_i) - \eta_S(x_i)|, |\vartheta_M(x_i) - \vartheta_S(x_i)|]]) \tag{6}$$

The $D_P^w(M, S)$ is weighted distance measure if it holds the following conditions:

(i) $0 \le D_P^w(M, S) \le 1$
(ii) $D_P^w(M, S) = 0 \text{iff}\, M = S$
(iii) $D_P^w(M, S) = D_P^w(S, M)$
(iv) For any $M, S, O \in PFSs(X)$ we have $D_P^w(M, O) \ge D_P^w(M, S)$ and $D_P^w(M, O) \ge D_P^w(S, O)$.

(c) Similarity Index Measure

Now, using above distance measures between two PFSs defined on $X = \{x_{1,}x_{2,}x_{3,\ldots,}x_n\}$, we can define a similarity index measure as:

$$I_p(M,S) = 1 - \frac{1}{3}\sum_{i=1}^{n} w_j(|\mu_M(x_i) - \mu_S(x_i)| + |\eta_M(x_i) - \eta_S(x_i)| + |\vartheta_M(x_i) - \vartheta_S(x_i)| + \max[|\mu_M(x_i) - \mu_S(x_i)|, |\eta_M(x_i) - \eta_S(x_i)|, |\vartheta_M(x_i) - \vartheta_S(x_i)|]) \quad (7)$$

Here, the weights of the "m" criteria are w_j holding condition that $\sum_{j=1}^{m} w_j = 1$. The $I_P(M,S)$ is a similarity measure if it holds the following conditions:

(v) $0 \leq I_P(M,S) \leq 1$
(vi) $I_P(M,S) = 0$iff $M = S$
(vii) $I_P(M,S) = I_P(S,M)$
(viii) For any $M, S, O \in PFSs(X)$, we have $I_P(M,O) \geq I_P(M,S)$ and $I_P(M,O) \geq I_P(S,O)$.

Step by Step Process of Maximum Entropy Minimum Variance OWA-Picture Fuzzy TOPSIS (MEMV-OWA-PF-TOPSIS).

Now, we have demonstrated a Multi-criteria decision-making approach, i.e. TOPSIS with picture fuzzy information. And the maximum entropy minimum variance OWA approach is used to find out the criteria weights. Consider, a discrete set of alternatives say $A = \{A_1, A_2, A_3, \ldots, A_n\}$ and "m" criteria $C = \{C_1, C_2, C_3, \ldots, C_m\}$ having weights say $W = \{w_1, w_2, w_3, \ldots, w_m\}$ such that $\sum_{j=1}^{m} w_j = 1$.

We also define Picture Fuzzy decision matrix say $M = [\Delta_{ij}]_{m\times n} = [(\mu_{ij}, \eta_{ij}, \vartheta_{ij})]_{m\times n}$, where $\mu_{ij}, \eta_{ij}, \vartheta_{ij}$ is the degree of acceptance, neutral and rejection respectively that alternatives A_i fulfils. The MCDM procedure is briefly explained below in order to make the best decision:

Step 1: Firstly build the hierarchical model for the selection of software.

Step 2: Obtain the MEMV-OWA weights $W=\{w_1, w_2, w_3, \ldots, w_m\}$ for the given set of criteria by solving non-linear programming problem given in **Definition** 5.

Step 3: Develop the matrix $M = [\Delta_{ij}]_{m\times n}$ which is the picture fuzzy decision matrix using decision maker's information.

Step 4: Identify the benefit criteria (B_1) and cost criteria (B_2). Later, we need to determine the Picture Fuzzy Positive Ideal Solution $\left(\Delta_p^+\right)$ and Picture Fuzzy Negative Ideal Solution $\left(\Delta_p^-\right)$ given as:

$$\Delta_p^+ = \begin{cases} [(\max_j(\mu_{ij}), \min_j(\eta_{ij}), \min_j(\vartheta_{ij})], & C_j \in B_1 \\ [(\max_j(\mu_{ij}), \max_j(\eta_{ij}), \min_j(\vartheta_{ij})], & C_j \in B_2, \end{cases} \quad (8)$$

$$\Delta_p^- = \begin{cases} [(\min_j(\mu_{ij}), \min_j(\eta_{ij}), \max_j(\vartheta_{ij})], & C_j \in B_1 \\ [(\min_j(\mu_{ij}), \max_j(\eta_{ij}), \max_j(\vartheta_{ij})], & C_j \in B_2, \end{cases} \tag{9}$$

Step 5: Find out the degree of weighted similarity index $\left(I_{p_i}^+\right)$ and $\left(I_{p_i}^-\right)$ between Δ_p^+ and Δ_p^- respectively where $1 \le i \le n$.

$$I_I^+\left(A_i, \Delta_p^+\right) = 1 - \frac{1}{3}\sum_{j=1}^{m} w_j\Big(\left|\mu_A(x_i) - \mu_{ij}^+\right| + \left|\eta_M(x_i) - \eta_{ij}^+\right| + \left|\vartheta_M(x_i) - \vartheta_{ij}^+\right| + \max\Big[\left|\mu_M(x_i) - \mu_{ij}^+\right|, \left|\eta_M(x_i) - \eta_{ij}^+\right|, \left|\vartheta_M(x_i) - \vartheta_{ij}^+\right|\Big]\Big) \tag{10}$$

$$I_I^-\left(A_i, \Delta_p^-\right) = 1 - \frac{1}{3}\sum_{j=1}^{m} w_j\Big(\left|\mu_A(x_i) - \mu_{ij}^-\right| + \left|\eta_M(x_i) - \eta_{ij}^-\right| + \left|\vartheta_M(x_i) - \vartheta_{ij}^-\right| + \max\Big[\left|\mu_M(x_i) - \mu_{ij}^-\right|, \left|\eta_M(x_i) - \eta_{ij}^-\right|, \left|\vartheta_M(x_i) - \vartheta_{ij}^-\right|\Big]\Big) \tag{1.11}$$

Step 6: Using the above equations, the degree of weighted similarity index $\left(I_{p_i}^+\right)$ and $\left(I_{p_i}^-\right)$ is calculated and later evaluate the Relative Closeness measure for a given set of alternatives with respect to Δ_p^+.

$$RelativeCloseness(RC_i) = \frac{I_{p_i}^+}{I_{p_i}^+ + I_{p_i}^-} \tag{12}$$

The higher the value of RC_i of the given alternatives with respect to Δ_p^+ which picture fuzzy positive ideal solution corresponds to best alternatives from A_i.

4 Numerical Illustrations

In this chapter, a numerical illustration related to software quality assessment of open source software (OSS) is provided in order to validate the application of MEMV-OWA-PF-TOPSIS. Several OSS is freely available online. The quality of OSS is questioned and has become the primary concern because OSS makes a significant influence on commercial marketing sector (Karakoidas et al. 2007).

In this research, four open-source-software (OSS) are assessed, which are freely available online. The assessment information is collected through a decision-maker who is the users of the software. The set of alternatives for OSS is denoted by

$S = \{S_1, S_2, S_3, S_4\}$. All these OSS software are assessed on the basis six criteria which comprise a set denoted by

$$\begin{aligned} C &= \{C_1, C_2, C_3, C_4, C_5, C_6\} \\ &= \{\mathit{TechnicalAspects}, \mathit{Cost}, \mathit{SystemReliability}, \mathit{compatibility}, \mathit{ImplemetationTime}, \mathit{Functionality}\}' \end{aligned}$$

where *Cost* and *Implementation Time* are the cost criteria and others are the benefit criteria. The steps for software quality assessment of four OSS on the basis of six criteria through MEMV-OWA-PF-TOPSIS are given as follows:

Step 1: The hierarchal model for software quality assessment is provided in Fig. 1.

Step 2: The orness (α) level is selected by the uncertain preferences of experts. If they are in moderating state, then $\alpha = 0.5$ and if they are maximally optimistic then $\alpha = 1$. In our study, OSS software experts have given moderate optimistic preferences; therefore, the level of orness (α) will be equal to 0.8. The weight vector (W_i) of MEMV-OWA averaging operator with respect to $n = 6$ and the particular level of orness (α) is provided in Table 1.

These W_i can be used to solve the information of performance and cost-related criteria of open source software (OSS). So, the weight vector concerning $\alpha = 0.8$ from Table 1 are as follows:

$$W_1 = 0.4352;\ W_2 = 0.2492;\ W_3 = 0.1441;\ W_4 = 0.0974;$$
$$W_5 = 0.0471;\ W_6 = 0.0270.$$

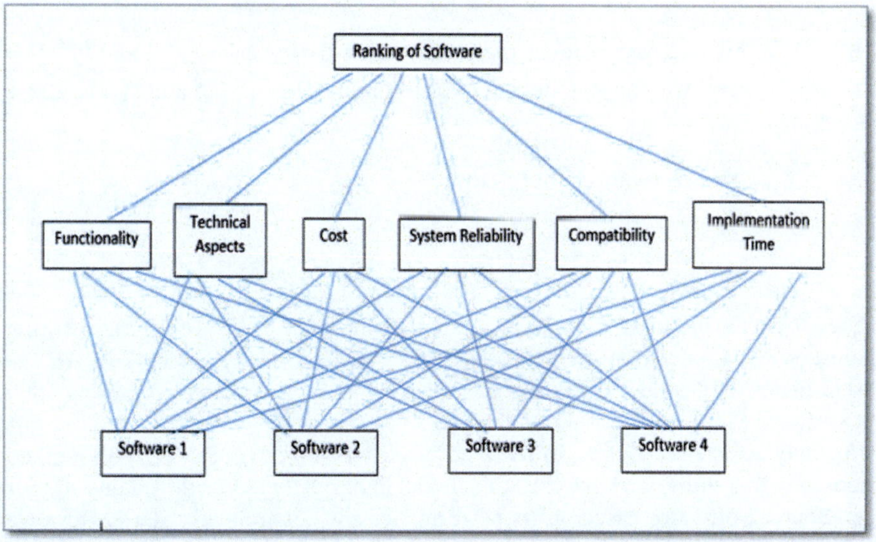

Fig. 1 Hierarchical model for quality assessment of differeent OSS

Table 1 Weight vector of MEMV-OWA averaging operator

W_i	Orness (α)					
	1	0.9	0.8	0.7	0.6	0.5
W_1	1	0.6227	0.4352	0.3096	0.2158	0.1667
W_2	0	0.2345	0.2492	0.2236	0.1852	0.1667
W_3	0	0.0894	0.1441	0.1614	0.1589	0.1667
W_4	0	0.0337	0.0974	0.1165	0.1364	0.1667
W_5	0	0.0129	0.0471	0.0842	0.1171	0.1667
W_6	0	0.0068	0.0270	0.1047	0.1866	0.1667

Step 3: The matrix $M = [\Delta_{ij}]_{m\times n}$ which is the picture fuzzy decision matrix using decision maker's information is in Table 2.

Step 4: Now, we calculate the Picture Fuzzy Positive Ideal Solution $\left(\Delta_p^+\right)$ and Picture Fuzzy Negative Ideal Solution $\left(\Delta_p^-\right)$ on the basis of Eqs. (8) and (9).

$$\Delta_p^+ = [(0.9, 0.0, 0.05), (0.75, 01, 0.1), (0.6, 0.0, 0.30), \\ (0.75, 0.0, 0.1), (0.5, 0.1, 0.4), (0.6, 0.0, 0.3)]$$

$$\Delta_p^- = [(0.5, 0.0, 0.4), (0.5, 01, 0.4), (0.3, 0.0, 0.6), \\ (0.3, 0.0, 0.6), (0.25, 0.1, 0.6), (0.25, 0.0, 0.6)]$$

Step 5: In this step, we calculate the weighted similarity index $\left(I_{p_i}^+\right)$ and $\left(I_{p_i}^-\right)$ by putting the criteria weights calculated in **Step 2** in the Eqs. (1.10) and (1.11) where $1 \le i \le n$.

$$I_{p_1}^+ = 0.9490; I_{p_2}^+ = 0.6399; I_{p_3}^+ = 0.7542; I = 0.6352.$$

$$I_{p_1}^- = 0.6819; I_{p_2}^- = 0.9028; I_{p_3}^- = 0.8347; I_{p_4}^- = 0.9167.$$

Step 6: On the basis of Eq. (1.12), we calculate the value of RC_i of the ith software, where $1 \le i \le 4$, such that: $RC_1 = 0.5819$; $RC_2 = 0.4148$; $RC_3 = 0.4746$; $RC_4 = 0.4093$ that provides the assessment order as: $S_1 > S_3 > S_2 > S_4$, represents that first software (S_1) is the best alternative.

To analyze the effectiveness of MCDM based MEMV-OWA-PF-TOPSIS method proposed in this study, we have used this model to basically assess the quality of four Open Source Softwares based on six performance and cost-related criteria. Now, we can observe from Fig. 2 that the first software S_1 is relatively more effective than other OSS software because the value of its relative closeness is much larger than the rest of the software.

Table 2 Picture fuzzy decision matrix

	C_1	C_2	C_3	C_4	C_5	C_6
S_1	(0.90,0.00,0.0)	(0.75,0.05,0.10)	(0.60,0.00,0.30)	(0.50,0.10,0.40)	(0.25,0.05,0.60)	(0.60,0.00,0.30)
S_2	(0.60,0.00,0.3)	(0.50,0.10,0.40)	(0.30,0.00,0.60)	(0.75,0.05,0.10)	(0.30,0.00,0.60)	(0.30,0.00,0.60)
S_3	(0.75,0.05,0.1)	(0.50,0.10,0.40)	(0.30,0.00,0.60)	(0.50,0.10,0.40)	(0.50,0.10,0.40)	(0.30,0.00,0.60)
S_4	(0.50,0.10,0.4)	(0.60,0.00,0.30)	(0.30,0.00,0.60)	(0.30,0.00,0.60)	(0.25,0.05,0.60)	(0.25,0.05,0.60)

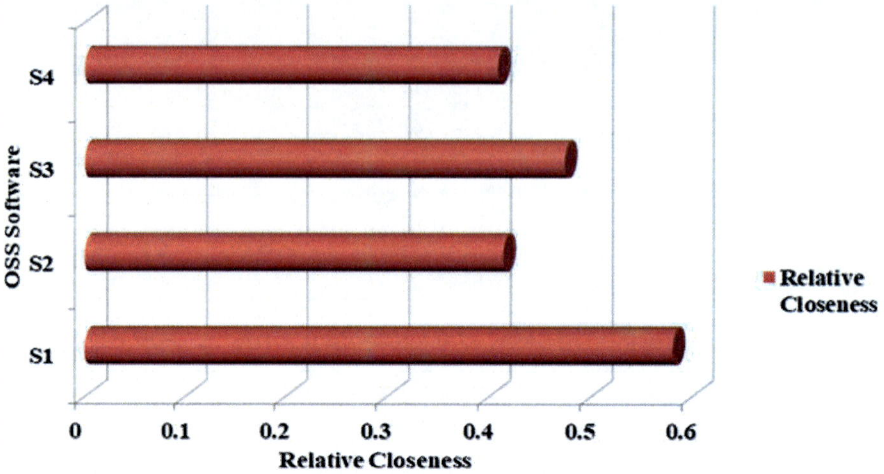

Fig. 2 OSS quality assessment by applying MEMV-OWA-PF-TOPSIS method

5 Conclusion and Future Scope

The current software environment has become active and influential and software customers' needs to think about the quality of the software before buying the license. The selection of software has become a serious activity as there various criteria for the quality. The software selection can influence various software companies, customers, developers in various aspects. Therefore, it is significant to select software for every user or customer before using any software. This selection process is quiet challenging task as it involves multiple criteria. As we know, MCDM approach is well known for ranking, selecting, evaluating these multiple criteria. So, in this chapter, the quality of open-source software (OSS) are assessed using the MCDM based approach: Maximum Entropy Minimum Variance OWA-Picture Fuzzy TOPSIS (MEMV-OWA-PF-TOPSIS). The criteria considered in this study to rank or select software are based on quality and cost. The following are the criteria taken: Technical Aspects, Cost, System Reliability, Compatibility, Implementation Time, and Functionality. The basic approach of TOPSIS technique is to first divide the criteria into cost and benefit criteria. So in our model, Cost and Implementation Time are the cost criteria as we need to minimize these whereas Technical Aspects, System Reliability, Compatibility, and Functionality are the benefit criteria which need to be maximized. The unknown weights of the criteria are evaluated using MEMV-OWA method which is a bi-objective non-linear programming approach. Another reason to use MEMV-OWA is that it handles the uncertainty while finding the weights of the criteria.

In the past studies, most researchers have applied Intuitionistic Fuzzy Sets (IFS) based MCDM technique for assessing the software reliability, but, IFSs cannot incorporate all the cases efficiently. For example, in the case of election voting, people thoughts consider more degrees as, refusal, no, neutral, yes. To overcome this, in this research, we have applied PFSs based MCDM technique which is the extension of IFSs. The Picture fuzzy decision matrix has been created to remove uncertainty to another level that is present in the real-life decision-making problem like software quality assessment. Hence, the uncertainty factor is handled twice in this model.

This chapter also adds to the literature on software quality assessment by providing an advanced Picture Fuzzy based MEMV-OWA operator technique which includes the decision maker's uncertain preferences. The proposed method has been demonstrated with a numerical illustration for validating their reliability and effectiveness. The MEMV-OWA-PF-TOPSIS method has been implemented in a numerical example and we have reported the results obtained graphically as well.

The future research direction should focus on the techniques that can be extended in the field of software using MCDM approach under the environment of multi-granular fuzzy linguistic, polygonal fuzzy sets and other unclear situations.

References

Ashraf S, Abdullah S, Qadir A (2018) Novel concept of cubic picture fuzzy sets. J New Theory 24:59–72

Atanassov KT (1999) Intuitionistic fuzzy sets. In: Intuitionistic fuzzy sets, Physica, Heidelberg. Springer, pp 1–137

Boehm BW (1978) Characteristics of software quality, vol 1. North-Holland

Brans J-P, Vincke P, Mareschal B (1986) How to select and how to rank projects: the PROMETHEE method. Eur J Oper Res 24(2):228–238

Charnes A, Cooper W (1961) Management models and industrial applications of linear programming. Wiley, New York

Chen C-T (2000) Extensions of the TOPSIS for group decision-making under fuzzy environment. Fuzzy Sets Syst 114(1):1–9

Coyle G (2004) The analytic hierarchy process (AHP). Practical strategy: structured tools and techniques. Open access material. Pearson Education Ltd., Glasgow

Cường BC (2014) Picture fuzzy sets. J Comput Sci Cybern 30(4):409

Cuong BC, Kreinovich V (2013) Picture fuzzy sets-a new concept for computational intelligence problems. In: 2013 third world congress on information and communication technologies (WICT 2013). IEEE, pp 1–6

Deng H (1999) Multicriteria analysis with fuzzy pairwise comparison. Int J Approx Reason 21(3):215–231

Dromey RG (1996) Cornering the chimera [software quality]. IEEE Softw 13(1):33–43

Efe B (2016) An integrated fuzzy multi criteria group decision making approach for ERP system selection. Appl Soft Comput 38:106–117

Fullér R, Majlender P (2001) An analytic approach for obtaining maximal entropy OWA operator weights. Fuzzy Sets Syst 124(1):53–57

Garg H (2017) Some picture fuzzy aggregation operators and their applications to multicriteria decision-making. Arab J Sci Eng 42(12):5275–5290

Grady RB (1992) Practical software metrics for project management and process improvement. Prentice-Hall, Inc.,

Hanine M, Boutkhoum O, Tikniouine A, Agouti T (2016) Application of an integrated multi-criteria decision making AHP-TOPSIS methodology for ETL software selection. Springerplus 5(1):263

Hwang C-L, Yoon K (1981) Methods for multiple attribute decision making. In: Multiple attribute decision making. Springer, pp 58–191
Jacobson I, Booch G, Rumbaugh J (1999) The unified software development process. Addison-Wesley Longman Publishing Co., Inc.,
Jana C, Senapati T, Pal M, Yager RR (2019) Picture fuzzy Dombi aggregation operators: application to MADM process. Appl Soft Comput 74:99–109
Kacprzak D (2019) A doubly extended TOPSIS method for group decision making based on ordered fuzzy numbers. Expert Syst Appl 116:243–254
Kara SS, Cheikhrouhou N (2014) A multi criteria group decision making approach for collaborative software selection problem. J Intell Fuzzy Syst 26(1):37–47
Karakoidas V, Vlachos V, Instit TE (2007) Software quality assessment of open source software. Current trends in informatics: 11th panhellenic conference on informatics, vol A. PCI, Athens, pp 303–315
Karayalcin II (1982) The analytic hierarchy process: planning, priority setting, resource allocation: Thomas L. SAATY McGraw-Hill, New York, 1980, xiii+ 287 p, £ 15.65. North-Holland,
Kuo T (2017) A modified TOPSIS with a different ranking index. Eur J Oper Res 260(1):152–160
Lee Y-C, Tang N-H, Sugumaran V (2014) Open source CRM software selection using the analytic hierarchy process. Inf Syst Manag 31(1):2–20
Linda RT, Shaw HJ (1998) Software metrics and reliability.
McCall JA (1977) Factors in software quality. US Rome air development center reports NY, Tech Rep RADC-TR-77–369, vol 1–2
Peng S-M (2017) Study on enterprise risk management assessment based on picture fuzzy multiple attribute decision-making method. J Intell Fuzzy Syst 33(6):3451–3458
Saaty TL (1996) Decision making with dependence and feedback: the analytic network process, vol 4922. RWS Publ.,
Shukla S, Mishra P, Jain R, Yadav H (2016) An integrated decision making approach for ERP system selection using SWARA and PROMETHEE method. Int J Intell Enterprise 3(2):120–147
Singh P (2015) Correlation coefficients for picture fuzzy sets. J Intell Fuzzy Syst 28(2):591–604
Son LH (2016) Generalized picture distance measure and applications to picture fuzzy clustering. Appl Soft Comput 46 (C):284–295
Tian Z-P, Zhang H-Y, Wang J-Q, Wang T-L (2018) Green supplier selection using improved TOPSIS and best-worst method under intuitionistic fuzzy environment. Informatica 29(4):773–800
Van Viet P, Van Hai P (2017) Picture inference system: a new fuzzy inference system on picture fuzzy set. Appl Intell 46(3):652–669
Wang L, Peng J-j, Wang J-q (2018) A multi-criteria decision-making framework for risk ranking of energy performance contracting project under picture fuzzy environment. J Clean Prod 191:105–118
Wang R, Li Y (2018) Picture hesitant fuzzy set and its application to multiple criteria decision-making. Symmetry 10(7):295
Wang X, Wang J, Zhang H (2019) Distance-based multicriteria group decision-making approach with probabilistic linguistic term sets. Expert Systems 36(2):e12352
Wei G (2016) Picture fuzzy cross-entropy for multiple attribute decision making problems. J Bus Econ Manag 17(4):491–502
Wei G (2017) Picture fuzzy aggregation operators and their application to multiple attribute decision making. J Intell Fuzzy Syst 33(2):713–724
Wei G, Wang J, Lu J, Wu J, Wei C, Alsaadi FE, Hayat T (2019) VIKOR method for multiple criteria group decision making under 2-tuple linguistic neutrosophic environment. Econ Res-Ekonomska Istraživanja 33(1):1–24
Wu S-J, Wei G-W (2017) Picture uncertain linguistic aggregation operators and their application to multiple attribute decision making. Int J Knowl-Based Intell Eng Syst 21(4):243–256
Yadav A, Khan R (2012) Reliability estimation framework-complexity perspective. Comput Sci Inform Technol (CS & IT) 2(5):97–104

Yager RR (1988) On ordered weighted averaging aggregation operators in multicriteria decision-making. IEEE Trans Syst Man Cybern 18(1):183–190
Yazgan HR, Boran S, Goztepe K (2009) An ERP software selection process with using artificial neural network based on analytic network process approach. Expert Syst Appl 36(5):9214–9222
Zadeh LA (1965) Fuzzy Sets. Inform Control 8(3):338–353
Zaidan A, Zaidan B, Hussain M, Haiqi A, Kiah MM, Abdulnabi M (2015) Multi-criteria analysis for OS-EMR software selection problem: a comparative study. Decis Support Syst 78:15–27
Zhang S, Gao H, Wei G, Wei Y, Wei C (2019) Evaluation based on distance from average solution method for multiple criteria group decision making under picture 2-tuple linguistic environment. Mathematics 7(3):243

Requirement Barriers to Implement the Software Projects in Agile Development

Deepak Kumar and Saru Dhir

Abstract The success of an organization is to deliver the good-quality products as per the client's needs. But few organizations are unable to deliver the successful product due to number of software barriers. The research is based on the study and analysis of different requirement barriers, which causes problem in agile implementation. Several authors identified the barriers for successful implementation of software, but none have found the barriers at the initial stage of requirements. The motivation behind this work is to classify the main requirement barriers to the effective implementation of software projects in agile development. For the study, interviews were conducted with developers and testers. The results recognized the key barriers and it will deliver the roadmap to managers to take suitable steps to overwhelm the major barriers to effective software implementation.

Keywords Software requirement · Agile project implementation · Interpretive structural modelling (ISM) · Requirement barriers · Agile methodology · Requirement engineering

1 Introduction

Software development is a teamwork where each member has different roles such as software development, testing, project analysis etc. A project quality, delivery time and cost of delivered product specify its success and failure rate. Agile software development is a leading approach in software organizations during last few years; to fulfil the client's need of producing quick, better and cost-effective solutions. In agile development, client's have direct interaction with team members of project to improve the communication among them. As the concept of quality is relative, the aim of this work is to comprehend the factors that affect the failure of software project and its quality with regards to Agile Software Development (ASD).

D. Kumar (✉) · S. Dhir
Amity University, Noida, Uttar Pradesh, India

A. G. Aggarwal et al. (eds.), *Optimization Models in Software Reliability*, Springer Series in Reliability Engineering,
https://doi.org/10.1007/978-3-030-78919-0_15

The popularity of agile development is increased during the last years; despite of that, agile methods are also criticized for successful delivery on functional requirements and on neglecting the quality requirements. Ignorance of quality requirements becomes the result of non-satisfaction of user's requirements.

In common practice, all the individuals or groups faces challenges during the implementation, which reduces the performance of the system (Boehm and Turner 2005). For a successful service workflow, requirement management should be considered as an initial point. Different success and failure factors were identified in the project implantation (Dhir et al. 2017, 2019). For a successful implementation of the software project, requirement elicitation and management are a significant task (Dhir and Kumar 2015; Dhir et al. 2019).

Software complexity and their issues are fully involved in requirement and design factors. Project requirements are chosen rendering to the client's end product's need (Rai and Dhir 2014; Rajagopal et al. 2005). An approach was planned to collect the requirements and appropriate steps were taken to eradicate the barriers (Rajagopal et al. 2005). It is necessary to remove the barriers and ensured about the software requirements should be according to the customer's requirements to maintain the complete software quality.

The topical survey of Standish Group (2014) represents the success reason of the project is: requirements statement, user contribution and management support. This survey report considered the standards of a project that are eventually based on the requirement management.

During agile implementation, requirements are adaptive in nature and it is also not easy to maintain the requirement specification documents. Requirements are continuously change in agile development, due to different reasons such as: missing requirements, customers lack knowledge, market change and bug fixing.

Barriers in software requirements would affect the budget and quality of product. Number of barriers means number of risk factors increases for the failure of software project. So, it is essential to recognize the different barriers during implementation to improve the functionality, including efficiency, performance, quality and security of the system.

Requirement documentation is the key deliverable of requirements for the implementation of software. Lack of documentation is also another barrier as the documentation is not possible in agile.

A survey was directed with the help of industry experts and identifies the barriers during software implementation. Industry experts were the software developers, testers and leaders who implemented the software using agile methodology (Dhir and Kumar 2015).

The research recognizes the most significant barriers that affects the effectiveness and quality of software implementation. This work signifies the ISM practice to classify the connection between the diverse barriers and find the most significant barrier that affect the software implementation using agile methodology. The research work presents a roadmap for managers of an organization to resolve the issues influencing during the agile development, so that the management or senior members can take appropriate actions to resolve these issues.

2 Literature Survey

There are different studies have been focused on the impact and acceptance of agile development in different organizations (Rai and Dhir 2014; Aggarwal and Dhir 2013). Authors discussed about the agile principles and the impact on current software development.

There are different uncertainties occurred during project planning such as uncertain estimates, requirements management and prioritization, ignorance of non-functional requirement. Other studies have been evaluated directing on how the acceptance of agile issues can be resolved (Misra et al. 2009). Qu et al. (2012) identified different risk factors for project management. Result analysis were analysed to evaluate the perplexing possible relationship between risk factors using interpretative structural modelling.

Alsaqaf et al. (2018) identified the nine challenges faced by the agile developers in large distributed projects that harmed the implementation of quality requirements. There are different challenges in software development such as: organization environment, communication and time differences in distributed environment.

Srinivasan and Lundqvist (2009) faced the challenges, that the agile team was not involved in the initial estimation of project due to which ambiguous requirements become poor in quality and schedule overruns. The literature lacks the obtainability of framework that can identify the barriers of requirement elicitation in agile software development, where the software requirements are changed very frequently (Srinivasan and Lundqvist 2009).

Conboy et al. (2011) conducted a study focusing on the challenges by the people in the agile development such as transparency by the team members, lack of business knowledge among team members etc. (Conboy et al. 2011). The selection of accurate requirement is a big challenge. Overall quality of the product depends upon the selection of requirements. A survey analysis was done by different practitioners to identify the prevalence's and challenges using agile software development. Statistical analysis was executed to identify the significant value of agile implementation over the traditional development (Dhir et al. 2017).

There are different decision-making techniques are applied for improvement of selection, such as Analytic Hierarchy Process (AHP), Paired Comparison Analysis, Game Theory, Multiple Criteria Decision Analysis and Interpretive structural modelling (ISM). ISM is an interpretive because it decides the findings of the group, how the variables are associated (Dhir et al. 2017).

Researcher studied and analysis among the barriers on a case study in 'just in time' production using ISM. Paper described the hidden barriers to 'just in time' production using ISM (Jadhav et al. 2015).

3 Methodology

ISM methodology has been introduced by prof. John Warfield with an objective to examine socio economics system issues and understand the complex relationship in different areas.

It has been demonstrated that ISM is a fixed decision-making tool that permits entities, groups and organizations to develop a connection for determining the complex situation and signifies the relationship through binary matrix (Huang et al. 2005; Warfield 1974). This method is used to understand composite structure with a simple geometric model articulating the complex relationship between numerous elements.

The usage of this method ranges for modelling systems interconnected to planning, decision making, competitive analysis, process re-engineering and many more. Statistical techniques provide the quantitative results using the large number of variables, whereas ISM provides the relationship among qualitative variables. The relationships between different variables are established with the repetition of questions, such as: 'Does target A supports to accomplish target B?', 'Does target A supports to accomplish target B?' for all pair of elements. According to the established relationship, a structure is created which is modelled through a digraph.

Steps for model development using ISM methodology are (Jadhav et al. 2015):

- Identify the barriers.
- Establish the relationship between barriers by conducting the interviews. Generate a self-structural interaction matrix (SSIM) of variables showing pairwise connection among barriers.
- Generate an initial reachability matrix (RM) and eliminate transitive relations in final reachability.
- Partition reachability matrix into various levels in different iterations.
- Plot a directed graph (digraph) in view of relations.
- Modify the digraph into an ISM model.
- Analyse an ISM model and examine hypothetical inconsistency.

Step 1: Recognize the Barriers

A wide analysis study is finished to recognize the barriers for the agile implementation in software development. The survey is directed during agile implementation to recognize the barriers for avoiding the effective and productive implementation. Survey data is collected by conducting the interviews with many industry experts. Interviews and discussions are done to control the barriers during the agile adoption and implementation. The main concern includes ever changing requirements, communication gap, undefined goals, incomplete requirements, lack of plan, shortage of expert members, requirement management, budgetary constraint, lack of documentation and ignorance of non-functional requirement. The barriers are listed as shown below in Table 1.

Table 1 Barriers to successful agile implementation

Barriers	Description
B_1	Lack of communication
B_2	Unclear goals
B_3	Incomplete requirement
B_4	Lack of planning
B_5	Ever changing requirement
B_6	Lack of expertise
B_7	Requirement management
B_8	Budgetary constraint
B_9	Lack of documentation
B_{10}	Ignorance of non-functional requirement

Step 2: Structural Self Interaction Matrix (SSIM)

A team of five members containing of industry experts especially developers, testers were interviewed to recognize the relationship between barriers. Following keywords signify the connection among barriers:

"V" signifies 'a' and helps to attain 'b'.
"A" signifies 'b' and helps to attain 'a'.
"X" signifies 'a' and 'b' helps to attain each other.
"O" signifies both 'a' and 'b' are not linked.

SSIM was implemented according to the known relationships in Table 2 and shows, 'a' and 'b' signifies rows and columns and it also represents the interrelationships between those 10 barriers. Table 3 is then used to generate reachability matrix.

Table 2 Structural self interaction matrix

Attribute (B_i)	B_1	B_2	B_3	B_4	B_5	B_6	B_7	B_8	B_9	B_{10}
B_1	1	V	A	A	O	A	V	V	V	V
B_2		1	A	V	O	V	V	V	V	A
B_3			1	V	A	V	V	V	V	V
B_4				1	A	V	V	V	V	V
B_5					1	V	V	V	V	V
B_6						1	V	V	V	V
B_7							1	V	A	V
B_8								1	A	V
B_9									1	X
B_{10}										1

Table 3 Initial reachability matrix

Attribute (B_i)	B_1	B_2	B_3	B_4	B_5	B_6	B_7	B_8	B_9	B_{10}
B_1	1	1	0	0	0	0	1	1	1	1
B_2	0	1	0	1	0	1	1	1	1	0
B_3	1	1	1	1	0	1	1	1	1	1
B_4	1	0	0	1	0	1	1	1	1	1
B_5	0	0	1	1	1	1	1	1	1	1
B_6	1	0	0	0	0	1	1	1	1	1
B_7	0	0	0	0	0	0	1	1	0	1
B_8	0	0	0	0	0	0	0	1	0	1
B_9	0	0	0	0	0	0	1	1	1	1
B_{10}	0	1	0	0	0	0	0	0	0	1

Step 3: Generate initial reachability matrix (IRM)

Generate IRM Table 3 from Table 2 SSIM.

If (a, b) in SSIM i.e. in Table 2 is "V"; then reachability matrix converts into 1 and (b, a) converts 0.

If (a, b) in SSIM i.e. in Table 2 is "A"; then reachability matrix converts into 0 and (b, a) converts 1.

If (a, b) in SSIM i.e. in Table 2 is "X"; then reachability matrix converts 1 and (b, a) converts 1.

If (a, b) in SSIM i.e. in Table 2 is "O"; then reachability matrix converts 0 and (b, a) converts 0.

Table 3 represents an IRM of 1's and 0's and currently it contains transitive relations.

The transitive relationship of Table 3 matrix is planned by squaring the matrix and will be transitive, if the resulting value of squared matrix has 1 which was earlier value 0. The transitive relation for reachability matrix is verified. Final reachability matrix (FRM) after verifying the transitive values is given in Table 4 and demonstrated in rows and columns, wherever rows specify driving power and columns indicate dependence power (Table 5).

Row wise barriers are the driving power to each barrier. Dependence power is sum of barriers (Tables 6, 7, 8, 9, 10 and 11).

Step 4: Dividing into levels

Both reachability and antecedent set to each barrier is estimated. Later, the connection of reachability and antecedent sets is estimated to all barriers. If the connection and reachability set are similar, then allot the level in the ISM Model. This process is iteratively estimated for whole barriers till the level for each barrier is recognized. Table 5 validates the early iteration with barriers B8 and B10 creating the first level. The whole levels to each of the barriers are signified in Table 12. Level I barriers have the lowermost whereas level VII has the uppermost driving power.

Table 4 Final reachability matrix

Attribute (B_i)	B_1	B_2	B_3	B_4	B_5	B_6	B_7	B_8	B_9	B_{10}	Drive power
B_1	1	1	0	1	0	1	1	1	1	1	8
B_2	1	1	0	1	0	1	1	1	1	1	8
B_3	1	1	1	1	0	1	1	1	1	1	9
B_4	1	1	0	1	0	1	1	1	1	1	8
B_5	1	1	1	1	1	1	1	1	1	1	10
B_6	1	1	0	0	0	1	1	1	1	1	7
B_7	0	0	0	0	0	0	1	1	0	1	3
B_8	0	0	0	0	0	0	0	1	0	1	2
B_9	0	1	0	0	0	0	1	1	1	1	5
B_{10}	0	1	0	1	0	1	1	1	1	1	7
Dependence power	6	8	2	6	1	7	9	10	8	10	67

[a]Entries are included to incorporate transitivity

Table 5 RM partitioning iteration 1

Attribute (B_i)	Reachability set	Antecedent set	Intersection set	Level
B_1	B_1, B_2, B_4, B_6, B_7, B_8, B_9, B_{10}	B_1, B_2, B_3, B_4, B_5, B_6, B_7	B_1, B_2, B_4, B_6, B_7	
B_2	B_1, B_2, B_4, B_6, B_7, B_8, B_9, B_{10}	B_1, B_2, B_3, B_4, B_5, B_6, B_7, B_8, B_9, B_{10}	B_1, B_2, B_4, B_6, B_7, B_9, B_{10}	
B_3	B_1, B_2, B_3, B_4, B_6, B_7, B_8, B_9, B_{10}	B_3, B_5	B_3	
B_4	B_1, B_2, B_4, B_6, B_7, B_8, B_9, B_{10}	B_1, B_2, B_3, B_4, B_5, B_{10}	B_1, B_2, B_4, B_{10}	
B_5	B_1, B_2, B_3, B_4, B_5, B_6, B_7, B_8, B_9, B_{10}	B_5	B_5	
B_6	B_1, B_2, B_6, B_7, B_8, B_9, B_{10}	B_1, B_2, B_3, B_4, B_5, B_6, B_{10}	B_1, B_2, B_6, B_{10}	
B_7	B_7, B_8, B_{10}	B_1, B_2, B_3, B_4, B_5, B_6, B_7, B_9, B_{10}	B_7, B_{10}	
B_8	B_8, B_{10}	B_1, B_2, B_3, B_4, B_5, B_6, B_7, B_8, B_9, B_{10}	B_8, B_{10}	I
B_9	B_2, B_7, B_8, B_9, B_{10}	B_1, B_2, B_3, B_4, B_5, B_6, B_8, B_9, B_{10}	B_2, B_9, B_{10}	
B_{10}	B_2, B_4, B_6, B_7, B_8, B_9, B_{10}	B_1, B_2, B_3, B_4, B_5, B_6, B_7, B_8, B_9, B_{10}	B_2, B_4, B_6, B_7, B_8, B_9, B_{10}	I

Table 6 RM partitioning iteration 2

Attribute (B_i)	Reachability set	Antecedent set	Intersection set	Level
B_1	B_1, B_2, B_4, B_6, B_7	$B_1, B_2, B_3, B_4, B_5, B_6, B_7$	B_1, B_2, B_4, B_6, B_7	
B_2	$B_1, B_2, B_4, B_6, B_7, B_9$	$B_1, B_2, B_3, B_4, B_5, B_6, B_7, B_9$	$B_1, B_2, B_4, B_6, B_7, B_9$	II
B_3	$B_1, B_2, B_3, B_4, B_6, B_7, B_9$	B_3, B_5	B_3	
B_4	$B_1, B_2, B_4, B_6, B_7, B_9$	B_1, B_2, B_3, B_4, B_5	B_1, B_2, B_4	
B_5	$B_1, B_2, B_3, B_4, B_5, B_6, B_7, B_9$	B_5	B_5	
B_6	B_1, B_2, B_6, B_7, B_9	$B_1, B_2, B_3, B_4, B_5, B_6$	B_1, B_2, B_6	
B_7	B_7	$B_1, B_2, B_3, B_4, B_5, B_6, B_7, B_9$	B_7	II
B_9	B_2, B_7, B_9	$B_1, B_2, B_3, B_4, B_5, B_6, B_9$	B_2, B_9	

Table 7 RM partitioning iteration 3

Attribute (B_i)	Reachability set	Antecedent set	Intersection set	Level
B_1	B_1, B_4, B_6	B_1, B_3, B_4, B_5, B_6	B_1, B_4, B_6	
B_3	B_1, B_3, B_4, B_6, B_9	B_3, B_5	B_3	
B_4	B_1, B_4, B_6, B_9	B_1, B_3, B_4, B_5	B_1, B_4	
B_5	$B_1, B_3, B_4, B_5, B_6, B_9$	B_5	B_5	
B_6	B_1, B_6, B_9	B_1, B_3, B_4, B_5, B_6	B_1, B_6	
B_9	B_9	$B_1, B_3, B_4, B_5, B_6, B_9$	B_9	III

Table 8 RM partitioning iteration 4

Attribute (B_i)	Reachability set	Antecedent set	Intersection set	Level
B_1	B_1, B_4, B_6	B_1, B_3, B_4, B_5, B_6	B_1, B_4, B_6	IV
B_3	B_1, B_3, B_4, B_6	B_3, B_5	B_3	
B_4	B_1, B_4, B_6	B_1, B_3, B_4, B_5	B_1, B_4	
B_5	B_1, B_3, B_4, B_5, B_6	B_5	B_5	
B_6	B_1, B_6	B_1, B_3, B_4, B_5, B_6	B_1, B_6	IV

Table 9 RM partitioning iteration 5

Attribute (B_i)	Reachability set	Antecedent set	Intersection set	Level
B_3	B_3, B_4	B_3, B_5	B_3	
B_4	B_4	B_3, B_4, B_5	B_4	V
B_5	B_3, B_4, B_5	B_5	B_5	

Table 10 RM partitioning iteration 6

Attribute (B_i)	Reachability set	Antecedent set	Intersection set	Level
B_3	B_3	B_3, B_5	B_3	VI
B_5	B_3, B_5	B_5	B_5	

Table 11 RM partitioning iteration 6

Attribute (B_i)	Reachability set	Antecedent set	Intersection set	Level
B_5	B_5	B_5	B_5	VII

Table 12 Level of requirement barriers

Attribute (B_i)	Reachability set	Antecedent set	Intersection set	Level
B_1	$B_1, B_2, B_4, B_6, B_7, B_8, B_9, B_{10}$	$B_1, B_2, B_3, B_4, B_5, B_6, B_7$	B_1, B_2, B_4, B_6, B_7	IV
B_2	$B_1, B_2, B_4, B_6, B_7, B_8, B_9, B_{10}$	$B_1, B_2, B_3, B_4, B_5, B_6, B_7, B_8, B_9, B_{10}$	$B_1, B_2, B_4, B_6, B_7, B_9, B_{10}$	II
B_3	$B_1, B_2, B_3, B_4, B_6, B_7, B_8, B_9, B_{10}$	B_3, B_5	B_3	VI
B_4	$B_1, B_2, B_4, B_6, B_7, B_8, B_9, B_{10}$	$B_1, B_2, B_3, B_4, B_5, B_{10}$	B_1, B_2, B_4, B_{10}	V
B_5	$B_1, B_2, B_3, B_4, B_5, B_6, B_7, B_8, B_9, B_{10}$	B_5	B_5	VII
B_6	$B_1, B_2, B_6, B_7, B_8, B_9, B_{10}$	$B_1, B_2, B_3, B_4, B_5, B_6, B_{10}$	B_1, B_2, B_6, B_{10}	IV
B_7	B_7, B_8, B_{10}	$B_1, B_2, B_3, B_4, B_5, B_6, B_7, B_9, B_{10}$	B_7, B_{10}	II
B_8	B_8, B_{10}	$B_1, B_2, B_3, B_4, B_5, B_6, B_7, B_8, B_9, B_{10}$	B_8, B_{10}	I
B_9	$B_2, B_7, B_8, B_9, B_{10}$	$B_1, B_2, B_3, B_4, B_5, B_6, B_8, B_9, B_{10}$	B_2, B_9, B_{10}	III
B_{10}	$B_2, B_4, B_6, B_7, B_8, B_9, B_{10}$	$B_1, B_2, B_3, B_4, B_5, B_6, B_7, B_8, B_9, B_{10}$	$B_2, B_4, B_6, B_7, B_8, B_9, B_{10}$	I

Barriers are classified into four classes:

1. Autonomous barriers.
2. Dependent Barriers.
3. Linkage Barriers.
4. Independent Barriers.

Step 5: Digraph

Digraph is created representing the directed link among various barriers. Here, level VII generates the root node, means B5 is root driving power besides controls further barriers.

Figure 1 illustrate the digraph having the partition into diverse levels through I to VII.

Fig. 1 Digraph representing inter-relationship among barriers

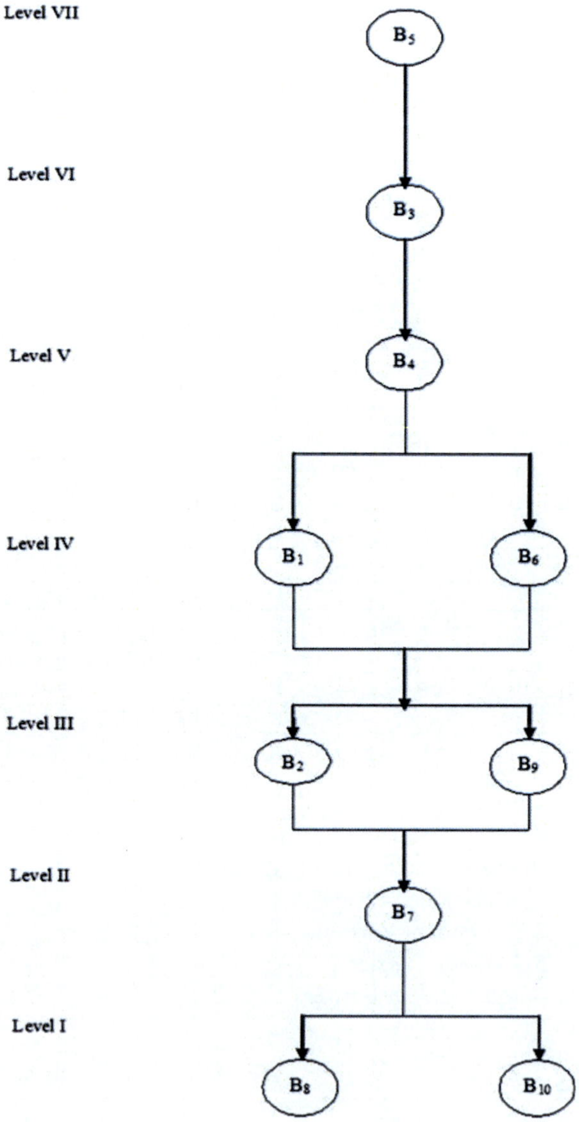

Step 6: Convert the digraph to ISM model

Figure 2 indicates the ISM model to successfully implement agile software. Figure 2 represents the ever-changing requirement is the major barrier followed by incomplete requirement, lack of planning etc.

Step 7: Analyse the ISM model

ISM model was finally reviewed by industry experts and approved the results.

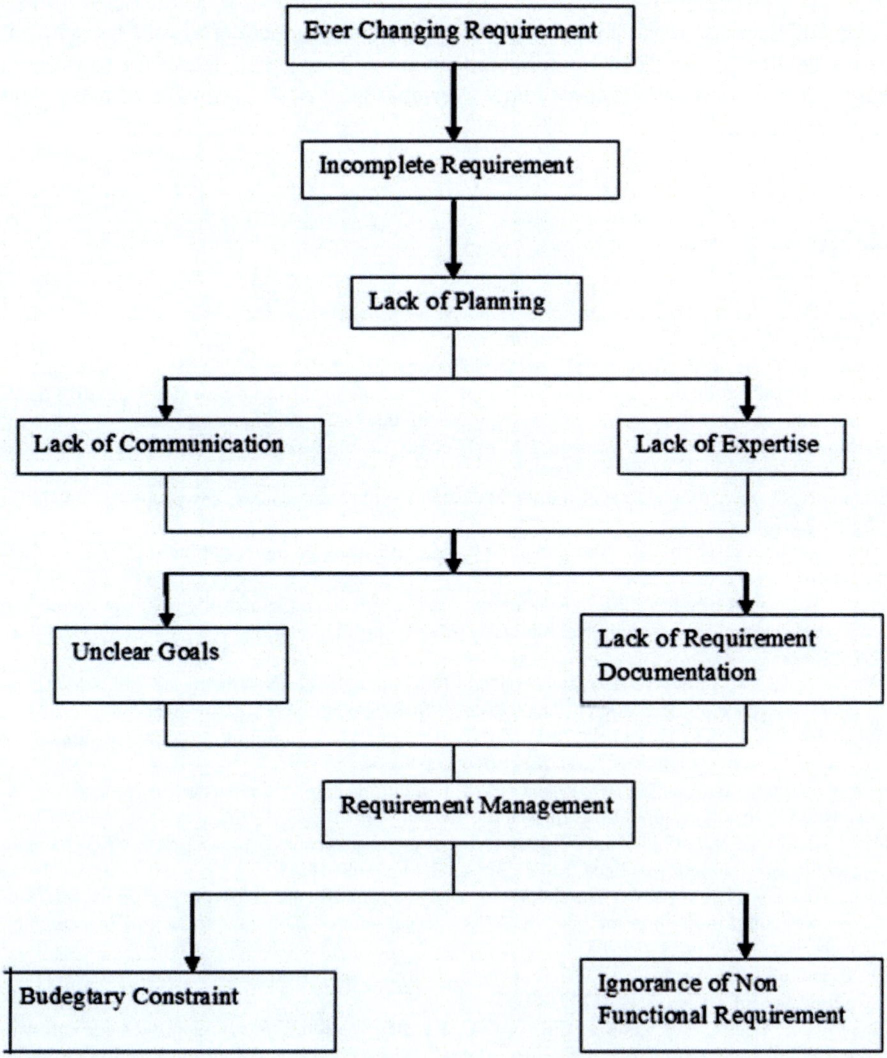

Fig. 2 ISM Model

4 Conclusion

The paper analyzed barriers in agile implementation and ISM model is represented by using Interpretive Structural Modelling (ISM) technique. It is concluded from the ISM model, that the ever-changing requirements (frequent change in project requirements from the customer side) forms the main barrier preventing the successful implementation of agile. Ever-changing requirement as barrier B5 the uppermost driving power besides the lowermost dependence power value creating the main barrier persuading altogether further barriers. Hence, it's vital for all project managers, developers and team to coordinate on the time, so that its complete effect can be reduced, and a quality product can be delivered on time. Thus, it is necessary to provide the resources to complete the client's ever-changing requirements and incomplete requirements on time.

References

Aggarwal S, Dhir S (2013) Ground axioms to achieve movables: methodology. Int J Comput Appl 69(14)

Alsaqaf W, Daneva M, Wieringa R (2018) Understanding challenging situations in agile quality requirements engineering and their solution strategies: insights from a case study. In: 2018 IEEE 26th international requirements engineering conference (RE). IEEE, pp 274–285

Boehm B, Turner R (2005) Management challenges to implement agile processes. Traditional development organizations

Conboy K, Coyle S, Wang X, Pikkarainen M (2011) People over process: key people challenges in agile development

Dhir S, Kumar D (2015) Factors persuading nuts and bolts of agile estimation. Adv Sci Lett 21(10):3118–3122

Dhir S, Kumar D, Singh V (2017) Requirement paradigms to implement the software projects in agile development using analytical hierarchy process. Int J Decis Support Syst Technol (IJDSST) 9(3):28–41

Dhir S, Kumar D, Singh V (2019) Success and failure factors that impact on project implementation using agile software development methodology. Software engineering. Springer, pp 647–654

Huang J-J, Tzeng G-H, Ong C-S (2005) Multidimensional data in multidimensional scaling using the analytic network process. Pattern Recogn Lett 26(6):755–767

Jadhav J, Mantha S, Rane S (2015) Analysis of interactions among the barriers to JIT production: interpretive structural modelling approach. J Ind Eng Int 11(3):331–352

Misra SC, Kumar V, Kumar U (2009) Identifying some important success factors in adopting agile software development practices. J Syst Softw 82(11):1869–1890

Qu Y, Yuan M-J, Liu F (2012) The risk factor analysis for software project based on the interpretative structural modelling method. In: 2012 international conference on machine learning and cybernetics. IEEE, pp 1019–1024

Rai P, Dhir S (2014) Impact of different methodologies in software development process. Int J Comput Sci Inf Technol 5(2):1112–1116

Rajagopal P, Lee R, Ahlswede T, Chiang C-C, Karolak D (2005) A new approach for software requirements elicitation. In: Sixth international conference on software engineering, artificial intelligence, networking and parallel/distributed computing and first ACIS international workshop on self-assembling wireless network. IEEE, pp 32–42

Srinivasan J, Lundqvist K (2009) Using agile methods in software product development: a case study. In: 2009 sixth international conference on information technology: new generations. IEEE, pp 1415–1420
The Standish Group (2014), The Standish group: the chaos report. Proj. Smart. 16
Warfield JN (1974) Developing interconnection matrices in structural modeling. IEEE Trans Syst Man Cybern 1:81–87

Ranking of Multi-release Software Reliability Growth Model Using Weighted Distance-Based Approach

Ritu Bibyan and Sameer Anand

Abstract Today's software systems and applications are expanded in almost all the firms and are indulged in various business units that need customer base. The methodology of the multi-release software reliability growth model (SRGM) deals with customer demand and market requirements. There are various multi-release SRGMs given by researchers, but it is challenging to select the optimal model. Traditionally, a multi-criteria decision-making approach has been used to resolve the problem of the ranking of models. In this chapter, the Weighted Distance-based Approach has been proposed to rank the multi-release SRGMs using the Maximum Deviation Method (MDM) and Distance-Based approach (DBA). The models are ranked based on selection criteria having different priority weights and composite distance values. The method MDM is a technique of Multi Criteria Decision Making (MCDM) in which non-linear programming is performed.

Keywords Software reliability · Multi release software reliability growth model · Maximum deviation method · Distance-based approach · Multi-criteria decision making · Ranking

1 Introduction

The use of software has become an essential part of our life, and it is growing exponentially. Almost in all the enterprises and even in our daily life, various applications run through software. In the information technology sector, humans are dependent on computers that use different software. Due to such advancement, software companies are majorly concerned about the software quality and requirements of the customers. Since the growth is exponential, the process of developing software has become

R. Bibyan
Department of Operational Research, University of Delhi, New-Delhi, India

S. Anand (✉)
S.S. College of Business Studies, University of Delhi, New-Delhi, India
e-mail: sameeranand@sscbsdu.ac.in

A. G. Aggarwal et al. (eds.), *Optimization Models in Software Reliability*, Springer Series in Reliability Engineering,
https://doi.org/10.1007/978-3-030-78919-0_16

costly and takes a lot of time, which makes the process more complicated. Therefore the main objective of a software developer is to create reliable software with less development cost. The performance of the software can be seen through its most crucial feature, i.e., Software Reliability. Software Reliability is defined as the failure-free operation under specifies environment and time. To estimate and make a future prediction about the software reliability, various software reliability growth models (SRGM) have been developed. Some of the models are proposed by Goel and Okumoto (1979), Yamada et al. (1983, 1984). Later Musa and Okumoto (1985) suggested a logarithmic Poisson Execution time, and Kapur and Garg (1992) proposed a flexible model that considered exponential and S-shaped growth. Three different types of faults were considered by Kapur et al. (1999) and modeled them as exponential, S-shaped, and three-stage Erlang model. These models are widely used to improve the quality of software by software industries, which later recommend them to top leading companies and research institutions. During the testing and operational phase of software, the fault detection aspect is explained by these SRGMs (Garmabaki et al. 2012). They provide the relationship in mathematical form between the time of testing and the cumulative number of faults. Some SRGMs were proposed with leaning processes and error generation for both perfect and imperfect debugging (Kapur et al. 2008, 2011). When software acquires desiderated reliability during the operational phase, then a firm comes up with a new version, which leads to an upgradation. The upgrade means the old product is supplanted by a new version of the same product, which has new features. The upgraded version gives better performance than the older version, but there is also a risk that it might contain some new faults and some faults from the previous version. These faults cannot be neglected by the testing team before releasing a new version. The fault removal of the older version is also acknowledged by testers during the testing of new code. Software firms intend to augment the reliability of a software by upgrading it. But there is also a possibility that upgradation makes the software/product worse. Due to which users might prefer to use older version. This is called software failure which can occur due to various reasons such as errors, incorrect code, misinterpretation, improper testing, or other issues. It is always a difficult job to update any software application which brings up risk related to multi release of software. There might be increase in failure rate when the upgradation takes place and it decreases by time as the testers fixes the faults. Testers are generally curious about demonstrating the glitch regarding the software which helps to understand the benefit of upgraded software. There are many advantages of multi releases for the software developing firms such as quick deliveries, increase in revenue and increment in market life of the software.

Currently, in the market, there are software products that are available with high potential in the first release. Such products include sufficient features when they start their life cycle to satisfy customers. However, later, with successive releases, they add on some new features or amplify the old features. Each release exists for a confined period as software products are not stagnant. There are various factors which call for change in the software when it is released in the market such as

a. Subsisting software release issues
b. Market competition
c. Policies of company
d. The necessity of additional functions
e. The requirement of new hardware supporting updated software
f. Customers demand.

These factors accrue together after reaching a point over time, which demands up-gradation in software.

Moreover, the cycle starts again as the new version comes in the market. Various models in the literature explain multiple releases for different scenarios. But in the past, no research has been done to rank multi-release SRGMs. In this chapter, our objective is to fill this research gap: to rank multi-release SRGMS. The study in this chapter uses weighted distance-based approach to rank these models based on performance criteria such as R^2, MSE, MAE, RMPSE, Bias, and PRR. The weights of the selection criteria are evaluated through the Maximum Deviation Method.

The chapter is summarized as follow: in the Sect. 2 we have provided a brief research background in the field of software reliability models. The Sect. 3 has notations and in Sect. 4 general multi-release SRGM has been explained. The selected multi-release SRGMs which are ranked are provided in Sect. 5. The Sect. 6 and 7 explains the performance criteria on the basis of which ranking has been done and the elaborated weighted ranking approach has been discussed. This approach is numerically applied on the 9 models in Sect. 8. The results, conclusion and future scope are discussed in the Sects. 9 and 10.

2 Literature Survey

Kapur et al. (2010) proposed an approach of multi-releasee of software in which they used the sigmoid curve. Garmabaki et al. (2011) categorized the faults into a simple and hard fault and proposed a new model for multi-release software, which considered faults having different severity levels. Garmabaki et al. (2012) proposed two- dimensional multi-release software reliability growth, model. Cobb Douglas production function was used to model the failure process, which studied the effect of the limitation of resources and schedule pressure. Kapur et al. (2013) modeled the extension of the Bass diffusion model for successive generations of software. Garmabaki et al. (2014) considered the effect of faults encountered during testing and operational phase together and modeled a SRGM for several versions using Kapur-Garg and Weibull fault removal model. Aggarwal et al. (2015) captured the effect of faults with upgrades and modeled a discrete SRGM for multiple releases. Furthermore, it suggested an optimal release policy, which minimizes the cost. Zhu and Pham (2018) proposed multi-release software reliability model with addition dependent fault detection process and validated on open source software project datasets. Anand et al. (2018) proposed 2-dimensional multi-release model in imperfect debugging scenario by considering constant fault reduction factor and rate of fault removal is given by Delayed S-shaped model.

All the SRGM works well to estimate software reliability with a particular data set but not on all types of data sets. To find out which SRGM model works well for a given data set becomes an important task. To find out the optimal SRGM, we make a comparison of different SRGMs. An initial attempt was made by Sukert (1979) and Schick and Wolverton (1978) to find the optimal model to estimate the reliability of the software for the failure data set. Knafl and Sacks (1991) made the comparison based on various parameters based on maximum likelihood. Asad et al. (2004) proposed an algorithm to select SRGM based on different criteria such as trend, expected output, life cycle, required input, nature of data, the structure of a project, testing, and development process. Liu and Gao (2009) solved the same using an automated tool called SRMSS to select a reliability model for software. The criteria which were used are Bias, Noise, and Goodness of fit. Sharma et al. (2010) developed a quantitative model to rank the SRGM model and select the optimal model using a distance-based approach (DBA). Since then, many researchers have investigated for selection of optimal SRGM. It has been accounted that multi-criteria decision-making methods (MCDM) have also been used such as DBA, Weighted criteria and greedy approach (Anjum et al. 2013; Cristescu et al. 2015; Khalid and Sharma 2015; Miglani and Rana 2011). An entropy distance-based approach (EDBA) was suggested by Gupta et al. (2018) to rank SRGMs using seven selection criteria. Kumar et al. (2018) used fuzzy data envelopment analysis to select the optimal SRGM.

3 Notations

$m(t)$	Expected number of faults removed by time t.
$\lambda(t)$	Failure Intensity
A	$=a_1 + a_2 + a_3 + a_4 =$ constant gives number of faults in the software at the beginning of testing
a_1	number of faults in the software at first release
a_2	number of faults added due to additional feature in release 2
a_3	number of faults added due to additional feature in release 3
a_4	number of faults added due to additional feature in release 4
$f(t)$	Probability density function
$F_i(t)$	Probability distribution function for *i*th release.
B	constant, fault detection rate

4 General Multi-release SRGM

In this section of the chapter, a fundamental review on multi-release SRGMs based on NHPP is provided along with the multi-release model.

The demeanor of the failure process of software is studied using SRGMs. The NHPP-based SRGM is broadly used mathematical models with the assumption that the occurrence of failure is a random process. Let $\{N(t), t \geq 0\}$ be a counting process that represents the cumulative number of software failures that are either detected or removed by time t.

Let the number of faults up to time t is represented by N(t).

The NHPP based counting process under the assumption that:

$$P\{N(t) = n\} = \frac{m(t)^n e^{-m(t)}}{n!} \tag{1}$$

The function m(t) is called the mean value function and describes the expected cumulative number of failures in (0, t]. Hence, m(t) is a handy descriptive measure of failure behavior.

Now the differential equation given below can be used to determine the expected number of faults removed:

$$\lambda(t) = \frac{dm(t)}{dt} = b(t)[a(t) - m(t)] \tag{2}$$

where $\lambda(t)$ is proportional to the residual fault content. a(t), b(t) defines different assumptions of the detection process. Where a(t), b(t) represents time-dependent initial fault content and fault detection rate, respectively.

$$\frac{dm(t)}{dt} = \frac{f(t)}{1 - F(t)}(a - m(t)) \tag{3}$$

Let 'a' denote the expected number of faults that would be detected given infinite testing time in the case of finite failure NHPP models. Then, the mean value function of the finite failure NHPP models can be written as

$$m(t) = \int_0^t \lambda(x)dx = a \cdot F(t) \tag{4}$$

.

Release 1

After developing a software most crucial task is the testing of the software before bringing it into the market, which is the operational phase. The testing team tests the software prudently and carefully to ensure that maximum faults are removed. Nevertheless, there are still possible chances that the faults still exist in the software because it is practically impossible to remove or detect all the faults. Therefore during the testing of the initial version of the software, there is a chance that testers might find a finite number of faults. Then these finite number of faults are removed, and it can be represented mathematically as:

$$m_1(t) = aF_1(t)\ 0 < t < t_1 \tag{5}$$

Release 2

As time passes, technological changes take place, customer demand increases, and even market competition increases. Due to which developer is compelled to add some new features to the software. The addition of a new feature can make the software more complex and fault content increases. During the testing of newly added feature code, testers might encounter faults from the previous version, which were not removed along with the additional faults due to new code. By testing the new version of the software, the overall software is improved. So during release two, we have two versions of the software. The leftover faults from the previous release i.e. $a_1(1 - F_1(t_1))$ Interact with the new detection rate due to which fractions of leftover faults from the previous release gets removed. Additionally, the fraction of faults due to addition of new feature gets removed with the detection rate $F_2(t - t_1)$. The expected number of faults removed is mathematically represented as:

$$m_2(t) = a_2F_2(t - t_1) + a_1(1 - F_1(t_1))F_2(t - t_1)\quad t_1 < t < t_2 \tag{6}$$

Release 3

When the new features are added twice, then some new codes are developed for the software. This new code with the existing code is then tested. During the testing of this version, testers might encounter leftover faults from release two and release one. The addition of the new feature helps in removing the maximum number of faults. Now there three versions of the software. The leftover faults from the previous release i.e. $a_2(1 - F_2(t_2))$ Interact with the new detection rate due to which fractions of leftover faults from the previous release gets removed. Also, the leftover fraction of faults from the release one i.e. $a_1(1 - F_1(t_1))(1 - F_2(t_2))$ interact with new detection rates. Additionally, the fraction of faults due to addition of new feature gets removed with the detection rate $F_3(t - t_2)$. The expected number of faults removed is mathematically represented as:

$$\begin{aligned} m_3(t) &= a_3F_3(t - t_2) + a_2(1 - F_2(t_2))F_3(t - t_2) \\ &\quad + a_1(1 - F_1(t_1))(1 - F_2(t_2))F_3(t - t_2),\ t_2 < t < t_3 \end{aligned} \tag{7}$$

Release 4

This process of adding new features is ongoing until the time software product exists in the market. It not only removes the maximum number of faults but also increases the reliability of the software. The mathematical equation for the expected number of faults removed during release 4 when the features were added for the third time is given by:

$$\begin{aligned} m_4(t) &= a_4F_4(t-t_3) + a_3(1-F_3(t_3))F_4(t-t_3) \\ &+ a_2(1-F_2(t_2))(1-F_3(t_3))F_4(t-t_3) \\ &+ a_1(1-F_1(t_1))(1-F_2(t_2))(1-F_3(t_3))F_4(t-t_3) \quad t_3 < t < t_4 \end{aligned} \tag{8}$$

5 Multi Release SRGM Models

In this chapter, we have considered various multi-release SRGMs as follow:

Model 1

Kapur et al. (2010) proposed that multi-release SRGM using S-shaped logistic distribution is given by him (1999). The mean value function m(t) is given as:

$$m(t) = \frac{a(1-\exp(-bt))}{1+\beta\exp(-bt)} = aF(t) \tag{9}$$

where $F(t) = \frac{(1-\exp(-bt))}{1+\beta\exp(-bt)}$

If we assume $\beta = 0$ in the above model, then it reduces to Goel Okumoto exponential model.

The software fault removed in each release is calculated using the mathematical structure given in Eqs. (5)–(8).

Model 2

Garmabaki et al. (2014) represented the fault removal phenomenon in a distributed environment where the "n" number of reused and "p" newly developed components were considered together. The explicit mean value function for both types of faults is given by:

$$\begin{aligned} m_i(t) &= a_ip_i\left(1-e^{-b_it}\right) + a_i(1-p_i)\left(1-\left(1+b_i^*t\right)e^{-b_i^*t}\right) \\ &= a_ip_iF_i(t) + a_i(1-p_i)F_i^*(t);\ i = 1, 2, 3, 4 \end{aligned} \tag{10}$$

where $F_i(t)$ and $F_i^*(t)$ are fault distribution function for reused and new components respectively for each release.

p_i the proportion of faults in both types of components.

Model 3

Pachauri et al. (2015) extended the model given by Hsu et al. (2011) having constant Fault reduction factors (FRF) with imperfect debugging. The mean value function for the next release model is based on the model given by Kapur et al. (2012). When a failure occurs, the fault which causes failure is instantly removed, and the new faults are brought in with some probability say γ. In this multi-release model, the fault of the current release and leftover faults of the previous release are considered. The mean value function for ith release is given as:

$$m_i(t) = (a_i + a_{i-1}(1 - F_{i-1}(t_{i-1})))F_i(t - t_{i-1});\ t_{i-1} \le t < t_i \tag{11}$$

$$F(t) = \frac{(1 - \exp(-rb(1-\gamma)t))}{(1-\gamma)} \tag{12}$$

where r is a constant detection rate, and b is a constant fault reduction factor.

Model 4

Garmabaki et al. (2015) proposed a multi-release SRGM model by using the Weibull model for estimating the mean number of faults. The mean value function for the ith release is given as:

$$m_i(t) = a_i F_i(t - t_i) + (a_{i-1} - m_{i-1}(t_{i-1} - t_{i-2}))F_i(t - t_i) \tag{13}$$

$$\text{Where, } F(t) = a\left(1 - e^{\left(-\frac{t}{\theta}\right)^{\beta}}\right);\ \theta > 0, \beta > 0 \tag{14}$$

And θ and β are the scale and shape parameters of the Weibul Distribution.

Model 5

Tandon et al. (2016) proposed a multi-release model with a change-point by considering the assumption that Fault Removal Rate (FRR) may change instead of staying constant during testing. This change in FRR occurs due to various changes in the change-point. For the fault removal process, S-shaped logistic distribution is used. The mean value function for the ith release is given as:

$$m_i(n) = \begin{cases} a_4 * (1 - \dfrac{(1+\beta_i)(1-b_{i1})^{n-t_{i-1}}}{(1+\beta_i(1-b_{i1})^{n-t_{i-1}}};\ 0 \le n < n_i \\ a_i(1 - \dfrac{(1+\beta_i)(1+\beta_i(1-b_{i2})^{n_i}(1-b_{i1})^{n_i}(1-b_{i2})^{n-(t_{i-1}+n_i)}}{(1+\beta_i(1-b_{i1})^{n_i}(1+\beta_i(1-b_{i2})^{n-t_{i-1}}};\ n > n_i;\ t_{1-1} < n \le t_1 \end{cases} \tag{15}$$

where n_i is the change point in ith release,

t_i is the ith release time,

b_{i1} is the fault removal rate before change point for ith release and

b_{i2} is the fault removal rate after change point for ith release.

Model 6, 7, 8

Mishra et al. (2017) bused 3 SRGMs to capture the behavior of testing phase namely

Model 6: Exponential SRGM by Goel and Okumoto

$$F^{ts}(t) = \left(1 - e^{-bt}\right) \tag{16}$$

Model 7: Two-phase S-shaped model by Yamada et al.

$$F^{ts}(t) = \left(1 - (1 + bt)e^{-bt}\right) \tag{17}$$

Model 8: Flexible SRGM by (Kapur and Garg)

$$F^{ts}(t) = \frac{1 - e^{-bt}}{1 + \beta e^{-bt}} \tag{18}$$

where b is error detection rate, and β is the learning parameter. The behavior of the operational phase is captured using Weibull Distribution with

$$F^{op}(t) = \left(1 - e^{-bt^k}\right) \tag{19}$$

The mean value function for the ith the release is given as:

$$\begin{aligned} m_i(t) = &\left[a_i + (1 - \alpha_{i-1}) \cdot a_{i-1}(1 - F_{i-1}(t_{i-1} - t_{i-2}))\right] \cdot F_i^{ts}(t - t_{i-1}) \\ &+ \left[\alpha_{i-1}.a_{i-1}(1 - F_{i-1}(t_{i-1} - t_{i-2}))\right]F_{i-1}^{op}(t - t_{i-1});\, t_{i-1} \le t < t_i \end{aligned} \tag{20}$$

where $(1 - \alpha_{i-1})$ represents the proportion of remaining bugs removed during the testing phase. (α_{i-1}), represents the proportion of remaining faults in the operational phase.

Model 9

Aggarwal et al. (2019) proposed multi-release SRGM with imperfect debugging time variable FRF.

The mean value function for the ith the release is given as:

$$m_i(t) = \left(a_i^* + a_{i-1}^*(1 - F_{i-1}(t_{i-1}))\right)F_i(t - t_{i-1});\, t_{i-10} \le t < t_i \tag{21}$$

where r_i is proportionality constant,

$$a_i^* = \frac{a_i}{(1 - \alpha_i)}$$

$$F_i(t) = \left(1 - (1 + b_{i-1}t)^{r_i(1-\alpha_{i-1})} e^{-b_{i-1}r_{i-1}(1-\alpha_{i-1})t}\right)$$

Table 1 Performance criteria

S. No.	Criteria	Description
1	R^2	It explains how successful the fit is in measuring the proportion of variation in the data $R^2 = 1 - \frac{\sum_{i=1}^{k}(m_i - \hat{m}(t_i))^2}{\sum_{i=1}^{k}\left(m_i - \sum_{j=1}^{k} m_j/n\right)^2}$
2	MSE	It calculates the deviation between expected and actual values $MSE = \frac{\sum_{i=1}^{k}(m_i - \hat{m}(t_i))^2}{k-p}$
3	MAE	It is mean absolute error which calculated deviation by using absolute values $MAE = \frac{\sum_{i=1}^{k}\lvert(m_i - \hat{m}(t_i))\rvert}{k-p}$
4	Bias	It gives the sum of the differences between actual and estimated values $Bias = \frac{\sum_{i=1}^{k}\lvert(\hat{m}(t_i) - m_i)\rvert}{k-p}$
5	Variance	It is the Standard Deviation of the prediction bias $Variance = \sqrt{\frac{\sum_{i=1}^{k}(m_i - \hat{m}(t_i) - Bias)^2}{k-1}}$
6	RMPSE	It stands for Root Mean Prediction Error, which calculates the closeness with which the model predicts the values $RMPSE = \sqrt{Variance^2 + Bias^2}$
7	PRR	It calculates the distance of model estimates from the actual values against the model estimate $PRR = \frac{\sum_{i=1}^{k} \hat{m}_i(t_i) - m_i}{\hat{m}_i(t_i)}$

6 Comparison Criteria

The parameters for multi-release SRGM are estimated through a statistical software SPSS. Later performance criteria such as R^2, MSE MAE, RMPSE, Bias, PRR, Variance are evaluated. The Table 1 describes performance criteria:

7 Weighted Distance-Based Approach

Sharma et al. (2010) proposed the distance-based approach to select the optimal SRGM, which is modified in this chapter. This approach is combined with the Maximum Deviation Method to evaluate the weights of all the criteria. Later weights are added with a Distance-based Approach to select optimal Multi-Release SRGM. The steps which are followed is explained below:

Step 1: Create Rating Matrix

The rating matrix contains the values of the performance measures for each model/alternative against the different criteria. The matrix R_{ij} with "m" criteria and "n" alternatives is given below:

$$[R_{ij}] = \begin{bmatrix} r_{11} & r_{12} & \ldots & r_{1m} \\ r_{21} & r_{22} & \ldots & r_{2m} \\ \vdots & \vdots & \vdots & \vdots \\ \vdots & \vdots & \vdots & \vdots \\ r_{n1} & r_{n1} & \ldots & r_{nm} \end{bmatrix} \quad (22)$$

where i = 1, 2, …, n and j = 1, 2, … m.

Step 2: Normalizing Rating matrix

The elements of the matrix R'_{ij} is are normalized to form Normalized Rating matrix $[R'_{ij}]$ below:

$$[R'_{ij}] = \begin{bmatrix} r'_{11} & r'_{12} & \ldots & r'_{1m} \\ r'_{21} & r'_{22} & \ldots & r'_{2m} \\ \vdots & \vdots & \vdots & \vdots \\ \vdots & \vdots & \vdots & \vdots \\ r'_{n1} & r'_{n1} & \ldots & r'_{nm} \end{bmatrix} \quad (23)$$

where $r'_{ij} = \frac{r_{ij}}{\sqrt{\sum_{j=1}^{m} r_{ij}}}$

Step 3: Maximum Deviation method for criteria weight calculation

The method is based on the fact that if for all the criterion values for a particular have small difference for all alternatives, then it has a small importance and is assigned small weight. Similarly, we can state this fact for large and no difference. The weights of these criteria are unknown to us (Yingming 1997). In this method, we solve a non-linear programming model to calculate the weights of "m" criteria. Let w = $(w_1, w_2, \ldots w_m)$ be the weight.

$$MaxD(w_j) = \sum_{j=1}^{m}\sum_{i=1}^{n}\sum_{k=1}^{n} d\left(r'_{ij}, r'_{kj}\right)w_j$$

$$subject\ to \sum_{j=1}^{m} w_j^2 = 1,\ w_j \in [0, 1] \quad (24)$$

Here the total weighted deviation for all the alternatives w.r.t. all criteria is expressed in the objective function. The deviation between the criteria value r'_{ij} and r'_{kj} for a criteria C_j is obtained by evaluating the distance between them. In this study, Euclidian distance is used, but any distance can be used like Hamming, Manhattan, and many more.

Solving the model we can obtain

$$w_j^* = \frac{\sum_{i=1}^{n} \sum_{k=1}^{n} d\left(r'_{ij}, r'_{kj}\right)}{\sum_{j=1}^{m} \sum_{i=1}^{n} \sum_{k=1}^{n} d\left(r'_{ij}, r'_{kj}\right)} \tag{25}$$

Step 4: Formation of Optimal matrix

The optimal value for each criterion is calculated where $\mathrm{r_{bj}} =$ best value for jth criteria

Case 1—When the smaller value of the criteria fits well for the model.

$\mathrm{r_{bj}} =$ minimum value of jth criteria

Case 2—When the larger value of the criteria fits well for the model.

$\mathrm{r_{bj}} =$ maximum value of jth criteria

The optimal matrix is given as:

$$\left[\mathrm{R}''_{\mathrm{ij}}\right] = \begin{bmatrix} r_{11} & r_{12} & \dots & r_{1m} \\ r_{21} & r_{22} & \dots & r_{2m} \\ \vdots & \vdots & \vdots & \vdots \\ r_{n1} & r_{n2} & \dots & r_{nm} \\ r_{b1} & r_{b1} & \dots & r_{bm} \end{bmatrix} \tag{26}$$

Step 5: Standardization of Optimal Matrix

The optimal matrix is standardized using the following-

$$Z_{ij} = \frac{r_{ij} - \overline{r_j}}{S_j} \tag{27}$$

$$\overline{r_j} = \frac{1}{n} \sum_{i=1}^{n} r_{ij} \tag{28}$$

$$S_j = \sqrt{\left[\frac{1}{n} \sum_{i=1}^{n} \left(r_{ij}^2 - \overline{r_j}\right)^2\right]} \tag{29}$$

$$[Z_j] = \begin{bmatrix} z_{11} & z_{12} & \cdots & z_{1m} \\ z_{21} & z_{21} & \cdots & z_{2m} \\ \cdots & \cdots & \vdots & \cdots \\ z_{n1} & z_{n2} & \cdots & z_{nm} \\ z_{opt1} & z_{opt2} & \cdots & z_{optm} \end{bmatrix} \tag{30}$$

Step 6: Formation of the Distance Matrix

The Distance Matrix is obtained by subtracting elements of a standardized optimal matrix from the elements of the optimal matrix.

$$[Z_d] = \begin{bmatrix} z_{opt1} - z_{11} & z_{opt2} - z_{12} & \cdots & z_{optm} - z_{1m} \\ z_{opt1} - z_{21} & z_{opt2} - z_{22} & \cdots & z_{optm} - z_{1m} \\ \vdots & \vdots & \cdots & \vdots \\ z_{opt1} - z_{n1} & z_{opt2} - z_{n2} & \cdots & z_{optm} - z_{nm} \end{bmatrix} \tag{31}$$

Later this matrix is multiplied with the criteria weights calculated using Maximum Deviation Method to get weighted distance matrix as

$$[Z_{w_j}] = \begin{bmatrix} w_1(z_{opt1} - z_{11}) & w_2(z_{opt2} - z_{12}) & \cdots & w_m(z_{optm} - z_{1m}) \\ w_1(z_{opt1} - z_{21}) & w_2(z_{opt2} - z_{22}) & \cdots & w_m(z_{optm} - z_{1m}) \\ \vdots & \vdots & \cdots & \vdots \\ w_1(z_{opt1} - z_{n1}) & w_2(z_{opt2} - z_{n2}) & \cdots & w_m(z_{optm} - z_{nm}) \end{bmatrix} \tag{32}$$

Finally, the Euclidian distance for each alternative is calculated as:

$$D_i = \left[\sum_{j=1}^{m} \left(w_j\left(z_{optj} - z_{ij}\right)\right)2 \right] 1/2 \tag{33}$$

.

8 Numerical Illustration for Multi-release SRGM Ranking

The aim of the research is to select multi-release SRGMs by evaluation of criteria and ranking the models on the basis of the selection criteria. The unknown parameters are evaluated using the Least Square Estimate technique for the failure dataset of Tandem Computers having four releases (Wood 1996). The selection of the model is made against seven criteria, namely R2, MSE, MAE, RMPSE, Bias, and Variation, shown in Table 2. Once these models are estimated, the weighted distance-based approach is implemented to rank the models for selection.

Table 2 Criteria for different multi-release SRGM

	R sq	MSE	MAE	RMPSE	Bias	PRR	Variance
Model 1	0.995	0.9411	0.7194	1.003	0.055	0.3423	0.694
Model 2	0.981	3.45316	0.702	1.90444	0.570971	0.204	1.816833
Model 3	0.97	4.474	1.878	2.34136	0.427	0.958	2.362
Model 4	0.995	0.8904	0.68736	0.970076	−0.0168	0.146288	0.96933
Model 5	0.995	0.9411	0.7194	1.00307	0.5526	0.342316	1.00154822
Model 6	0.969	5.759	0.852	2.889	0.738	0.251	2.794
Model 7	0.995	0.98	0.723	1.017	−0.001	0.66	1.017
Model 8	0.995	0.939	0.691	1.002	0.055	0.573	1.001
Model 9	0.99	1.421	1.0242	1.225277	0.01052	0.64512	1.225

The steps discussed in the section Weighted Distance-Based Approach are applied and demonstrated below:

1. *Creation of Rating Matrix*: The matrix $\left[R_{ij}\right]$ is created on the basis of the criteria values given in Table 2.

$$\left[R_{ij}\right] = \begin{bmatrix} 0.995 & 0.9411 & 0.7194 & 1.003 & 0.055 & 0.3423 & 0.694 \\ 0.981 & 3.453 & 0.702 & 1.9044 & 0.5709 & 0.204 & 1.81683 \\ 0.97 & 4.474 & 1.878 & 2.34136 & 0.427 & 0.958 & 2.362 \\ 0.995 & 0.8904 & 0.68736 & 0.9700 & -0.0168 & 0.1462 & 0.9633 \\ 0.995 & 0.9411 & 0.7194 & 1.00307 & 0.5526 & 0.3423 & 1.0015 \\ 0.969 & 5.759 & 0.852 & 2.889 & 0.738 & 0.251 & 2.794 \\ 0.995 & 0.98 & 0.723 & 1.017 & -0.001 & 0.66 & 1.017 \\ 0.995 & 0.939 & 0.691 & 1.002 & 0.055 & 0.573 & 1.001 \\ 0.990 & 1.421 & 1.0242 & 1.225277 & 0.01052 & 0.64512 & 1.225 \end{bmatrix}$$

2. *Creation of Normalized Rating Matrix:* The elements of Rating Matrix is normalized to obtain Normalized Rating Matrix $\left[R'_{ij}\right]$.

$$\left[R'_{ij}\right] = \begin{bmatrix} 0.333806174 & 0.21150312 & 0.25440419 & 0.27445766 & 0.03556694 & 0.1685977 & 0.1933702 \\ 0.329109403 & 0.7760643 & 0.24825096 & 0.52112478 & 0.368923.073 & 0.10047891 & 0.50622675 \\ 0.325419084 & 1.00548821 & 0.66412437 & 0.64068215 & 0.27612877 & 0.47185684 & 0.6581274 \\ 0.333806174 & 0.20010879 & 0.24307376 & 0.26544845 & -0.0108641 & 0.07205323 & 0.27008579 \\ 0.33806174 & 0.21150312 & 0.25440419 & 0.27447682 & 0.35735072 & 0.16860558 & 0.2790628 \\ 0.3250836 & 1.29427953 & 0.30129604 & 0.79053658 & 0.47724364 & 0.1362849 & 0.77849617 \\ 0.333806174 & 0.22024552 & 0.25567727 & 0.27828858 & -0.0006467 & 0.32507882 & 0.28336815 \\ 0.333806174 & 0.211031117 & 0.24436099 & 0.2782888 & -0.0006467 & 0.32507882 & 0.28336815 \\ 0.332128756 & 0.319356 & 0.36219179 & 0.33528082 & 0.00680299 & 0.31774977 & 0.34312348 \end{bmatrix}$$

3. *Criteria Weights:* The criteria weights are calculated by the Maximum Deviation method using Euclidian Distance in Table 3.

Table 3 Criteria weights

Criteria	Weight
R^2	0.009673
MSE	0.315834
MAE	0.089097
RMPSE	0.151789
Bias	0.157886
PRR	0.112717
Variance	0.163005

4. *Formation of Optimal Matrix*: This matrix contains the value of each criterion against different models, and the last row is filled with the optimal value for each criterion as shown in the matrix $[Z_s]$:

$$[Z_s] = \begin{bmatrix} 2.2388 & -2.1417 & -1.3905 & -2.1297 & -2.2203 & -1.3788 & -3.2324 \\ -1.7910 & 2.1324 & -1.5336 & 1.8623 & 3.2169 & -3.0270 & 1.6909 \\ -4.9574 & 3.8694 & 8.13756 & 3.7973 & 1.6997 & 5.9585 & 4.0814 \\ 2.2388 & -2.2280 & -1.654 & -2.2755 & -2.9769 & -3.7147 & -2.0251 \\ 2.2388 & -2.1417 & -1.390 & -2.1294 & 3.0233 & -1.3786 & -1.8839 \\ -5.2453 & 6.0559 & -0.3004 & 6.2225 & 4.9770 & -2.4669 & 5.9757 \\ 2.2388 & -2.0756 & -1.3609 & -2.0677 & -2.8104 & 2.4072 & -1.8161 \\ 2.2388 & -2.1453 & -1.6240 & 2.1342 & -2.2203 & 1.3704 & -1.8863 \\ 0.7996 & -1.3252 & 1.1160 & -1.1454 & -2.6890 & 2.2299 & -0.9041 \\ \mathbf{2.2388} & \mathbf{-2.2281} & \mathbf{-1.654} & \mathbf{-2.2755} & \mathbf{-2.9769} & \mathbf{-3.7147} & \mathbf{-3.2324} \end{bmatrix}$$

The max value from the R2 values of all the models is considered to be optimal, whereas, for other criteria, the minimum value is considered to be optimal.

5. *Formation of Distance Matrix:* This matrix is created after the calculating weights for each criterion and formation of an optimal matrix. It is

$$[Z_d] = \begin{bmatrix} 0 & 0.0007 & 0.0005 & 0.0005 & 0.0142 & 0.0693 & 0 \\ 0.0015 & 1.896 & 0.0001 & 0.3945 & 0.9563 & 0.0060 & 0.6440 \\ 0.0049 & 3.708 & 0.7610 & 0.8497 & 0.5452 & 1.1888 & 1.4213 \\ 0 & 0 & 0 & 0 & 0 & 0 & 0.0387 \\ 0 & 0.0007 & 0.0005 & 0.0005 & 0.8974 & 0.0069 & 0.0483 \\ 0.005 & 6.8452 & 0.0145 & 1.6640 & 1.5770 & 0.0197 & 2.2529 \\ 0 & 0.0023 & 0.00068 & 0.0009 & 0.0069 & 0.4761 & 0.0532 \\ 0 & 0.0006 & 7.11E-06 & 0.0004 & 0.0142 & 0.3285 & 0.0481 \\ 0.0001 & 0.0813 & 0.060914 & 0.0294 & 0.0002 & 0.4489 & 0.1440 \end{bmatrix}$$

Table 4 Euclidian distance and ranking of multi-release SRGMs

Model	Distance	Rank
Model 1	0.292197	2
Mode 2	1.974653	7
Model 3	2.912006	8
Model 4	0.196792	1
Model 5	1.008423	6
Model 6	3.518351	9
Model 7	0.730863	4
Model 8	0.626192	3
Model 9	0.875753	5

6. Finally, the Euclidian Distance for each multi-release SRGM is evaluated. These models are ranked on the basis of Euclidian Distance values, as shown in Table 4.

9 Results

The main motive of this chapter is to handle the problem of Multi release SRGM selection using MCDM. An interspersed approach with the amalgamation of the Maximum deviation method and DBA is applied to obtain the ranking of the Multi-Release SRGM model. The findings from this research are discussed below:

1. The primary task is to identify the selection criteria for each model and calculate the weights using the Maximum Deviation Method, as discussed in the section Weighted Distance Approach. The selection criteria MSE has the highest weight value, i.e., 0.315834, and R^2 has the least weight, i.e., 0.009673, as shown in Fig. 1.
2. The results shown in the Table 4 shows that Model 4 is ranked one as it has the lowest weighted Euclidian distance, i.e., 0.196792. This model was proposed by Garmabaki et al. (2015) using the Weibull model for estimating the mean number of faults.
   On the other hand, Model 6 is ranked last amongst the other model as it has the most considerable weighted Euclidian distance i.e. 3.518351. This model was proposed by Mishra et al. (2017) for using Exponential SRGM (Goel and Okumoto) for the testing phase and Weibull distribution for the operational phase.
3. The weighted distance-based approach has smooth and straightforward calculations, which makes it simple to use. There is no convoluted programming in this approach because it can easily be performed in EXCEL. It also deals with selection criteria weights, which helps to select the optimal multi-releasee SRGM. Even if the number of criteria and alternatives are increased, there is no impact on the approach.

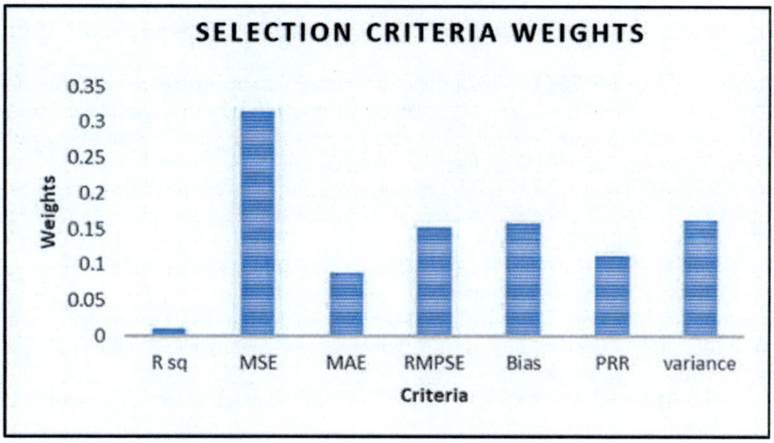

Fig. 1 Selection Criteria Weights

10 Conclusion and Future Scope

The research in this chapter was done to select optimal multi-release SRGM and rank the models. An interspersed MCDM approach has been created, which assembles the weights and distance-based approach. The weights are evaluated through the Maximum Deviation Method and combined with DBA to rank the model. The distance which has been used in this approach is Euclidian distance. The proposed approach contains simple mathematical matrix calculations making it less complicated and saves implementation time. This research can be extended in the following aspects:

(i) Other distances such as Hamming, Mahalanobis, Absolute, and Lee and many more can be used instead of Euclidian distance.
(ii) It can be further extended by considering more selection criteria and alternatives.
(iii) This approach can be compared with other approaches, such as the Simple Matrix method, AHP, and TOPSIS.
(iv) Sensitivity Analysis can be performed to handle the complexity of the criteria.

References

Aggarwal AG, Gandhi N, Verma V, Tandon A (2019) Multi-release software reliability growth assessment: an approach incorporating fault reduction factor and imperfect debugging. Int J Math Oper Res 15(4):446–463

Aggarwal AG, Nijhawan N, Kapur P (2015) A discrete SRGM for multi-release software system with imperfect debugging and related optimal release policy. In: 2015 international conference on futuristic trends on computational analysis and knowledge management (ABLAZE). IEEE, pp 186–192

Anand S, Verma V, Aggarwal AG (2018) 2-dimensional multi-release software reliability modelling considering fault reduction factor under imperfect debugging. Ingeniería Solidaria 14(25):1–12

Anjum M, Haque MA, Ahmad N (2013) Analysis and ranking of software reliability models based on weighted criteria value. IJ Inf Technol Comput Sci 2:1–14

Asad CA, Ullah MI, Rehman M-U (2004) An approach for software reliability model selection. In: Proceedings of the 28th annual international computer software and applications conference. COMPSAC 2004. IEEE, pp 534–539

Cristescu MP, Stoica EA, Ciovică LV (2015) The comparison of software reliability assessment models. Procedia Econ Finance 27:669–675

Garmabaki AH, Aggarwal AG, Kapur P (2011) Multi up-gradation software reliability growth model with faults of different severity. In: 2011 IEEE international conference on industrial engineering and engineering management. IEEE, pp 1539–1543

Garmabaki AH, Aggarwal AG, Kapur P, Yadavali V (2012) Modeling two-dimensional software multi-upgradation and related release problem (a multi-attribute utility approach). Int J Reliab Qual Safe Eng 19(03):1250012

Garmabaki AH, Kapur P, Aggarwal AG, Yadavali V (2014) The impact of bugs reported from operational phase on successive software releases. Int J Prod Qual Manage 14(4):423–440

Garmabaki AHS, Barabadi A, Yuan F, Lu J, Ayele YZ (2015) Reliability modeling of successive release of software using NHPP. In: 2015 IEEE international conference on industrial engineering and engineering management (IEEM). IEEE, pp 761–766

Goel AL, Okumoto K (1979) Time-dependent error-detection rate model for software reliability and other performance measures. IEEE Trans Reliab 28(3):206–211

Gupta A, Gupta N, Garg R, Kumar R (2018) Evaluation, selection and ranking of software reliability growth models using multi criteria decision making approach. In: 2018 4th international conference on computing communication and automation (ICCCA). IEEE, pp 1–8

Hsu C-J, Huang C-Y, Chang J-R (2011) Enhancing software reliability modeling and prediction through the introduction of time-variable fault reduction factor. Appl Math Model 35(1):506–521

Kapur P, Aggarwal AG, Garmabaki AHS, Singh G (2013) Modelling diffusion of successive generations of technology: a general framework. Int J Oper Res 16(4):465–484

Kapur P, Garg R (1992) A software reliability growth model for an error-removal phenomenon. Softw Eng J 7(4):291–294

Kapur P, Goswami D, Bardhan A, Singh O (2008) Flexible software reliability growth model with testing effort dependent learning process. Appl Math Model 32(7):1298–1307

Kapur P, Pham H, Aggarwal AG, Kaur G (2012) Two dimensional multi-release software reliability modeling and optimal release planning. IEEE Trans Reliab 61(3):758–768

Kapur P, Pham H, Anand S, Yadav K (2011) A unified approach for developing software reliability growth models in the presence of imperfect debugging and error generation. IEEE Trans Reliab 60(1):331–340

Kapur P, Tandon A, Kaur G (2010) Multi up-gradation software reliability model. In: 2010 2nd international conference on reliability, safety and hazard-risk-based technologies and physics-of-failure methods (ICRESH). IEEE, pp 468–474

Kapur PK, Kumar S, Garg R (1999) Contributions to hardware and software reliability, vol 3. World Scientific

Khalid B, Sharma K (2015) Ranking of software reliability growth models using Bacterial Foraging Optimization Algorithm. In: 2015 2nd international conference on computing for sustainable global development (INDIACom). IEEE, pp 1643–1648

Knafl GJ, Sacks J (1991) Software reliability model selection. In: [1991] Proceedings the fifteenth annual international computer software & applications conference. IEEE, pp 597–601

Kumar V, Singh V, Garg A, Kumar G (2018) Selection of optimal software reliability growth models: a fuzzy DEA ranking approach. In: Quality, IT and business operations. Springer, pp 347–357

Liu Y, Gao Y (2009) Automation software reliability model selection based on unascertained set. In: 2009 international conference on information management, innovation management and industrial engineering. IEEE, pp 643–646
Miglani N, Rana P (2011) Ranking of software reliability growth models using Greedy approach. Global J Bus Manage Inf Technol 1(11)
Mishra G, Kapur P, Shrivastava A (2017) Multi Release Cost Model—A New Perspective. Int J Reliab Qual Safe Eng 24(06):1740007
Musa J, Okumoto K (1985) Applications of basic and logarithmic Poisson execution model in software reliability measure. In: The challenge of advanced computing technology to system design methods
Pachauri B, Dhar J, Kumar A (2015) Incorporating inflection S-shaped fault reduction factor to enhance software reliability growth. Appl Math Model 39(5–6):1463–1469
Schick GJ, Wolverton RW (1978) An analysis of competing software reliability models. IEEE Trans Softw Eng 2:104–120
Sharma K, Garg R, Nagpal C, Garg R (2010) Selection of optimal software reliability growth models using a distance based approach. IEEE Trans Reliab 59(2):266–276
Sukert AN (1979) Empirical validation of three software error prediction models. IEEE Trans Reliab 28(3):199–205
Tandon A, Aggarwal AG, Nijhawan N (2016) An NHPP SRGM with change point and multiple releases. Int J Inf Syst Serv Sector (IJISSS) 8(4):57–68
Wood A (1996) Predicting software reliability. Computer 29(11):69–77
Yamada S, Ohba M, Osaki S (1983) S-shaped reliability growth modeling for software error detection. IEEE Trans Reliab 32(5):475–484
Yamada S, Ohba M, Osaki S (1984) S-shaped software reliability growth models and their applications. IEEE Trans Reliab 33(4):289–292
Yingming W (1997) Using the method of maximizing deviation to make decision for multiindices. J Syst Eng Electron 8(3):21–26
Zhu M, Pham H (2018) A multi-release software reliability modeling for open source software incorporating dependent fault detection process. Ann Oper Res 269(1–2):773–790